Philip L. Penfold

Jan M. Provis

Editors

**Macular Degeneration**

Philip L. Penfold
Jan M. Provis
Editors

# Macular Degeneration

With 70 Figures in 193 Separate Illustrations
and 14 Tables

**Dr. Philip L. Penfold**
Chief Scientist
Regenera Limited
PO Box 9
2 Brindabella Circuit
Canberra Airport
ACT 2609
Australia

**Dr. Jan M. Provis**
Research School of Biological Sciences
The Australian National University
GPO Box 475
Canberra
ACT 2601
Australia

Library of Congress Control Number: 2004105923

ISBN 3-540-20058-4 Springer Berlin Heidelberg New York

This work is subject to copyright. All rights are reserved, whether the whole or part of the material is concerned, specifically the rights of translation, reprinting, reuse of illustrations, recitation, broadcasting, reproduction on microfilm or in any other way, and storage in data banks. Duplication of this publication or parts thereof is permitted only under the provisions of the German Copyright Law of September 9, 1965, in its current version, and permission for use must always be obtained from Springer. Violations are liable to prosecution under the German Copyright Law.

**Springer is a part of Springer Science+Business Media**
springeronline.com
© Springer-Verlag Berlin Heidelberg 2005
Printed in Germany

The use of general descriptive names, registered names, trademarks, etc. in this publication does not imply, even in the absence of a specific statement, that such names are exempt from the relevant protective laws and regulations and therefore free for general use.

Product liability: the publishers cannot guarantee the accuracy of any information about dosage and application contained in this book. In every individual case the user must check such information by consulting the relevant literature.

Editor: Marion Philipp, Heidelberg, Gemrany
Desk Editor: Martina Himberger, Heidelberg, Gemrany
Production: ProEdit GmbH, 69126 Heidelberg, Germany
Cover: E. Kirchner, Heidelberg, Germany
Typesetting: K. Detzner, 67346 Speyer, Germany

Printed on acid-free paper   24/3150 ML   5 4 3 2 1 0

# Preface

This book provides a unique overview of current thinking on the pathogenesis, incidence and treatment of age-related macular degeneration (AMD). It includes, for the first time, a synthesis of the views of the world's leading scientists and clinical practitioners regarding retinal biology and the basic mechanisms, clinical and pathogenetic processes and rational approaches to the treatment of AMD.

Although the fovea is less than a millimetre in diameter, disorders of the fovea and its immediately surrounding area (the *macula*) are responsible for the majority of cases of untreatable blindness in the developed world. The basis for the vulnerability of the macula region in these degenerative changes is beginning to emerge. The fovea has a number of features that distinguish it from other parts of the retina and reflect its specialization for high visual acuity, principally a high density of photoreceptors and a lack of retinal vessels. Chapter 1, written by Anita Hendrickson, provides an overview of the anatomy of the primate macula. The fovea is a characteristic feature of the primate retina, lies on the temporal side of the optic disc and regards the central visual field. A sound understanding of macular anatomy is essential for understanding the impact of AMD on the patient.

In Chap. 2, we summarise the evidence suggesting a critical dependence of the central retina on vascular supply. The interrelationships between the physiological and immunological function of the blood-retinal barrier and the consequences of barrier breakdown are described. Increasing evidence is presented for the involvement of both resident microglia and choroidal leukocytes. New observations concerning the significance of drusen, the involvement of the retinal vasculature and the measurement of inflammation in AMD are presented for the first time. Taken together, the data lead to the conclusion that immunity plays both a primary and secondary role in the pathogenesis of AMD.

The link between photoreceptor dysfunction and the risk of neovascularization in Bruch's membrane is explored by Jackson and colleagues in Chap. 3. Because the RPE is polarized, problems pertaining to the re-supply of photoreceptors on the apical aspect of the RPE (leading to photoreceptor death) should be conceptually separated from problems pertaining to waste removal on the basal aspect of the RPE (leading to Bruch's membrane damage and neovascularization), at least for the purposes of designing mechanistic experiments. These processes are governed by different proteins and pathways at the cellular level and will be reflected in different risk factors and genetic predispositions at the population level. A rigorous test of a nutrient-deficiency hypothesis of AMD-associated photoreceptor death awaits more information about normal nutrient delivery mechanisms across the RPE/Bruch's membrane complex, intra-retinal contributions to photoreceptor nutrition, changes in these mechanisms with age and pathology, and differential effects on rods and cones.

In Chap. 4, AMD is considered as a complex genetic disease in which environmental risk factors impact on a genetic background. Finding the genes that determine susceptibility or modify the disease process is one of today's challenges, but also offers a chance for understanding underlying disease processes and for the development of preventive strategies and treatments. This chapter explores our current knowledge about the genetic influences on AMD and indicates possible directions for future study.

Until recently, most of the information about the natural history of AMD has come from clinical and histopathological studies. Most such studies have previously been of short duration involving select groups of patients attending ophthalmology clinics or participating in trials in which severe disease may be overrepresented. In the past 15 years data from population-based studies have resulted in a better understanding of the epidemiology of this disease. Chapter 5 examines the epidemiology of AMD, focusing on data from several recent population-based studies.

In Chap. 6, the racial/ethnic differences in the incidence and prevalence of AMD in China are examined. In China AMD is considered one of the most important causes of blindness in those over the age of 50. With improvement of economic conditions in China, the most common causes of blindness such as cataract, corneal diseases, trachoma and glaucoma have been largely brought under control, while AMD has increased in prevalence, now fourth on the list of causes of blindness in the age group of those 60 years and over. Considering the large population of China, it has been estimated that AMD currently affects at least 20 million individuals.

Experimental models of age-related macular degeneration capture only selected features of the human disease. Animal models that encompass both atrophic and exudative aspects of retinal degenerations are needed to better understand disease progression and to predict and assess potential therapeutic approaches. Recent insights into the pathogenesis of macular degeneration, along with the combination of rapid screening techniques with transgenic and other methods, are giving rise to several promising experimental systems outlined by Ray Gariano in Chap. 7.

Normal function is dependent upon a balance between the generation of free radicals and oxidative species and the availability of antioxidants and free radical scavengers. David Pow and colleagues investigate the role of transporters and oxidative stress in AMD in Chap. 8. In Chap. 9, Jonathan Stone and colleagues describe a widespread degenerative phenomena observed at the edge of the 'normal' retina. The observations suggest that the edge of the retina is subject to localized stress throughout life, inducing a progressive degenerative process. These edge-specific changes are part of the life history of the normal retina and form part of the baseline against which retinal degeneration takes place.

In the final chapter Scott Cousins and colleagues address – in the context of the scientific information described in the other sections – current clinical research strategies to provide a concept-based overview of the status of current and future treatments for AMD. A brief review of key scientific definitions and pathogenic theories is followed by the rationale for current treatments and ongoing trials. Space is also set aside for a bit of "educated speculation" about the potential future directions of clinical research based upon scientific discoveries described in other chapters.

**Acknowledgments**

We are indebted to the contributing authors and many other colleagues for their help in preparing this book. Special thanks are due to Emily Bell for her assistance in formatting the initial submissions and to Diana van Driel, both for her contributions as an author and fellow electron microscopist and for her diligence as proofreader *par excellence*.

**Philip L. Penfold, MPhil., PhD.**
**Jan M. Provis, PhD.**

Australian National University
Canberra, Australia

September 2004

# Contents

**1 Organization of the Adult Primate Fovea** . . . . . . . . . . . . . . 1
Anita Hendrickson

**1.1** Anatomy of the Human Fovea . . . . . . . . . . . . . . . . . . . . 1
1.1.1 General Anatomy . . . . . . . . . . . . . . . . . . . . . . . 1
1.1.2 Photoreceptor Distribution, Types, and Numbers
in the Human Retina . . . . . . . . . . . . . . . . . . . . . 3
1.1.3 Inner Retinal Neurons Associated with Cones
in Central Primate Retina . . . . . . . . . . . . . . . . . . . 9
1.1.4 Vascular and Glial Specializations of the Fovea . . . . . . . . 12
1.1.5 Pigment Epithelium Numerical Relationships
with Foveal Photoreceptors . . . . . . . . . . . . . . . . . . 13

**1.2** Anatomy of the Old and New World Monkey Fovea:
What Are the Differences with Human Foveas? . . . . . . . . . . . 14

**1.3** What Are the Anatomical Requirements to Create a Fovea? . . . . . 17
1.3.1 Midget Ganglion Cells . . . . . . . . . . . . . . . . . . . . . 17
1.3.2 High Cone Density and Types of Foveal Cones . . . . . . . . 18
1.3.3 Absence of Rods . . . . . . . . . . . . . . . . . . . . . . . . 18
1.3.4 Striate Cortex Expansion . . . . . . . . . . . . . . . . . . . 19
1.3.5 Vascular Specializations . . . . . . . . . . . . . . . . . . . . 19

References . . . . . . . . . . . . . . . . . . . . . . . . . . . 20

**2 Immunology and Age-Related Macular Degeneration** . . . . . . . . 25
Philip L. Penfold, James Wong, Diana van Driel, Jan M. Provis,
Michele C. Madigan

**2.1** Introduction: Why the Macula? . . . . . . . . . . . . . . . . . . . 25
**2.2** The Immune Status of the Retina . . . . . . . . . . . . . . . . . . 26
2.2.1 The Blood-Retinal Barrier . . . . . . . . . . . . . . . . . . . 26
2.2.2 Microglia . . . . . . . . . . . . . . . . . . . . . . . . . . . 29
**2.3** Immune Mechanisms in AMD . . . . . . . . . . . . . . . . . . . 30
2.3.1 Blood-Retinal Barrier Breakdown . . . . . . . . . . . . . . . 30

| | | |
|---|---|---|
| 2.3.2 | Pigmentary Disturbance | 30 |
| 2.3.3 | Drusen | 31 |
| 2.3.4 | Cell-Mediated Immunity and Inflammation | 33 |
| 2.3.5 | Humoral Immunity | 35 |
| 2.4 | Clinical Significance of Drusen | 37 |
| 2.5 | Atrophic ("Dry") Macular Degeneration | 37 |
| 2.6 | Neovascular ("Wet") Macular Degeneration | 37 |
| 2.7 | Involvement of the Retinal Vasculature in AMD | 38 |
| 2.8 | Leucocyte Common Antigen (CD45) Expression in AMD: A Measure of Inflammation | 40 |
| 2.9 | Conclusion | 41 |
| | References | 41 |

## 3  Photoreceptor Degeneration in Aging and Age-Related Maculopathy ... 45

Gregory R. Jackson, Christine A. Curcio, Kenneth R. Sloan, Cynthia Owsley

| | | |
|---|---|---|
| 3.1 | Introduction to Age-Related Maculopathy | 45 |
| 3.2 | Photoreceptor Loss | 47 |
| 3.3 | Photoreceptor Dysfunction | 51 |
| 3.3.1 | Topography of Loss and Dysfunction | 52 |
| 3.4 | Photoreceptor Function as a Bioassay of RPE and Bruch's Membrane Health | 54 |
| 3.5 | Impairment of Transport Between RPE and Photoreceptors | 56 |
| 3.6 | Summary | 57 |
| | References | 58 |

## 4  Genes and Age-Related Macular Degeneration ... 63

Robyn H. Guymer, Niro Narendran, Paul N. Baird

| | | |
|---|---|---|
| 4.1 | Age-Related Macular Degeneration, a Complex Genetic Disease | 63 |
| 4.2 | Genetic Basis of Disease | 64 |
| 4.2.1 | Family Studies | 64 |
| 4.2.2 | Sibling Studies | 64 |
| 4.2.3 | Twin Studies | 65 |
| 4.3 | Approaches to Genetic Investigation of AMD | 65 |
| 4.3.1 | Linkage Analysis | 65 |
| 4.3.2 | Candidate-Gene Screening | 67 |

| | | |
|---|---|---|
| 4.4 | Future Directions | 73 |
| 4.4.1 | Single-Nucleotide Polymorphisms | 73 |
| 4.4.2 | Microarrays | 73 |
| 4.4.3 | Proteomics | 74 |
| 4.5 | Conclusions | 74 |
| | References | 75 |

| | | |
|---|---|---|
| **5** | **Epidemiology of Age-Related Macular Degeneration** | 79 |
| | Ronald Klein | |
| 5.1 | Prevalence and Incidence of Age-Related Macular Degeneration | 79 |
| 5.1.1 | Introduction | 79 |
| 5.1.2 | Prevalence of Age-Related Macular Degeneration | 80 |
| 5.1.3 | Prevalence: Race/Ethnicity | 81 |
| 5.1.4 | Incidence of Age-Related Macular Degeneration | 82 |
| 5.2 | Risk Factors for Age-Related Macular Degeneration | 82 |
| 5.2.1 | Introduction | 82 |
| 5.2.2 | Familial Factors | 83 |
| 5.2.3 | Systemic Factors | 83 |
| 5.2.4 | Lifestyle Behavior | 87 |
| 5.2.5 | Environmental Factors | 89 |
| 5.2.6 | Ocular Factors | 90 |
| 5.2.7 | Socioeconomic Factors and Work Exposures | 92 |
| 5.3 | Age-Related Macular Degeneration and Survival | 93 |
| 5.4 | Public Health Issues | 93 |
| 5.5 | Conclusions | 93 |
| | References | 94 |

| | | |
|---|---|---|
| **6** | **Prevalence and Risk Factors for Age-Related Macular Degeneration in China** | 103 |
| | Zheng Qin Yin, Meidong Zhu, Wen Shan Jiang | |
| 6.1 | Introduction | 103 |
| 6.2 | Regional Prevalence and Relationship of Morbidity to Age and Gender | 104 |
| 6.2.1 | Population-Based Studies | 105 |
| 6.2.2 | Clinical Review Studies | 106 |
| 6.2.3 | The Relationship of Morbidity to Age and Gender | 107 |
| 6.3 | Prevalence of 'Wet' and 'Dry' AMD | 107 |
| 6.4 | Visual Acuity in AMD | 107 |

| 6.5 | **Risk Factors of AMD in China** | 107 |
| 6.5.1 | Race/Ethnicity | 109 |
| 6.5.2 | Light Exposures/Occupation | 109 |
| 6.5.3 | Smoking | 109 |
| 6.5.4 | Drinking | 110 |
| 6.5.5 | Systemic Diseases | 111 |
| 6.6 | **Conclusions** | 111 |
| | References | 111 |

| 7 | **Experimental Models of Macular Degeneration** | 113 |
|---|---|---|
| | Ray F. Gariano | |
| 7.1 | Introduction | 113 |
| 7.2 | Aging | 113 |
| 7.3 | Laser-Induced CNV | 114 |
| 7.4 | Growth Factor-Induced CNV | 114 |
| 7.5 | Genetically Defined Animal Models | 116 |
| 7.6 | Conclusions | 118 |
| | References | 118 |

| 8 | **Transporters and Oxidative Stress in AMD** | 123 |
|---|---|---|
| | David V. Pow, Robert K.P. Sullivan, Susan M. Williams, Elizabeth WoldeMussie | |
| 8.1 | Redox Reactions, Health, Oxidative Damage and Disease | 123 |
| 8.2 | Does Epidemiology Support a Role for Oxidation in AMD? | 124 |
| 8.3 | Exogenous Factors Influencing AMD Incidence | 125 |
| 8.4 | Hallmark Features of Oxidative Damage and Free Radical Damage | 125 |
| 8.4.1 | Oxidized Proteins | 125 |
| 8.4.2 | Oxidized Lipids | 125 |
| 8.4.3 | DNA Damage | 125 |
| 8.5 | Oxidative Challenges in the Eye | 125 |
| 8.5.1 | Endogenous Antioxidants, Enzymes and Related Molecules in the Retina | 126 |
| 8.5.2 | Classes of Antioxidants | 128 |
| 8.5.3 | The Glutathione System | 128 |
| 8.5.4 | GSH in the Retina: Intrinsic Synthesis and Possible Interplay with the RPE | 128 |
| 8.5.5 | Synthesis of GSH | 128 |

| | | |
|---|---|---|
| 8.5.6 | The GSH-GSSG Cycle | 129 |
| 8.5.7 | Elimination of Superoxides by Superoxide Dismutases | 131 |
| 8.5.8 | Ascorbic Acid | 132 |
| 8.5.9 | Vitamin E | 132 |
| 8.5.10 | Taurine | 133 |
| 8.5.11 | Metallothionein | 136 |
| 8.5.12 | Zinc, Copper, Selenium and Manganese | 136 |
| 8.5.13 | Selenium Compounds | 137 |
| 8.5.14 | Manganese | 137 |
| 8.5.15 | Macular Pigments | 138 |
| 8.5.16 | Ascorbate and Recycling of Carotenoids | 138 |
| 8.5.17 | A Summary of Mechanisms for Antioxidant Protection | 138 |
| **8.6** | **Cellular Interactions and Photoreceptor Death** | **138** |
| 8.6.1 | Light-Mediated Damage: Experimental Lesions and AMD | 139 |
| 8.6.2 | Mechanisms of Cell Death | 140 |
| 8.6.3 | Cell Death in Response to Oxidative Damage | 140 |
| **8.7** | **A Role for Glial cells in AMD?** | **141** |
| 8.7.1 | Metabolic Functions of Müller Cells in the AMD Retina | 141 |
| 8.7.2 | Cell Death and Glutamate Toxicity | 144 |
| **8.8** | **Conclusions** | **144** |
| | **References** | **145** |

## 9 Photoreceptor Stability and Degeneration in Mammalian Retina: Lessons from the Edge ... 149

Jonathan Stone, Kyle Mervin, Natalie Walsh, Krisztina Valter,
Jan M. Provis, Philip L. Penfold

| | | |
|---|---|---|
| **9.1** | **Introduction** | **149** |
| **9.2** | **Approach** | **150** |
| **9.3** | **Stages in Photoreceptor Degeneration at the Edge of the Retina** | **150** |
| 9.3.1 | Postnatal Development of the Edge: Site of Early Stress and Degeneration | 150 |
| 9.3.2 | The Edge of the Retina Is Functionally Degraded | 154 |
| 9.3.3 | The Edge of the Retina Is Highly Stable in the Face of Acute Stress | 155 |
| 9.3.4 | The Adult Retina Aged 2 Years: Observations in the Marmoset | 155 |
| 9.3.5 | The Adult Retina Aged 20 Years: Observations in the Baboon | 155 |
| 9.3.6 | The Adult Retina Aged 30–70 Years: Observations in the Human | 158 |
| 9.3.7 | Evidence of Progression in Edge Degeneration in Humans | 161 |
| **9.4** | **Discussion** | **161** |
| 9.4.1 | The Edge of the Retina as a Model of Retinal Degeneration | 162 |
| 9.4.2 | Why Is the Protection of Photoreceptors Stress-Inducible? | 162 |

9.4.3 Vulnerability to Oxidative Stress in Mice and Humans . . . . . . . . . 163
9.4.4 The Link to AMD . . . . . . . . . . . . . . . . . . . . . . . . . . 163
   **References** . . . . . . . . . . . . . . . . . . . . . . . . . . . . . 164

# 10 Clinical Strategies for Diagnosis and Treatment of AMD: Implications from Research . . . . . . . . . . . . . . . . . . . . . 167
Scott W. Cousins, Karl G. Csaky, Diego G. Espinosa-Heidmann

**10.1 Dry AMD** . . . . . . . . . . . . . . . . . . . . . . . . . . . . . . 167
10.1.1 Definition of Dry AMD . . . . . . . . . . . . . . . . . . . . . . . 167
10.1.2 Pathogenic Mechanisms for Drusen Formation . . . . . . . . . . . 168
10.1.3 Established or Evaluated Treatments . . . . . . . . . . . . . . . . 170
10.1.4 Ongoing Trials: Multicenter Investigation of Rheopheresis
       for AMD (MIRA-1) Study Group . . . . . . . . . . . . . . . . . . 172
10.1.5 Future Research . . . . . . . . . . . . . . . . . . . . . . . . . . 172
**10.2 Wet AMD** . . . . . . . . . . . . . . . . . . . . . . . . . . . . . . 178
10.2.1 Definition of Neovascular AMD . . . . . . . . . . . . . . . . . . 178
10.2.2 Pathogenic Mechanisms for CNV Formation . . . . . . . . . . . . 180
10.2.3 Established or Evaluated Treatments for Neovascular AMD . . . . . 184
10.2.4 Therapies Currently in Clinical Trial . . . . . . . . . . . . . . . . 185
10.2.5 Future Research . . . . . . . . . . . . . . . . . . . . . . . . . . 189
   **References** . . . . . . . . . . . . . . . . . . . . . . . . . . . . . 191

**Subject Index** . . . . . . . . . . . . . . . . . . . . . . . . . . . . . 201

# List of Contributors

## Chapter 1

**Anita Hendrickson**

Department of Biological Structure, University of Washington, 1959 NE Pacific Street, Box 357420, Seattle, WA 98195-7420, USA

Department of Ophthalmology, University of Washington, 1959 NE Pacific Street, Box 357420, Seattle, WA 98195-7420, USA
e-mail: anitah@u.washington.edu, Tel.: +1-206-6852273, Fax: +1-206-5431524

## Chapter 2

**Philip L. Penfold, James Wong, Diana van Driel, Jan M. Provis, Michele C. Madigan**

Save Sight Institute, University of Sydney, GPO 4337 Sydney 2001, NSW, Australia

**Jan M. Provis**

Department of Anatomy and Histology, University of Sydney 2006, NSW, Australia

*Current address*: Philip L. Penfold, Jan M. Provis, Research School of Biological Sciences, The Australian National University, GPO Box 475, Canberra, ACT 2601, ACT, Australia
e-mail: philip.penfold@regenera.com.au, Tel.: +61-2-62574155, Fax: +61-2-62574355
e-mail: jan.provis@anu.edu.au, Tel.: +61-2-61254242, Fax: +61-2-61250758

## Chapter 3

**Gregory R. Jackson, Christine A. Curcio, Cynthia Owsley**

Department of Ophthalmology, University of Alabama School of Medicine, Birmingham, AL 35294-0009, USA
e-mail: jackson@uab.edu, Tel.: +1-205-325-8674, Fax: +1-205-325-8692

**Kenneth R. Sloan**

Department of Computer and Information Sciences, University of Alabama at Birmingham, AL, USA

## Chapter 4

**Robyn H. Guymer, Paul N. Baird**

Department of Ophthalmology, University of Melbourne, 1st Floor Royal Victorian
Eye and Ear Hospital, 32 Gisborne Street. East Melbourne 3002
e-mail: rhg@unimelb.edu.au, Tel.:+613 9929-8360, Fax:+613-9662-3859

**Robyn H. Guymer, Niro Narendran, Paul N. Baird**

Centre For Eye Research Australia, University of Melbourne,
1st Floor Royal Victorian Eye and Ear Hospital, 32 Gisborne Street,
East Melbourne 3002, Australia

## Chapter 5

**Ronald Klein**

Department of Ophthalmology and Visual Sciences,
University of Wisconsin Medical School, Madison, WI, USA
e-mail: kleinr@epi.ophth.wisc.edu, Tel.: +1-608-263-0280, Fax: +1-608-263-0279

## Chapter 6

**Zheng Qin Yin, Meidong Zhu, Wen Shan Jiang**

Southwest Eye Hospital, Southwest Hospital, Third Military Medical University,
40038, Chongqing, Peoples Republic of China

**Meidong Zhu**

Save Sight Institute, University of Sydney, NSW 2006, Sydney, Australia
e-mail: zqyin@mail.tmmu.com.cn, Tel.: +86-23-6875-4401, Fax: +86-23-6875-3686

## Chapter 7

**Ray F. Gariano**

Department of Ophthalmology, Stanford University School of Medicine
330 Pasteur Dr., Palo Alto, CA 94305, USA

Santa Clara Valley Medical Center, Santa Clara, CA, USA
e-mail: Ray.Gariano@hhs.co.santa-clara.ca.us, Tel.: +1-408-885-6775,
Fax: +1-408-885-7166

## Chapter 8

**David V. Pow, Robert K.P. Sullivan, Susan M. Williams**

School of Biomedical Sciences, University of Queensland, Australia
e-mail: d.pow@mailbox.uq.edu.au, Tel.: +7-33654086, Fax: +7-33651766

**Elizabeth WoldeMussie**

Department of Biological Sciences, Allergan Inc., Irvine, CA, USA

## Chapter 9

**Jonathan Stone, Krisztina Valter, Jan Provis, Philip Penfold**

Biological Sciences, Australian National University, Canberra, Australia
e-mail: stone@rsbs.anu.edu.au, Tel.: +61-26125-3841, Fax: +61-26125-0758

**Jonathan Stone, Kyle Mervin, Natalie Walsh, Krisztina Valter, Jan Provis**

Department of Anatomy and Histology and Institute for Biomedical Research, University of Sydney, Australia

**Jan Provis, Philip Penfold**

Save Sight Institute, Department of Clinical Ophthalmology, University of Sydney

## Chapter 10

**Scott W. Cousins, Diego G. Espinosa-Heidmann**

Department of Ophthalmology, Bascom Palmer Eye Institute, University of Miami School of Medicine, Miami, FL, USA

**Karl G. Csaky**

National Eye Institute, National Institutes of Health, Bethesda, MD, USA
e-mail: scousins@med.miami.edu, despinosa@med.miami.edu,
Tel.: +1-305-326-6329, Fax: +1-305-326-6536

Chapter 1

# Organization of the Adult Primate Fovea

Anita Hendrickson

## Contents

1.1 Anatomy of the Human Fovea 1
1.1.1 General Anatomy 1
1.1.2 Photoreceptor Distribution, Types, and Numbers in the Human Retina 3
1.1.3 Inner Retinal Neurons Associated with Cones in Central Primate Retina 9
1.1.4 Vascular and Glial Specializations of the Fovea 12
1.1.5 Pigment Epithelium Numerical Relationships with Foveal Photoreceptors 13

1.2 Anatomy of the Old and New World Monkey Fovea: What Are the Differences with Human Foveas? 14

1.3 What Are the Anatomical Requirements to Create a Fovea? 17
1.3.1 Midget Ganglion Cells 17
1.3.2 High Cone Density and Types of Foveal Cones 18
1.3.3 Absence of Rods 18
1.3.4 Striate Cortex Expansion 19
1.3.5 Vascular Specializations 19

References 20

## 1.1 Anatomy of the Human Fovea

### 1.1.1 General Anatomy

The fovea centralis is a characteristic feature of all primate retinas except the nocturnal New World owl monkey *Aotes*. It lies on the temporal side of the optic disc within the area centralis or macular region which is outlined by the superior and inferior temporal branches of the central retinal artery and vein (Fig. 1.1). The fovea lies on the visual axis of the eye such that a beam of light passing perpendicularly through the center of the lens will fall on the fovea. The fovea has long been recognized as the site of maximal visual acuity (Helmholz 1924). Although the fovea only occupies 0.02% of the total retinal area and contains 0.3% of the total cones, it contains 25% of the ganglion cells (Curcio and Allen 1990; Curcio et al. 1990), illustrating its importance in primate vision. A large portion of the central visual centers is devoted to processing foveal information. For instance, 40% of the primary visual cortex processes the central 5° of the visual field which is approximately the retinal area of the foveal pit (Tootell et al. 1988; Curcio et al. 1990). A major function of the superior colliculus, pretectum, cranial nerve nuclei of III, IV, and VI and associated centers is to integrate cortical information and retinal input so that eye movements keep a point in space focused on the fovea of both eyes to facilitate binocular high-acuity vision (Goldberg 2000).

The macular region contains a yellow pigment which is particularly intense around the fovea (Delori et al. 2001). The fovea also is identified by the foveal pit which indents the vitreal surface of the retina (Fig. 1.2). This pit and the concentration of yellow pigment around it ("yellow spot" or "macula lutea") were recognized in human and monkey retina by anatomists such as Buzzi, Soemmerring, and Fragonard in the late eighteenth century (cited in Polyak 1941). Although Ramon y Cajal made

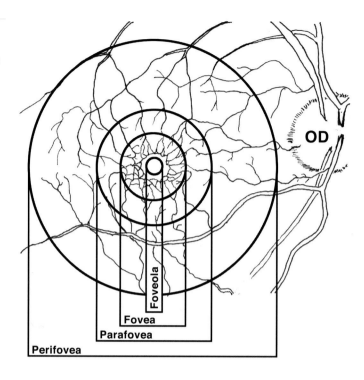

Fig. 1.1.
The central human retina or macula, with an idealized view of its blood vessels. The optic disc (*OD*) is to the left. The macula is subdivided into the central-most foveola, which is surrounded in turn by the fovea, parafovea and wider perifovea. Each is indicated by *concentric circles*. Blood vessels form a ring outlining the foveal avascular zone, which marks the inner limits of the foveal pit

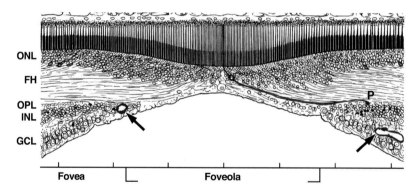

Fig. 1.2. Modification of a drawing of the macaque monkey foveola and foveal slope from Polyak (1941). A cone with an outer and inner segment in the foveola is filled in to show the length of the cone axon and displacement of the synaptic pedicle (*P*) from the cell body. The *arrow on the right* indicates the capillaries forming the wider inner portion of the foveal avascular zone and the *arrow on the left* the narrower outer ring of capillaries of the foveal avascular zone. The markings on the *lower scale* each indicate 100 µm

significant advances in our understanding of general retinal cellular anatomy and connectivity using the Golgi impregnation method (Cajal 1892), it was the work of Stefan Polyak (1941) and Brian Boycott and John Dowling (1969), also using Golgi impregnation, which revealed the intricate relationships of primate retinal neurons, including those forming the fovea. The high-resolution anatomical relationship between neurons and glia within the human and monkey fovea also has been described using electron microscopy (Dowling 1965; Missotten 1965, 1974; Yamada 1969; Borwein et al. 1980; Schein 1988; Krebs and Krebs 1991).

In the adult human eye (Fig. 1.1), the fovea centralis lies 4 mm temporal and 0.8 mm inferior to the center of the optic disc (Hogan et al. 1971). The human foveal pit generally is round, approximately 1000 µm at its widest, and 200–240 µm deep (Fig. 1.2). This is about half the thickness of adjacent retina, as can be seen in Fig. 1.3A. All of these dimensions can vary slightly between individuals (Polyak 1941) and also are subject to the methods used to preserve and examine the fovea. A study of many species of New World monkeys finds that, despite a fivefold variation in eye size and retinal area, the dimensions of the fovea remain constant (Franco et al. 2000). This strongly suggests that the same basic mechanism(s) must operate in all primates to create this stable structure.

The central retina or macula is divided into four concentric zones (Fig. 1.1), with the foveola in the center surrounded in turn by the fovea, parafovea, and perifovea. The centralmost region of the pit is called the *foveola* (Figs. 1.2, 1.3A,B,G). Here the center of the pit is slightly flattened and is overlain by the highest cone density in the retina. The foveola is 250–350 µm in diameter and represents the central 1° 20′ of the visual field. At its thinnest point, the cellular portion of the foveola is slightly more than 100 µm thick and is formed only by cone cell bodies surrounded by Müller glial cell processes (Yamada 1969; Burris et al. 2002). However the foveola has the longest outer and inner segments in the retina and these indent the cellular layers, forming an inward curve called the fovea externa (Fig. 1.3A,G). The inner retinal layers of the foveola, including the outer plexiform layer, are moved laterally onto the foveal slope, although occasional neurons are found on the foveal floor (Röhrenbeck et al. 1989; Curcio and Allen 1990). The *fovea* (Figs. 1.1, 1.2, 1.3C,F) includes the adjacent 750 µm around the foveola, making it 1.85 mm wide or 5.5° of the central visual field. The fovea contains all of the layers of the retina, including the thickest part, which is called the foveal slope. This region is characterized histologically by having a layer of ganglion cells up to eight deep and a very thick outer plexiform layer. Most of this layer is made up by the elongated cone axons called the fibers of Henle, but it also contains the synaptic pedicles of the foveolar cones (Figs. 1.2, 1.3C,F). Rods first appear in the fovea and also have elongated axons (Fig. 1.3D,E). These long photoreceptor axons are diagrammatically shown for a single foveolar cone in Fig. 1.2 and indicated by arrows in Fig. 1.3E. They are formed during pit development to allow foveal photoreceptors to retain their synaptic connections to bipolar and horizontal cells during the peripheral-ward displacement of inner retinal neurons and the central-ward displacement of photoreceptors (Hendrickson 1992; Provis et al. 1998). Capillaries are present in the inner retina up to the foveal slope, where they form a foveal avascular zone (FAZ, see Fig. 1.4A) about 400–500 µm wide (Figs. 1.1, 1.2, 1.4). Two more peripheral zones are identified around the fovea but these do not have sharp boundaries. The region next to the fovea is the *parafovea*, a zone about 500 µm wide (Figs. 1.1, 1.3D). It also has a thick outer plexiform layer and rods are more numerous. The ganglion cell layer is still eight cells thick near the fovea but decreases to four cells thick at the peripheral edge of the parafovea. The remainder of the central retina is called the *perifovea*, which is 1.5 mm wide with the peripheral perifoveal edge close to the nasal side of the optic disc (Figs. 1.1, 1.3E). The perifovea ganglion cell layer decreases to one cell thick at its peripheral edge. The perifovea contains many more rods than cones in its outer nuclear layer.

## 1.1.2
**Photoreceptor Distribution, Types, and Numbers in the Human Retina**

The fovea is the center of photoreceptor topography in the primate retina. The first quantitative analysis of the human photoreceptor layer was done by Osterberg (1935), using sections from a single celloidin-embedded 16-year-old eye. The advantages of immunocytochemical labeling of specific photoreceptor types combined with recent advances in optical imaging and computerized counting of retinal wholemounts have been used by Curcio and col-

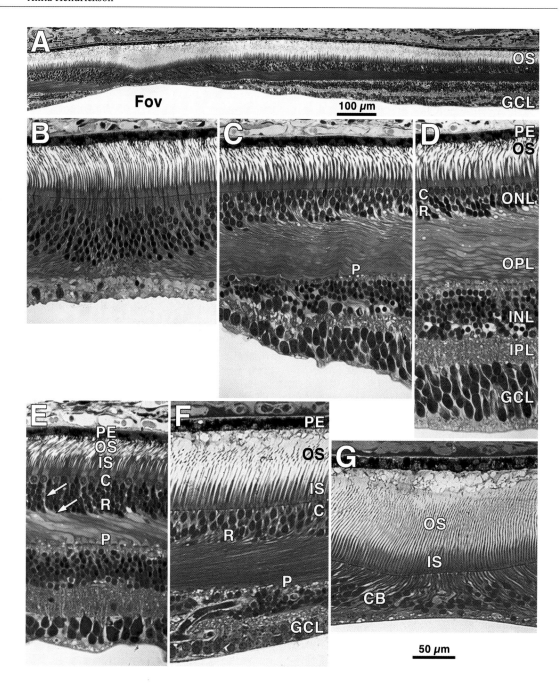

leagues (Curcio et al. 1990, 1991) to considerably extend that pioneering effort.

The mean number of photoreceptors per eye based on a sample of seven human retinas between 27 and 44 years old (Curcio et al. 1990) is 4.6 million cones and 92 million rods, which is slightly lower than the 6 million cones found by Osterberg (1935). Cones can be differentiated from rods on the basis of size, allowing the creation of color-coded computer-generated maps for cone (Fig. 1.5A) and rod (Fig. 1.5B) distribution in human retina. Different types of photoreceptors have been identified using immunocytochemical labels. Each photoreceptor

Fig. 1.3A–G. Photomicrographs of human central retina: A, F, G are from a 72-year-old retina and B–E, a 13-year-old retina, as seen in semithin plastic sections of glutaraldehyde-fixed retina. (*PE* pigment epithelium; *OS* outer segments; *IS* inner segments; *C* cone; *R* rod; *ONL* outer nuclear layer; *P* synaptic pedicle; *OPL* outer plexiform layer; *INL* inner nuclear layer, *IPL* inner plexiform layer, *GCL* ganglion cell layer.) A A low-power view of the fovea and foveal slope. The foveola (*fov*) is the thinnest part of this region, but has the longest OS. The outer foveola is indented in this region, forming the fovea externa. Layers are marked on the left side of D for B–D. B This 13-year-old foveola contains tightly-packed cone cell bodies, long and very thin IS and OS, as well as Müller glial processes. C The inner foveal slope has all retinal layers, although the GCL is still thinner than its maximum on the peak of the slope (not shown). The OPL is formed mainly of long fibers of Henle and foveolar cone P are first found at this location. The first R cell bodies also are seen at the bottom of the ONL. The increase in thickness of cone IS is obvious. D The parafovea has the thickest GCL. R cell bodies have increased sharply in number in the ONL, but C cell bodies still form a single uniform layer at the outer edge of the ONL. E The perifovea is thinner overall, due to shorter OS and a much thinner GCL. R are now the predominant cell in the ONL. The fiber of Henle cone axon (arrows) can be traced from the cone cell body into the OPL. F, G The fovea (F) and foveola (G) of this 72-year-old retina have longer and thinner cone IS and OS than the young retina shown in B, but the GCL is thinner throughout. Note that the foveola (G) contains only long thin OS and IS, some cone cell bodies (*CB*) surrounded by pale Müller cell processes. *Scale bar* in G for B–G

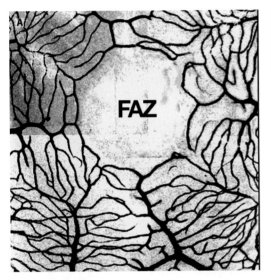
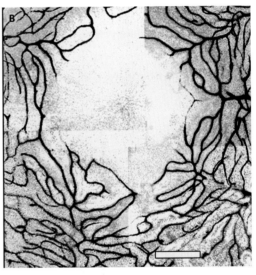
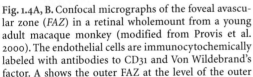

Fig. 1.4A, B. Confocal micrographs of the foveal avascular zone (*FAZ*) in a retinal wholemount from a young adult macaque monkey (modified from Provis et al. 2000). The endothelial cells are immunocytochemically labeled with antibodies to CD31 and Von Wildebrand's factor. A shows the outer FAZ at the level of the outer plexiform layer and B the inner wider FAZ at the level of the ganglion cells. Apparent blunt endings of some vessels actually are anastomoses between the two layers of capillaries which form a cone around the inner walls of the foveal slope. *Scale bar* 200 µm

type has a characteristic opsin protein which, combined with 11-*cis*-retinal forms a unique phototransduction molecule in the outer segment. The amino acid sequence of each opsin determines the basis of wavelength selectivity of the three cone types and the single rod type (Nathans et al. 1986; Jacobs 1998; Sharpe et al. 1999). The human retina contains a mixture of short- (S), medium- (M), and long-wavelength (L) selective cones and rods. S, M, and L cones also are called blue, green, and red cones, respectively, for the color perception which they support. The amino acid sequence in S opsin and rod opsin are sufficiently different from each other and from the amino acid sequences of M and L opsins so that both riboprobes and antibodies have been generated which specifically label rods, S cones, and M/L cones in histo-

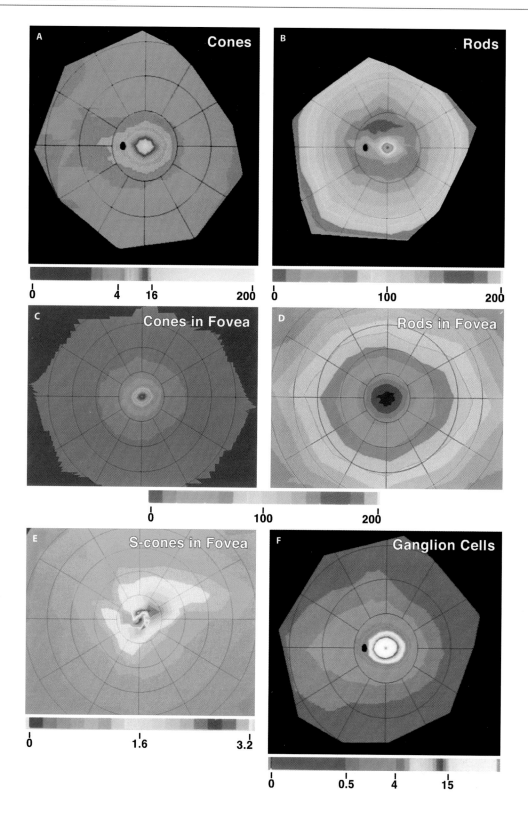

logical preparations such as retinal wholemounts. A map of S cones determined by these methods is shown in Fig. 1.5E. However, M and L opsins differ by only a few amino acids so thus far it is not possible to generate riboprobes or antibodies which can distinguish between M and L cones (reviewed in Bumsted et al. 1997; Xiao and Hendrickson 2000). Recently in vivo studies in humans using adaptive optics (Roorda and Williams 1999; Roorda et al. 2001) or electroretinographic flicker photometry (Carroll et al. 2002) has made it possible to directly observe living cones to determine M and L ratios. It also is possible to determine the ratio of L to M cones in vivo (Carroll et al. 2002) or in vitro (Hagstrom et al. 1998) using quantitative Rt-PCR molecular techniques.

Cone density in the human periphery is 2000–4000/mm$^2$ (Fig. 1.5A). M and L cones form the majority, with S cones only being 6–12% of total cones (Curcio et al. 1991). Cone density remains higher along the nasal horizontal meridian so that the same eccentricity in the nasal retina will have 2–3 times as many cones as the eccentricity in temporal retina. Individual S cones are typically separated from one another by intervening photoreceptors, forming a mosaic in the peripheral retina. Analysis of this mosaic indicates that it is randomly arranged (Curcio et al. 1991; Martin et al. 2000; Roorda et al. 2001). Although direct visualization has not been done for peripheral M and L cones (see above), their arrangement in more central retina suggests that M and L cones each form small clusters which also are randomly arranged (Roorda et al. 2001). Quantitative PCR finds that the majority of human retinas show a decrease in M cones with increasing eccentricity, with many human retinas having very few M cones in the far periphery (Hagstrom et al. 1998). Rods are the predominant photoreceptor outside the fovea (Fig. 1.5B), outnumbering cones 20:1 on average (Curcio and Hendrickson 1991). Rod density is relatively uniform across most of the retina, but it is highest in a ring at approximately 20° (Osterberg 1935; Curcio et al. 1990). The highest rod density is found superior to the optic disc in most retinas, where this rod "hot spot" contains 176,000 rods/mm$^2$ (Fig. 1.5B).

In the macula, cone density rises (Figs. 1.5C, 1.6A) and rod density drops (Figs. 1.5D, 1.6A), until, within the central 500–600 µm of the fovea, photoreceptor topography changes radically. At 500 µm, the ratio of cones to rods is 1:1, with an equal density of 40,000/mm$^2$ (Fig. 1.6A,E). The last significant number of rods is found at 300 µm (Figs. 1.5D, 1.6A), with a few as close to the foveolar center as 100 µm. S cones continue in significant numbers up to 100 µm, but are absent from the central foveola (Figs. 1.5E, 1.9C,D). This S cone-free region often is not centered on the spot of highest cone density, and its shape, size, and the number of S cones adjacent to the foveola varies between individuals (Curcio et al. 1991; Bumsted and Hendrickson 1999). Peak S cone density is 5100/mm$^2$ near the foveola, where S cones constitute only ~3% of all cones. Because S cone density drops more slowly than M and L cone density in the fovea, S cone percentage actually rises to 6–7% in the parafovea. The central 100 µm of the foveola (Figs. 1.5C, 1.6A–C) is centered over the deepest part of the foveal pit and contains only M and L cones. Direct visualization of normal adult foveal and perifoveal human cone populations using adaptive optics has shown a remarkable range of M-to-L ratios in the fovea. The average is 2L:1M, but individuals with normal color vision can range between 8M:1L and 8L:1M (Roorda and Williams 1999; Carroll et al. 2002). The central foveola contains the highest

Fig. 1.5A–F. Computer-generated color-coded maps of photoreceptor density (A–E) and ganglion cell density (F) in the human retina. These maps display the left eye with the nasal retina to the left. A, B, F The entire retina from fovea in the center to the ora serrata at the edge, with the black oval representing the optic disc. Lines of isoeccentricity are spaced at 5.94 mm. C–E only Foveal photoreceptor densities. Lines of isoeccentricity are spaced at 0.4 mm. Color-coded scales indicating cell density_1000/mm$^2$ are located *below* each map with the highest density in each scale to the *right*. (A–D from Curcio et al. 1990; E from Curcio et al. 1991; F from Curcio and Allen 1990)

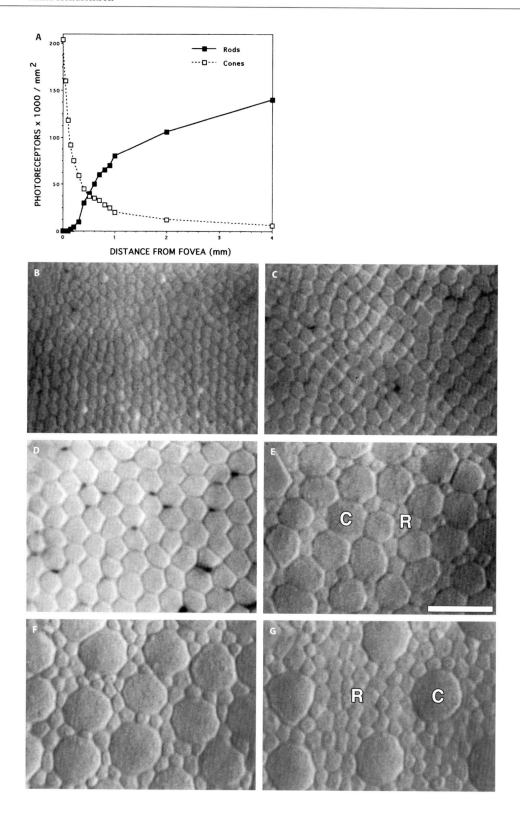

cone density in the retina (Figs. 1.5C, 1.6A–C) which ranges in normal young adults between 99,000–324,000/mm$^2$, with a mean of 199,000/mm$^2$ (Curcio et al. 1991) or 208,000/mm$^2$ (Curcio and Allen 1990). However, the actual number of cones forming this high density is a small fraction of the total, with only 7,000–10,000 in the centralmost 300 µm and 90,000 in the central 2 mm of human retina (Curcio et al. 1990; Wässle et al. 1990). Therefore our highest visual acuity depends on the functional integrity of less than 100,000 cones out of almost 5 million, emphasizing how critical optimal health of the fovea is for good vision throughout life.

Foveal cone diameter changes dramatically over the central 5 mm, which is well illustrated in Fig. 1.6B–G. Foveolar cones are the longest and thinnest in the retina, as well as being the most tightly packed (Figs. 1.2, 1.3B,G, 1.6B,C). Their outer segments lie in the foveola, but their cell bodies are displaced into more lateral fovea. Foveolar cones have a very long axon or fiber of Henle which extends to the synaptic pedicle, well into the fovea. This axon is formed during pit development to retain synaptic contact with inner retinal neurons as these are displaced more peripherally (Hendrickson 1992; Provis et al. 1998). The overall total length of foveal cones is 500–700 µm. Adult foveal cones measured in plastic sections have outer segments 1.2 µm wide and 40–50 µm long, while the slightly tapered inner segments are 3 µm at the outer limiting membrane, 2 µm at their tip, and 30–35 µm long (Yuodelis and Hendrickson 1986). Cone length decreases and cone diameter increases rapidly with increasing distance from the foveal center. For instance, 400 µm outside the foveola center, inner segments are 6 µm wide and 24 µm long, while outer-segment width is the same but length has decreased to 27 µm (Yuodelis and Hendrickson 1986). This change in size also is reflected in the drop in cone density from 200,000mm$^2$ at the foveola (Fig. 1.6B,C) to 100,000/mm$^2$ just 150 µm away (Fig. 1.6D) and 15,000/mm$^2$ at 1.5 mm (Fig. 1.6F).

These studies show that human foveal photoreceptor topography is a series of narrow concentric rings. Only M and L cones form the highest cone-packing density in the center. The next ring starts at 100 µm, where S cones are added and finally rods are added at 300 µm. Throughout these rings, M, L, and S cone density drops and rod density increases so that a rod-to-cone ratio of 1:1 is reached at 500 µm. During foveal development, a similar arrangement of concentric rings can be identified at fetal week 15–16, when all photoreceptor opsins are expressed (Bumsted and Hendrickson 1999; Xiao and Hendrickson 2000). Almost nothing is known about the molecular controls underlying the development of this complex topography.

## 1.1.3
### Inner Retinal Neurons Associated with Cones in Central Primate Retina

There are almost no inner retinal neurons in the foveola, because they have been moved peripherally during development (Fig. 1.2) to form the foveal pit. Given the high density of foveal cones, there must be a similar high density of bipolar and ganglion cells to transmit infor-

Fig. 1.6A–G. Changes in the rod-to-cone ratios across the central human retina. A This graph shows the rapid decrease in cone density (*open squares*) and rapid increase in rod density (*solid squares*) across the foveola and fovea. A 1:1 ratio is reached at 500 µm from the foveolar center. *Below the graph* are DIC images of human photoreceptors in a retinal wholemount. This type of image was used to calculate the color-coded computer graphs in Fig. 1.5, and the densities in the graph above. B, C The difference between normal young individuals in peak cone-packing in the center of the foveola. D Increase in cone diameter and drop in packing density just 125 µm from the foveolar center. Although a few rods are present at this eccentricity, none are shown. E From 660 µm eccentricity where the rod (R)-to-cone (C) ratio is slightly more than 1:1, and cone diameter has increased markedly. F At 1.4 mm or the outer edge of the parafovea where the rod-to-cone ratio is 6:1. Note that cones are now well separated by intervening rods. G From 5 mm, where the peak density of rods is found. (Modified from Curcio et al. 1990)

mation received by each individual cone. The combination of high inner-neuron density and pit formation creates the characteristic foveal slope which has the thickest ganglion cell layer in the retina and also contains the synaptic pedicles of the foveolar cones (Figs. 1.2, 1.3C,F). Each photoreceptor type has a specific subset of inner retinal neurons associated with it which forms the M or L cone, S cone, or rod circuits. Although a detailed description of these circuits is beyond the scope of this chapter (see Boycott and Dowling 1969; Schein 1988; Wässle and Boycott 1991; Dacey 1999), a brief discussion will help understand why the fovea supports our highest visual acuity.

Polyak (1941) has pointed out that there are classes of bipolar and ganglion cells around the fovea which have very tiny dendritic fields, and he infers that they receive synapses from a single cone or bipolar. These cells are termed the *midget system* (Fig. 1.7A), and form the basis of subsequent work that shows that individual foveal cones have a "private line" synaptic circuit (Kolb 1970). Foveal M and L cones have a large synaptic pedicle 7–9 µm in diameter, and each pedicle contains 21 basal membrane infoldings marked by a synaptic ribbon structure (Chun et al. 1996; Haverkamp et al. 2000). Each infolding is presynaptic to two main cell types, a central dendritic process from an invaginating midget bipolar cell and two lateral processes from horizontal cells (Fig. 1.7A, midget bipolar, MBp, on right). On the base of the pedicle, more conventional synapses are made onto the flat-

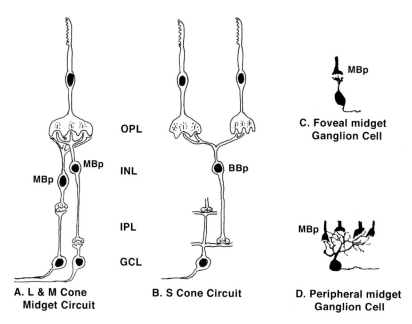

Fig. 1.7A–D. The L and M cone and S cone circuits found in primate central retina. A Individual L and M cones are synaptically connected in the outer plexiform layer (*OPL*) to an invaginating ON midget bipolar (*MBp*) on the *right* and a flat OFF MBp on the *left*. Each MBp only receives input from a single cone. In the inner plexiform layer (*IPL*), each MBp in turn provides all of the bipolar synaptic input to a midget OFF (*left*) and midget ON (*right*) ganglion cell. Note that this input is stratified, with the OFF bipolar/ganglion cell synapses in the outer portion, and the ON in the inner portion. B S cones have a blue ON bipolar (*BBp*) which provides invaginating dendrites to two to four S cones, so several S cones share a single blue bipolar. The BBp synapses on a small, bistratified ganglion cell in the inner part of the IPL. This same ganglion cell receives mixed L and M cone input to its dendrites in the outer IPL. C, D Midget human ganglion cells at the same scale near the fovea and in the peripheral retina. Note that the peripheral midget ganglion cell has a much larger dendritic tree and receives synaptic input from four midget bipolar cells

tened dendritic tips of a flat midget bipolar cell (Fig. 1.7A, MBp on left). Each midget BP type provides dendrites to a single cone and each cone contacts one of each midget type (Kolb 1970). The axon of the invaginating midget bipolar terminates deep in the inner plexiform layer, while the flat midget bipolar terminates in the outer part of the inner plexiform layer. The invaginating bipolar provides almost all of the bipolar synaptic input to the tiny dendritic tree of a single midget ganglion cell which lies in the inner part of the inner plexiform layer. The flat midget bipolar has the same relationship, but terminates on a tiny dendritic tree in the outer part of the inner plexiform layer. The pioneering work of Kolb and colleagues (reviewed in Kolb 1994) has shown that this stratification reflects processing of circuits stimulated when light goes on in a point in space (inner part of inner plexiform layer) and when light goes off (outer part of inner plexiform layer). Thus a more modern view of the foveal midget system is that it conveys both on and off information from a single cone through two different midget bipolar cells, each of which synapse onto two different midget ganglion cells (Schein 1988; Wässle and Boycott 1991; Dacey 1999). This 1:2:2 circuit provides a faithful capture of visual information in a spot the size of a foveal cone, which underlies the high visual acuity of the primate fovea. Because each cone driving this circuit contains only M or L opsin, current anatomical and electrophysiological evidence also indicates that the midget system transmits color vision for red and green (Dacey 1999). Acuity declines with eccentricity because cones are larger and spaced farther apart, and midget ganglion cell dendritic fields receive input from more than one midget bipolar which further increases the effective capture area (compare Fig. 1.7C with D).

S cones are absent from the central 100 µm but then rise to their highest density in the parafovea (Curcio et al. 1991). S cone pedicles are smaller than L or M cones (Ahnelt et al. 1990) but show the same basic synaptic structures. However, S cones have their own specialized circuit (Fig. 1.7B) which transmits blue color information (reviewed in Martin 1998; Dacey 1999; Calkins 2001). The invaginating or "on" "blue" bipolar is a specialized type that is postsynaptic only to S cones (Fig. 1.7B, BBp). Even near the fovea these blue bipolars contact more than a single S cone, which reduces visual acuity in the blue system. The blue bipolars synapse in the inner part of the inner plexiform layer with the inner dendrites of a specialized blue bistratified ganglion cell. Its dendrites in the outer part of the inner plexiform layer receive mixed M and L input from other types of cone bipolars.

The retinal ganglion cell is the output cell of the retina. Young human retinas contain a mean of 1.07 million ganglion cells with a range of 710,000–1.54 million (Curcio and Allen 1990). This range appears to be due to individual variation in absolute ganglion cell number and retinal area. The distribution of ganglion cells across human retina is shown in Fig. 1.5F, where it can be seen that it fairly closely matches cone distribution (Fig. 1.5A). The thickness of the ganglion cell layer helps define the zones in central retina (Polyak 1941; Hogan et al. 1971). The thickest ganglion cell layer begins on the peak of the foveal slope where it is up to eight cells deep, and this thick region continues into the parafovea (Fig. 1.3C). The peak density of 35,000/mm$^2$ occurs about 1 mm from the foveal center (Curcio and Allen 1990). In the perifovea (Fig. 1.3D) the ganglion cell layer drops from four thick at the central edge to one thick at the peripheral edge. At this point ganglion cell density has dropped to >3000/mm$^2$ at the nasal and >2000/mm$^2$ at the temporal perifoveal edge, reflecting the higher cone density in nasal retina. About 50% of all ganglion cells are found within 4.5 mm of the foveal center, which accounts for the large amount of central visual centers devoted to foveal processing.

There has been considerable quantitative work done to test the hypothesis that foveal acuity depends on a 1:2:2 midget system. If this is true, then there has to be a minimum of two ganglion cells for every foveal cone. Counts comparing cone and ganglion cell density in human retina (Curcio and Allen 1990) find that a 1:2 ratio is certainly supported, and a ratio of 1:3 is possible for the fovea. This would allow additional foveal cone output to other types of ganglion cells. Counts in *Macaca* (Schein 1988;

Wässle et al. 1990) and marmoset retina (Wilder et al. 1996) also support a 1:3 ratio out to the perifovea.

## 1.1.4
### Vascular and Glial Specializations of the Fovea

The primate fovea is characterized by the absence of blood vessels in its center, forming the foveal avascular zone (FAZ). The FAZ outlines the foveal slope (Figs. 1.1, 1.2, 1.4) and is 500–600 μm in diameter in humans (Mansour et al. 1993) and monkeys (Weinhaus et al. 1995; Provis et al. 2000). The blood vessels on the foveal slope are mainly capillaries, and they form a cone around the foveal pit (Snodderly et al. 1992; Provis et al. 2000, also see Fig. 1.4). Most of the primate central retina contains four main laminar plexuses (Iwasaki and Inomata 1986; Snodderly and Weinhaus 1990; Snodderly et al. 1992; Gariano et al. 1994; Provis 2001). Starting from the vitreal surface of the retina, the innermost plexus lies within the nerve fiber layer and the next within the ganglion cell layer. The two outer plexuses are found at the inner and outer borders of the inner nuclear layer. The outer nuclear layer is avascular across the retina. Near the foveal slope, there are no capillaries in the nerve fiber or ganglion cell layer; instead these are replaced by a single capillary plexus at the border of the ganglion cell and inner plexiform layers (Figs. 1.2, 1.4B). This plexus is the major visible component of the FAZ (Snodderly et al. 1992; Provis et al. 2000). The outer capillaries around the fovea are concentrated along the outer plexiform layer (Figs. 1.2, 1.4A). Snodderly et al. (1992) show that vascular coverage or "screening of light" of the photoreceptor layer by blood vessels is 0 in the FAZ, but then rises rapidly to 30% at 1 mm and 45% at 2 mm. One functional reason for the FAZ and foveal pit may be to provide an optically clear path to the foveolar cones.

Developmental studies in *Macaca* monkey show that the fovea is avascular throughout development (Engerman 1976; Gariano et al. 1994), and that the FAZ forms coincident with initial pit formation (Provis et al. 2000). Throughout foveal development the FAZ remains at or near adult diameter, suggesting that it remodels as the foveal pit matures and widens. Computer modeling of pit development (Springer and Hendrickson 2003) indicates that this neurovascular relationship may be crucial for pit formation (see below).

The adult primate peripheral retina has a large population of astrocytes which are mainly found in the nerve fiber and ganglion cell layers, where they often are seen adjacent to ganglion cell axons or blood vessels (Distler et al. 1993; Gariano et al. 1996a, 1996b). In both fetal and adult retina, astrocytes can be identified by immunocytochemical labeling for glial fibrillary acidic protein (GFAP; Distler et al. 1993; Gariano et al. 1996a, 1996b). Astrocytes originate from stem cells near and within the optic disc (reviewed in Chan-Ling 1994; Provis 2001), but also proliferate within the fetal retina itself (Sandercoe et al. 1999). Fetal astrocytes are closely related to immature blood vessels and, especially at the front of the vascular advance, label for the angiogenic factor vascular endothelial growth factor (Provis et al. 1997). Although astrocytes are present throughout vascular development around the fovea (Distler and Kirby 1996; Provis et al. 2000), they rapidly disappear after birth. This disappearance does not seem to involve elevated cell death in the astrocyte population, but may reflect a migration out of the fovea, a downregulation of GFAP, or the disappearance of astrocyte-promoting factors (Distler et al. 2000). Given the wide range of functions now ascribed to astrocytes, this raises a question as to whether their absence from the fovea may make this region more vulnerable to insults.

Microglia are marrow-derived glial cells which are present within all layers of the adult primate retina. Several types are present which may be associated with neurons or with blood vessels, and some of these are antigen-presenting cells (reviewed in Provis et al. 1996; Penfold et al. 2001; Provis 2001). Whether or not there is a unique distribution or population of microglia around the human fovea has not been determined.

Müller cells originate from the same progenitor stem cells which give rise to all of the retinal neurons (reviewed in Fischer and Reh 2003). Muller cells are found throughout the retina and they are the only cell type which spans all retinal layers. Their cell bodies form a single layer in the inner nuclear layer between the amacrine and bipolar cell bodies. Inner processes extend like pillars to form the end feet abutting the internal limiting membrane and thinner, multiple processes form the outer limiting membrane made up of complex junctions between Müller cells and photoreceptors (Polyak 1941; Yamada 1969; Krebs and Krebs 1991; Distler and Dreher 1996). Müller cells have been implicated in multiple roles including neurotransmitter recycling, glycogen metabolism, $K^+$-ion buffering and growth-factor production. Recent evidence indicates that a subset may be retinal stem cells which can give rise to neurons during retinal regeneration (Fischer and Reh 2003).

Müller cells are particularly important in the fovea because the foveola is formed only by cones and Müller cell processes (Yamada 1969). Quantitative EM finds that the number of foveal cones is matched 1:1 by Müller cell trunks (Burris et al. 2002), suggesting that the fovea could be especially vulnerable to Müller cell injury. Processes from several Müller cells form a glial basket around each cone pedicle which channels glutamate, the cone neurotransmitter, toward the synaptic contacts at its base and also effectively isolates each pedicle from glutamate released from neighboring cones. Müller cells in peripheral retina lose GFAP labeling by birth and only regain it under pathological or stressful conditions (Milam et al. 1998). In contrast, Müller cell processes within the fovea retain some GFAP immunogenicity throughout life (Gariano et al. 1996b; Distler et al. 2000; Provis et al. 2000), suggesting that their cellular environment is different.

## 1.1.5
### Pigment Epithelium Numerical Relationships with Foveal Photoreceptors

The pigment epithelium (PE) over the fovea does not differ obviously from that of the more peripheral retina. It is a single layer of cuboidal cells which contain melanin pigment granules. EM reconstructions (Borwein et al. 1980) show that cone outer segments do not reach the PE. Instead, PE microvilli lacking pigment granules surround the outer third of the outer segment while its inner portion is covered by microvilli from the cone inner segment and from surrounding Müller cells. This close relationship facilities the phagocytosis of cone outer segments by the PE, recycling of visual pigments, inactivation of free radicals produced during phototransduction, and bidirectional transport of metabolites (Bird 2003; Thompson and Gal 2003). There is some evidence that the complex interphotoreceptor matrix between photoreceptors and PE has a different molecular composition in the fovea and peripheral retina (Hollyfield et al. 2001).

Several studies have determined the numerical relationship between foveal photoreceptors and the overlying PE. It has been noted (Streeten 1969; Robb 1985) that central PE cells tended to get smaller and more closely packed during human retinal development, similar to the packing pattern found in foveal cones during development (Yuodelis and Hendrickson 1986). A striking change in the cone/PE relationship over time has been confirmed in a quantitative developmental study in macaque monkeys (Robinson and Hendrickson 1995) which counted both PE and the underlying photoreceptor density. These authors have found that in general PE cell density over the fovea rises, while peripheral PE density falls during development. For instance, foveal PE cell densities rise from around $2500/mm^2$ at midgestation to over $5500/mm^2$ in the adult. At midgestation, one perifoveal PE cell and one foveal PE cell each covers 5 cones. After birth the number of cones/PE cell changes rapidly, so that in the adult fovea each foveal PE cell covers 30–35 cones, a PE cell in the perifovea still covers about 5 cones, and one in the periphery covers 1–2 cones. A recent study (Snodderly et al. 2002) which counted PE cell density but not photoreceptors, estimates that each foveal PE cell covers 20 cones, while peripheral cells cover 1.5 cones. A study in humans which counted both PE and photoreceptors (Panda-Jones et al. 1996) has found a similar foveal 1:20 ratio, although it should be noted that their earlier photoreceptor counts for human fovea are significantly lower than Curcio et al. (1991) or Osterberg (1935). All of these studies indicate that each adult foveal PE cell has to support a much higher number of metabolically active cones than does an individual peripheral PE cell. However, if the cones + rods per PEcell are considered, perifoveal PE cells also have a total of 35 photoreceptors, mainly rods. This may produce differential stresses on the PE which vary with the regional differences in photoreceptor ratio.

## 1.2
## Anatomy of the Old and New World Monkey Fovea: What Are the Differences with Human Foveas?

Experimental approaches to find a cure for human blinding eye diseases such as adult macular dystrophy require a model for preliminary testing of new therapeutic approaches. The Old World macaque (*Macaca*) monkey has long been used as a model for human retina (Polyak 1941; Boycott and Dowling 1969; Wässle and Boycott 1991; Dacey 1999), while others have suggested that the New World marmoset is also appropriate (Troilo et al. 1993; Wilder et al. 1996; Jacobs 1998; Martin 1998). The fovea of a young adult *Macaca* monkey is shown in Fig. 1.8A and the fovea of a young adult marmoset in Fig. 1.8B.

A major difference between New and Old World monkeys is that most New World monkeys, including marmosets, only have one type of M or L cone combined with S cones, making them dichromats. Similar to humans, all known Old World monkeys have both M and L cones combined with S cones, giving them trichromatic vision (Jacobs 1998). However, this difference does not seem to have affected foveal anatomy in any obvious manner as shown in Fig. 1.8A,B. This marked similarity in foveal organization between monkeys strongly argues that the fovea had evolved for high visual acuity well before New and Old World monkeys were separated 30 million years ago. This also argues that the presence of one or two M/L genes does not affect foveal formation, because this separation occurred well before Old World monkeys developed trichromatic color vision 10–15 million years ago (Jacobs 1998; Kremers et al. 1999; Ahnelt and Kolb 2000). However, detailed analysis using cell-specific markers and cell counts show that there are subtle differences between these three primate groups. Peak cone density averages around 200,000/mm$^2$ in *Macaca* (Packer et al. 1989; Wikler and Rakic 1990; Robinson and Hendrickson 1995) and marmoset (Troilo et al. 1993; Wilder et al. 1996), as well as human (Curcio et al. 1990), but marmoset has a much higher peripheral cone density than either (Martin and Grünert 1999). A few rods are present within 50 µm of the foveal center in *Macaca* (Packer et al. 1989), which has not been seen in normal humans. No detailed foveal studies have been reported for the great apes.

The most striking difference in both peripheral and foveal retina between primate species is for S cone distribution. Humans, *Macaca* monkey and New World *Cebus* monkey have an S cone mosaic in which single S cones are consistently separated by other photoreceptors (Szél et al. 1988; Wikler and Rakic 1990; Curcio et al. 1991; Bumsted et al. 1997; Martin and Grunert 1999; Calkins 2001). Statistical measures of whether this mosaic is ordered or not have shown that the *Macaca* monkey S cone mosaic is somewhat ordered, while that in the human is random (Curcio et al. 1991; Martin et al. 2000; Roorda et al. 2001). On the other hand, the marmoset monkey has a clustered pattern of S cone distribution in which S cones can be immediate neighbors and the clusters have random order (Martin et al. 2000). Humans consistently lack a significant number of S cones within the central 100 µm of the foveola (Figs. 1.5E, 1.9C,D), although the size and shape of the S-cone-free region varies between individuals (Curcio et al. 1991; Bumsted and Hendrickson 1999). Preliminary studies in our laboratory find that chimpanzees and orangutans have a small S cone-free zone, similar to humans. In New World monkeys, the *Cebus* monkey also has an S-cone-free foveolar center, while the same study found that marmoset has S cones across the fovea (Martin and Grünert 1999). Several groups have studied *Macaca* monkey foveas with

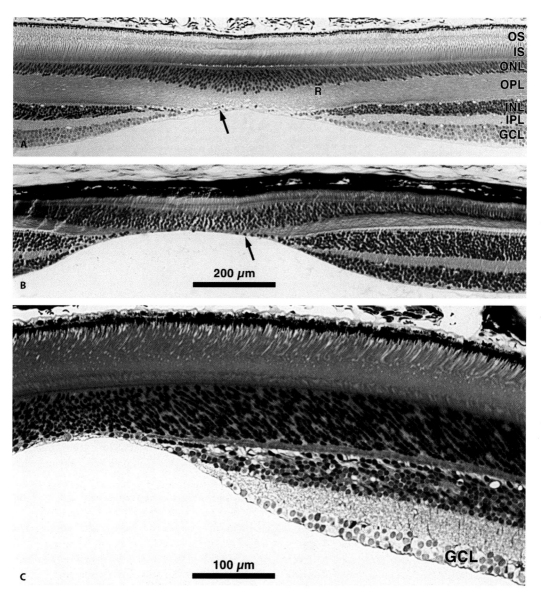

Fig. 1.8A–C. Comparison of primate foveal morphology in young adult primates. A is a semithin plastic section from a glutaraldehyde-fixed Old World *Macaca* monkey. B is a paraffin section from a Carnoy-fixed New World marmoset monkey. C is a semithin plastic section from a glutaraldehyde-fixed tarsier retina. The retinal layers are indicated on the *right* in A. The first rod cell bodies (*R*) in the *Macaca* fovea (A) are found at the same point where the foveal slope begins. *Arrows* in A and B indicate neurons in the floor of the foveola in both monkeys. Note that the tarsier fovea in C has relatively little packing of the outer nuclear layer over the deepest part of the pit and the OPL is thin. The tarsier GCL is no more than three deep on the slope, its thickest point

slightly different results. Using immunocytochemical labeling with different antibodies to S opsin, Szél et al. (1988) and Wikler and Rakic (1990) report that a few S cones are missing from the foveal center but do not give a dimension; while Martin and Grunert (1999) report a 50-μm-wide S-cone-free zone which is similar to the rod-free zone (Packer et al. 1989). Bum-

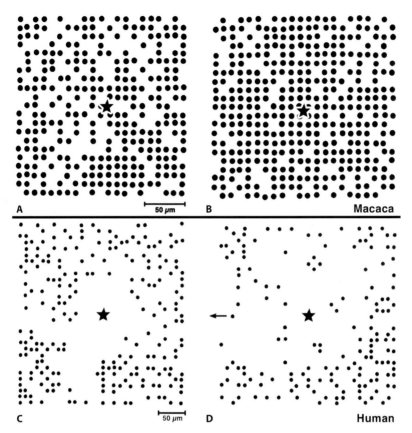

**Fig. 1.9A–D.** Comparison of S cone distribution in two different *Macaca* monkeys (A, B) and human (C, D). Adjacent counting fields 10 × 10 µm were sampled systematically across the fovea after the S cones were labeled with an antibody to S opsin. Each *dot* represents a field containing one to three S cones. The *star* indicates the point of highest cone density. Note that S cones are found across the foveal centers of both macaques. Both humans have a clear S cone-free area around the *star*, but the individual on the *right* has an S cone-free region which is much less regular than the one on the *left*. (Modified from Bumsted and Hendrickson 1999)

sted and Hendrickson (1999) have used both immunocytochemistry and in situ hybridization and find no S-cone-free zone in either fetal or adult *Macaca* monkeys compared with humans (Fig. 1.9A,B). All of these authors agree that the highest density of S cones is found adjacent to the fovea in both humans and macaques, where their percentage ranges between 5 and 15%. The highest S cone density in marmosets is in the foveola, where it reaches 10,000/mm$^2$ (Martin and Grünert 1999).

L and M cones have been identified in several monkey species by their selective wavelength absorption in vitro (Mollon and Bowmaker 1992; Packer et al. 1996) and more recently in vivo using adaptive optics (Roorda et al. 2001). These studies find that M and L cones are clustered in a random pattern, similar to humans. However, a marked difference in L-to-M ratio between macaque and human was noted in these studies, and also in studies using a rtPCR molecular approach (Deeb et al. 2000; McMahon et al. 2001). *Macaca* and talapoin Old World monkeys have a 1M:1L ratio near the fovea, and molecular studies of a larger sample of *Macaca* and baboon retinas show this is slightly M biased. A 1:1 ratio is in marked contrast to the typical 2:1 L-biased human central ratio determined by the same methods. The other difference is that this 1:1 ratio remains stable across the *Macaca* retina (McMahon et al. 2001), while the human periphery contains mainly L cones (Hagstrom et al. 1998).

The midget bipolar/ganglion cell system has been extensively analyzed in marmoset and *Macaca* retina using anatomical and electrophysiological methods (reviewed in Dacey 1999; Martin 1998). Although details are less well documented for human (Dacey 1999), current concepts and the original Golgi analysis of Polyak (1941) suggest that the midget system synaptic connections are very similar near the fovea in all three primates.

Detailed histological studies following labeling or section of one optic tract find that there is a slight vertical overlap of visual fields at the *Macaca* fovea (Stone et al. 1973; Fukuda et al. 1989). Some of the central nasal ganglion cells do not cross and some in temporal retina do cross, while the majority take the opposite course. This creates overlap of the visual fields within the central 1°. In recent psychophysical experiments on humans with visual field defects, a similar overlap was found (Reinhard and Trauzettel-Klosinski 2003). This probably is the basis of "macular sparing" described in human visual field studies.

## 1.3
## What Are the Anatomical Requirements to Create a Fovea?

The following anatomical and neuronal aspects of foveal organization seem to be distinctive for the primate retina compared with the "area centralis" of many mammals (Ahnelt and Kolb 2000). Undoubtedly there are molecular markers which are equally characteristic, and must underlie the differences between foveal and peripheral retina, but no molecules unique to the creation of a fovea have yet been identified. However, the fact that a fovea with a thick ganglion cell layer and a photoreceptor layer free of rods can be identified in humans at fetal week 11 and in monkeys at fetal day 50 (Hendrickson 1992) strongly suggests that these molecules must act at a very early stage of development to establish the characteristic cone-dominated circuits within the fovea.

It also is significant that foveal dimensions, peak cone density, and relative position in the temporal retina remain constant, despite a wide range of eye and retinal size (Packer at al. 1989; Curcio and Hendrickson 1991; Martin and Grünert 1999; Franco et al. 2000). This strongly suggests that the forces that create a fovea are localized to a small region. A constant dimension also suggests that optimal visual acuity is achieved using a relatively fixed number of neurons. Given the large amount of visual thalamus and cortex that is devoted to existing primate foveas, having larger foveas may not be possible without enlarging the brain beyond its cranial capacity.

### 1.3.1
### Midget Ganglion Cells

This ganglion cell type is generally considered to be found only in primate retinas (Fig. 1.7A,C,D) and, in its most characteristic form, only around the fovea where it makes up 80% of the ganglion cell layer (Boycott and Dowling 1969; Wässle and Boycott 1991; Dacey 1999; Kremers et al. 1999). In central retina the midget ganglion cell dendritic tree is slightly larger than a single bipolar axon terminal and it receives synapses from a single midget bipolar (Fig. 1.7C). One midget bipolar in turn receives input from a single M or L cone. This arrangement guarantees that a midget ganglion cell receives input from a single cone and also that it can convey the M or L wavelength-selectivity of that cone. In more peripheral retina (Fig. 1.7D), the dendritic field is larger and it receives input from several bipolar axons, but the 1:1 relationship between midget bipolar cell and M or L cone changes very little until the far periphery (Kolb 1970). Because of the random nature of M and L cone distribution (Roorda et al. 2001), a peripheral midget ganglion cell will receive input from several cones which are both larger and more widely spaced, and this input will also be a variable mixture of M and L. This degradation of the midget system with increasing eccentricity explains both lower visual acuity and poorer red/green color vision in the periphery (Dacey 1999, but see Martin et al. 2001).

In the afoveate owl monkey retina, there are ganglion cells which fit the description of the midget type, but even in the most central portion they have a large dendritic field and are more like peripheral midget ganglion cells found in macaque or *Cebus* retina (Silveira et al. 1994; Kremers et al. 1999). This suggests that both a true midget ganglion cell as well as a high central cone density is missing from the afoveate owl monkey retina. Interestingly, another nocturnal primate, the tarsier, has a fovea with a relatively high cone density, but a low ganglion cell density (Fig. 1.8C). This suggests that the appearance of the midget ganglion cell may be an essential step in evolution of the primate fovea.

### 1.3.2
### High Cone Density and Types of Foveal Cones

Most mammalian retinas show an uneven distribution of cones, with some specialized region, the "area centralis" containing a peak of cone density (reviewed in Ahnelt and Kolb 2000). Human (Curcio et al. 1990), macaque monkey (Packer et al. 1989; Wikler and Rakic 1990), and marmoset monkey (Martin and Grünert 1999) retinas have a foveal cone peak density around 200,000/mm$^2$, which is 2 orders of magnitude higher than peripheral cone density. However, the actual area that has the highest density contains no more than 10,000 cones out of a total of 4.6 million in a human retina (Wässle et al. 1990; Curcio et al. 1990). This small number emphasizes the critical role played by foveolar cones and that the loss of even a small number can significantly affect central vision.

The nocturnal tarsier (*Tarsius spectrum*) has a fovea with a narrow pit (Fig. 1.8C) and a peak cone density between 50,000 and 85,000/mm$^2$ (Hendrickson et al. 2000; Hendrickson, unpublished work), the lowest yet found in primate foveas. The nocturnal owl monkey *Aotes* lacks a fovea and has a peak cone density of 7000/mm$^2$ (Wikler and Rakic 1990). It also should be noted that peripheral cone density in all primates ranges between 2000 and 4000/mm$^2$, which is little different from many other mammals, including mice (Ahnelt and Kolb 2000). It therefore appears that a major driving factor toward a fovea is a drastic increase in cone density confined to a small central region.

Most of the high foveal cone density is formed by M or L cones, although peak S cone densities are close to the foveal center as well in macaque and human and at its center in marmoset (Curcio et al. 1991; Martin and Grünert 1999; Calkins 2001). Human foveolas lack S cones, as does the tarsier (Hendrickson et al. 2000), but monkey foveolas in general have some to many S cones (Martin and Grunert 1999; Bumsted and Hendrickson 1999). Thus the presence or absence of S cones does not seem to affect foveal formation, although it should be pointed out that the afoveate owl monkey totally lacks S cones in its retina (Jacobs 1998). Likewise, both trichromatic (Fig. 1.8A) and dichromatic (Fig. 1.8B) monkeys have virtually identical foveas, so the number of opsins seem irrelevant (Jacobs 1998; Kremers et al. 1999). However, across primate species, the greatest pressure to develop the characteristic high foveal cone density appears to be within the M and L cone population(s).

### 1.3.3
### Absence of Rods

In all primate retinas in which rod topography has been described, rods are absent from the center of the fovea (Fig. 1.5D), although the rod-free zone is slightly smaller in monkeys than humans (Packer et al. 1989; Curcio et al. 1990). A rod-free zone in both monkey and human retinas is present from the earliest stages of development in which rods can be identified by molecular markers (Swain et al. 2001; Bumsted-O'Brien et al. 2003). In contrast, in human albinos, rods are found throughout the fovea, cones are large, loosely packed and immature in shape, and the foveal pit is poorly formed or absent (Fulton et al. 1978; Mietz et al. 1992). This suggests that the molecular changes causing albinism allow rods to form within the fovea which could interfere in some unknown way with foveal pit development. A note of caution should be inserted as to whether the presence of rods directly interferes with pit formation. The nocturnal tarsier has a very high density of rods in central retina (Hendrickson et al. 2000; Hendrickson, unpublished work). Our preliminary studies of the tarsier fovea show that it has a narrow but deep foveal pit and minimal evidence of cone-packing into a foveola (Fig. 1.8C). Although cone density may reach 85,000/mm$^2$ over the pit, rods are still present at significant density throughout the foveal cone mosaic.

### 1.3.4
### Striate Cortex Expansion

The overall expansion in the area of primary visual cortex in primates is due in large part to the amount of cortex devoted to the high density of ganglion cells around the fovea. A similar expansion is seen in the lateral geniculate nucleus, the thalamic relay to striate cortex (Wässle et al. 1990; Kremers et al. 1999). In human retinas the peak ganglion cell density of 35,000/mm$^2$ occurs about 1 mm from the foveal center and 50% of all ganglion cells are found in the central 4.5 mm (Curcio and Allen 1990). In macaque monkeys, 40% of V1 contains the representation of the central 5°, or the 1.5 mm of retina surrounding the foveal pit (Tootell et al. 1988). It is possible that other mammals started to develop foveas but failure of their striate cortex to expand sufficiently to process the input from this large number of neurons negatively affected subsequent foveal evolution.

### 1.3.5
### Vascular Specializations

The primate retina shows a unique vascular developmental pattern (Gariano et al. 1994, 1996a, 1996b; Provis et al. 2000; Provis 2001). As blood vessels growing across the inner retina from the optic disc approach the site of the future fovea, they diverge to surround it, leaving a FAZ (Figs. 1.1, 1.2, 1.4). It is important to emphasize that blood vessels invade the foveal region at a time when it contains all retinal layers and has not yet started to form a pit (Provis et al. 2000).

The diameter of the fetal FAZ is similar to the adult FAZ, suggesting that it remodels in a dynamic fashion as the foveal pit develops. Because the adult FAZ outlines the foveal slope, this also strongly suggests that some molecule(s) within the fetal fovea mark out the future foveal zone and repel astrocytes, blood vessels, and ganglion cell axons from this region (Provis et al. 2000; Provis 2001). Computer modeling based on actual histological sections of developing macaque fovea shows that, because the FAZ is more elastic than surrounding vascularized retina, it is more easily deformed by retinal growth. These models indicate that unequal deformation within the FAZ combined with eye growth-induced retinal stretch causes the foveal pit to form (Springer and Hendrickson 2004). If blood vessels fill the fovea, no pit is formed. In turn, tensile forces generated within the retinal layers around the developing pit help pack cones into the foveal center (Springer 1999). These models conclude that the FAZ is a major factor in foveal pit formation and may also be important for cone packing. All retinas that have a foveal pit have a FAZ (Wolin and Massopust 1967). Human albino retinas and hypoplastic foveal syndromes have a poorly formed pit and a low cone-packing density, and these patients often have aberrant blood vessels in the fovea (Spedick and Beauchamp 1986; Oliver et al. 1987; Barbosa-Carneiro et al. 2000). If these vessels were present during foveal development, this could account for both the poor pit formation and low cone density. However, because these foveas are in the proper position on the retina but contain inappropriate cells such as rods, some earlier molecular mechanisms must be disturbed while others are maintained.

Thus it is likely that no single gene, developmental event, or set of neurons gives rise to a fovea. A number of genes must work together to first fix the foveal position on the retina and then generate a large number of cones with their specialized inner retinal neurons in this position. During this phase it is critical that a large number of midget bipolar and ganglion cells are generated to create the 1:2:2 high-acuity circuit. Other genes probably cause central visual centers to generate additional neurons to handle foveal processing. Next it is probable that unknown molecules must remain to exclude astrocytes and blood vessels, and probably ganglion cell axons, to form the FAZ and begin pit formation. Finally a combination of molecular changes within the foveal neurons and pit formation causes the central cones to remodel so that they become very thin and elongated, which facilitates packing into the foveola. All of these steps are necessary, but almost none of them are understood in any molecular or functional detail. However, a better understanding of foveal dynamics might aid in unraveling changes in aging which lead to foveal degeneration.

## References

Ahnelt P, Kolb H (2000) The mammalian photoreceptor mosaic-adaptive design. Prog Retina Eye Res 19: 711–777

Ahnelt P, Keri C, Kolb H (1990) Identification of pedicles of putative blue-sensitive cones in the human retina. J Comp Neurol 293: 39–53

Barbosa-Carneiro L, Peret M, Bichara A, Coscarelli G, Peret P (2000) Isolated foveal hypoplasia. Invest Ophthalmol Vis Sci 41: S571

Bird AC (2003) The Bowman Lecture. Towards an understanding of age-related macular disease. Eye 17: 457–466

Borwein B, Borwein D, Medeiros J, McGowan JW (1980) The ultrastructure of monkey foveal photoreceptors, with special reference to the structure, shape, size and spacing of the foveal cones. Am J Anat 159: 125–146

Boycott BB, Dowling JE (1969) Organization of the primate retina: light microscopy. Philos Trans R Soc Lond B Biol Sci 255: 109–184

Bumsted K, Hendrickson A (1999) Distribution and development of short-wavelength cones differ between *Macaca* monkey and human fovea. J Comp Neurol 403: 502–516

Bumsted K, Jasoni CL, Szél A, Hendrickson A (1997) Spatial and temporal expression of cone opsins during monkey retinal development. J Comp Neurol 378: 117–134

Bumsted-O'Brien K, Schulte D, Hendrickson A (2003) Expression of photoreceptor-associated molecules during human fetal eye development. Mol Vis 9: 401–409

Burris C, Klug K, Ngo IT, Sterling P, Schein S (2002) How Müller glial cells in macaque fovea coat and isolate the synaptic terminals of cone photoreceptors. J Comp Neurol 453: 100–111

Cajal S Ramon Y (1892) La rétine des vertébrés. Cellule 9: 121–255

Calkins DJ (2001) Seeing with S cones. Prog Retin Eye Res 20: 255–287

Carroll J, Neitz J, Neitz M (2002) Estimates of L:M cone ratio from ERG flicker photometry and genetics. J Vis 2: 531–542

Chan-Ling T (1994) Glial, neuronal and vascular interactions in the mammalian retina. Prog Retin Eye Res 13: 357–389

Chun M-H, Grünert U, Martin PR, Wässle H (1996) The synaptic complex for cones in the fovea and in the periphery of the macaque monkey retina. Vis Res 36: 3383–3395

Curcio CA, Allen KA (1990) Topography of ganglion cells in human retina. J Comp Neurol 300: 5–25

Curcio CA, Hendrickson AE (1991) Organization and development of the primate photoreceptor mosaic. Prog Retin Res 10: 89–120

Curcio CA, Sloan KR, Kalina RE, Hendrickson AE (1990) Human photoreceptor topography. J Comp Neurol 292: 497–523

Curcio CA, Allen KA, Sloan KR, Lerea CL, Hurley JB, Klock IB, Milam AH (1991) Distribution and morphology of human cone photoreceptors stained with anti-blue opsin. J Comp Neurol 312: 610–624

Dacey DM (1999) Primate retina: cell types, circuits and color opponency. Prog Retin Eye Res 18: 737–763

Deeb SS, Diller LC, Williams DR, Dacey DM (2000) Interindividual and topographical variation of L:M cone ratios in monkey retinas. J Opt Soc Am A Opt Imag Sci Vis 17: 538–544

Delori FC, Goger DG, Hammond BR, Snodderly DM, Burns SA (2001) Macular pigment density measured by autofluorescence spectrometry: comparison with reflectometry and heterochromatic flicker photometry. J Opt Soc Am A Opt Imag Sci Vis 18: 1212–1230

Distler C, Dreher Z (1996) Glia cells of the monkey retina. II. Müller cells. Vis Res 36: 2381–2394

Distler C, Kirby MA (1996) Transience of astrocytes in the newborn macaque monkey retina. Eur J Neurosci 8: 847–851

Distler C, Weigel H, Hoffman K-P (1993) Glia cells of the monkey retina. I. Astrocytes. J Comp Neurol 333: 134–147

Distler C, Kopatz K, Telkes I (2000) Developmental changes in astrocyte density in the macaque perifoveal region. Eur J Neurosci 12: 1331–1341

Dowling JE (1965) Foveal receptors of the monkey retina: fine structure. Science 147: 57–59

Engerman RL (1976) Development of the macular circulation. Invest Ophthalmol 15: 835–840

Fischer AJ, Reh TA (2003) Potential of Muller glia to become neurogenic retinal progenitor cells. Glia 43: 70–76

Franco ECS, Finlay BL, Silveira LCL, Yamada ES, Crowley JC (2000) Conservation of absolute foveal area in New World monkeys. A constraint on eye size and conformation. Brain Behav Evol 56: 276–286.

Fukada Y, Sawai H, Watanabe M, Wakakuwa K, Morigiwa K (1989) Nasotemporal overlap of crossed and uncrossed retinal ganglion cell projections in the Japanese monkey (*Macaca fuscata*). J Neurosci 9: 2353–2373

Fulton AB, Albert DM, Craft JL (1978) Human albinism: light and electron microscopy. Arch Ophthalmol 96:305–310

Gariano RF, Iruela-Arispe ML, Hendrickson AE (1994) Vascular development in primate retina: comparison of laminar plexus formation in monkey and human. Invest Ophthalmol Vis Sci 35:3442–3455

Gariano RF, Iruela-Arispe ML, Sage EH, Hendrickson AE (1996a) Immunohistochemical characterization of developing and mature primate retinal blood vessels. Invest Ophthalmol Vis Sci 37:93–103

Gariano RF, Sage EH, Kaplan HJ, Hendrickson AE (1996b) Development of astrocytes and their relation to blood vessels in fetal monkey retina. Invest Ophthalmol Vis Sci 37:2367–2375

Goldberg ME (2000) The control of gaze. In: Kandel ER, Schwartz JH, Jessell TM (eds) Principles of neural science, 4th edn. McGraw-Hill, New York, pp 782–800

Hagstrom SA, Neitz J, Neitz M (1998) Variations in cone populations for red-green color vision examined by analysis of mRNA. Neuroreport 9:1963–1967

Haverkamp S, Grünert U, Wässle H (2000) The cone pedicle, a complex synapse in the retina. Neuron 27:85–95

Helmholtz H (1924) Treatise on physiological optics, the sensation of vision (JPC Southall, translator). Optical Society of America

Hendrickson A (1992) A morphological comparison of foveal development in man and monkey. Eye 6:136–144

Hendrickson A, Djajadi HR, Nakaura L, Possin DE, Sajuthi D (2000) Nocturnal tarsier retina has both short and long/medium-wavelength cones in an unusual topography. J Comp Neurol 424:718–730

Hogan, MJ, Alvarado JA, Weddell JE (1971) Histology of the Human Eye. Saunders, Philadelphia

Hollyfield JG, Rayborn ME, Nihiyama K, Shadrach KG, Miyagi M, Crabb JW, Rodriquez IR (2001) Interphotoreceptor matrix in the fovea and peripheral retina of the primate *Macaca mulatta*; distribution and glyoforms of SPACR and SPACRCAN. Exp Eye Res 72:49–61

Iwasaki M, Inomata H (1986) Relation between superficial capillaries and foveal structures in the human retina. Invest Ophthalmol Vis Sci 27:1698–1705

Jacobs GH (1998) Photopigments and seeing – lessons from natural experiments. Invest Ophthalmol Vis Sci 39:2205–2216

Kolb H (1970) Organization of the outer plexiform layer of the primate retina: electron microscopy of Golgi-impregnated cells. Philos Trans R Soc London B Biol Sci 258:261–283

Kolb H (1994) The architecture of functional neural circuits in the vertebrate retina. The Proctor Lecture 1994. Invest Ophthalmol Vis Sci 35:2385–404.

Krebs W, Krebs IP (1989) Quantitative morphology of the central fovea in the primate retina. Am J Anat 184:225–236

Krebs W, Krebs I (1991) Primate retina and choroid: atlas of fine structure in man and monkey. Springer-Verlag, New York

Kremers J, Silveira LCL, Yamada ES, Lee BB (1999) The ecology and evolution of primate color vision. In: Gegenfurtner KR, Sharpe LT (eds) Color vision: from genes to perception. Cambridge University Press, Cambridge, UK, pp 123–142

Mansour AM, Schachat A, Bodiford G, Haymond R (1993) Foveal avascular zone in diabetes mellitus. Retina 13:125–128

Martin PR (1998) Colour processing in the primate retina: recent progress. J Physiol (Lond) 513:631–638

Martin PR, Grünert U (1999) Analysis of the short wavelength-sensitive "blue" cone mosaic in the primate retina: comparison of new world and old world monkeys. J Comp Neurol 406:1–14

Martin PR, Grünert U, Chan TL, Bumsted K (2000) Spatial order in short-wavelength-sensitive cone photoreceptors: a comparative study of the primate retina. J Opt Soc Am A Opt Imag Sci Vis 17:557–567

Martin PR, Lee BB, White AJ, Solomon SG, Ruttiger L (2001) Chromatic sensitivity of ganglion cells in the peripheral primate retina. Nature 410:933–936

McMahon C, Neitz J, Dacey DM, Neitz M (2001) Reversed L:M cone ratio in baboon indicates that pigment gene order does not determine relative cone number. Invest Ophthalmol Vis Sci 44:72S

Mietz H, Green WR, Wolff SM, Abundo GP (1992) Foveal hypoplasia in complete oculocutaneous albinism. Retina 12:254–260

Milam AH, Li ZY, Fariss RN (1998) Histopathology of the human retina in retinitis pigmentosa. Prog Retin Eye Res 17:175–205

Missotten L (1965) The ultrastructure of the human retina. Arsia, Brussels, Belgium

Missotten L (1974) Estimation of the ratio of the cones and neurons in the fovea of the human retina. Invest Ophthalmol 13:1045–1049

Mollon JD, Bowmaker JK (1992) The spatial arrangement of cones in the primate fovea. Nature 360:677–679

Nathans J, Thomas D, Hogness DS (1986) Molecular genetics of human color vision: the genes encoding blue, green and red pigments. Science 232:193–202

Oliver MD, Dotan SA, Chemke J, Abraham FA (1987) Isolated foveal hypoplasia. Br J Ophthalmol 71:926–930

Osterberg G (1935) Topography of the layer of rods and cones in the human retina. Acta Ophthalmol [Suppl VI]:1–102

Packer O, Hendrickson AE, Curcio CA (1989) Photoreceptor topography of the retina in the adult pigtail macaque (*Macaca nemestrina*). J Comp Neurol 288:165–183

Packer OS, Williams DR, Bensinger DG (1996) Photopigment transmittance imaging of the primate photoreceptor mosaic. J Neurosci 16:2251–2260

Panda-Jones S, Jonas JB, Jakobczyk-Zmija M (1996) Retinal pigment epithelial cell count, distribution and correlations in normal human eyes. Am J Ophthalmol 121:181–189

Penfold PL, Madigan MC, Gillies MC, Provis JM (2001) Immunological and aetiological aspects of macular degeneration. Prog Retin Eye Res 20:385–414

Polyak SL (1941) The retina. University of Chicago Press, Chicago, IL

Provis JM (2001) Development of the primate retinal vasculature. Prog Retin Eye Res 20:799–821

Provis JM, Diaz CM, Penfold PL (1996) Microglia in human retina: a heterogeneous population with distinct ontogenies. Perspect Dev Neurobiol 3:213–221

Provis JM, Leech J, Diaz CM, Penfold PL, Stone J, Keshet E (1997) Development of the human retinal vasculature: cellular relations and VEGF expression. Exp Eye Res 65:555–568

Provis JM, Diaz CM, Dreher B (1998) Ontogeny of the primate fovea: a central issue in retinal development. Prog Neurobiol 54:549–580

Provis JM, Sandercoe T, Hendrickson AE (2000) Astrocytes and blood vessels define the foveal rim during primate retinal development. Invest Ophthalmol Vis Sci 41:2827–2836

Reinhard J, Trauzettel-Klosinski S (2003) Nasotemporal overlap of retinal ganglion cells in humans: a functional study. Invest Ophthalmol Vis Sci 44:1568–1572

Robb RM (1985) Regional changes in retinal pigment epithelial cell density during ocular development. Invest Ophthalmol Vis Sci 26:614–620

Robinson SR, Hendrickson A (1995) Shifting relationships between photoreceptors and pigment epithelial cells in monkey retina: implications for the development of retinal topography. Vis Neurosci 12:767–778

Röhrenbeck J, Wässle H, Boycott BB (1989) Horizontal cells in the monkey retina: immunocytochemical staining with antibodies against calcium binding proteins. Eur J Neurosci 1:407–420

Roorda A, Williams DR (1999) The arrangement of the three cone classes in the living human eye. Nature 397:520–522

Roorda A, Metha AB, Lennie P, Williams DR (2001) Packing arrangement of the three cone classes in primate retina. Vis Res 41:1291–1306

Sandercoe TM, Madigan MC, Billson FA, Penfold PL, Provis JM (1999) Astrocyte proliferation during development of the human retinal vasculature. Exp Eye Res 69:511–523

Schein SJ (1988) Anatomy of macaque fovea and spatial densities of neurons in foveal representation. J Comp Neurol 269:479–505

Sharpe LT, Stockman A, Jägle H, Nathans J (1999) Opsin genes, cone photopigments, color vision and color blindness. In: KR Gegenfurtner, Sharpe LT (eds) Color vision – from genes to perception. Cambridge University Press, Cambridge, UK, pp 3–51

Silveira LCL, Yamada ES, Perry VH, Picanco-Diniz CW (1994) M and P retinal ganglion cells of diurnal and nocturnal New World monkeys. Neuroreport 5:2077–2081

Silveira LCL, Lee BB, Yamada ES, Kremers J, Hunt DM (1998) Postreceptoral mechanisms of colour vision in new world primates. Vis Res 38:3329–3337

Snodderly DM, Weinhaus RS (1990) Retinal vasculature of the fovea of the squirrel monkey *Saimiri sciureus*: three-dimensional architecture, visual screening and relationships to the neuronal layers. J Comp Neurol 297:145–163

Snodderly DM, Weinhaus RS, Choi JC (1992) Neural-vascular relationships in central retina of macaque monkeys (*Macaca fascicularis*). J Neurosci 12:1169–1193

Snodderly DM, Sandstrom MM, Leung IY-F, Zucker CL, Neuringer M (2002) Retinal pigment epithelial cell distribution in central retina of rhesus monkeys. Invest Ophthalmol Vis Sci 43:2815–2818

Spedick MJ, Beauchamp GR (1986) Retinal vascular and optic nerve abnormalities in albinism. J Pediatr Ophthalmol Strabismus 23:58–63

Springer AD (1999) New role for the primate fovea: a retinal excavation determines photoreceptor deployment and shape. Vis Neurosci 16:629–636.

Springer A, Hendrickson A (2004) Development of the primate fovea. 1. Use of finite element analysis models to identify mechanical variables affecting pit formation. Vis Neurosci 21:53–62

Stone J, Leicester J, Sherman M (1973) The naso-temporal division of the monkey's retina. J Comp Neurol 150:333–348

Streeten BW (1969) Development of the human retinal pigment epithelium and the posterior segment. Arch Ophthalmol 81:383–394

Swain PK, Hicks D, Mears AJ, Apel IJ, Smith JE, John SK, Hendrickson AE, Milam AH, Swaroop A (2001) Multiple phosphorylated isoforms of NRL are specifically expressed in photoreceptors. J Biol Chem 276:36824–36830

Szél A, Diamanstein T, Röhlich P (1988) Identification of the blue sensitive cones in the mammalian retina by anti-visual pigment antibody. J Comp Neurol 273:593–602

Thompson DA, Gal A (2003) Vitamin A metabolism in the retinal pigment epithelium: genes, mutations and diseases. Prog Retin Eye Res 22:683–703

Tootell RBH, Switkes E, Silverman MS, Hamilton SL (1988) Functional anatomy of macaque striate cortex. II. Retinotopic organization. J Neurosci 8:1531–1568

Troilo D, Howland HC, Judge SJ (1993) Visual optics and retinal cone topography in the common marmoset (*Callithrix jacchus*). Vis Res 33:1301–1310

Wässle H, Boycott BB (1991) Functional architecture of the mammalian retina. Physiol Rev 71:447–480

Wässle H, Grünert U, Röhrenbeck J, Boycott BB (1990) Retinal ganglion cell density and cortical magnification factor in the primate. Vis Res 30:1897–1911

Weinhaus RS, Burke JM, Delori FC, Snodderly DM (1995) Comparison of fluorescein angiography with microvascular anatomy of macaque retinas. Exp Eye Res 61:1–16

Wikler KC, Rakic P (1990) Distribution of photoreceptor subtypes in the retina of diurnal and nocturnal primates. J Neurosci 10:3390-3401

Wilder HC, Grünert U, Lee BB, Martin PR (1996) Topography of ganglion cells and photoreceptors in the retina of a New World monkey: the marmoset *Callithrix jacchus*. Vis Neurosci 13:335-352

Wolin LR, Massopust LC Jr (1967) Characteristics of the ocular fundus of the primate. J Anat 101:693-699

Xiao M, Hendrickson A (2000) Spatial and temporal expression of short, long/medium or both opsins in human fetal cones. J Comp Neurol 425:545-559

Yamada E (1969) Some structural features of the fovea centralis in the human retina. Arch Ophthalmol 82:151-155

Yuodelis C, Hendrickson A (1986) A qualitative and quantitative analysis of the human fovea during development. Vis Res 26:847-855

## Chapter 2

# Immunology and Age-Related Macular Degeneration

Philip L. Penfold, James Wong,
Diana van Driel, Jan M. Provis,
Michele C. Madigan

### Contents

2.1  Introduction: Why the Macula?  25
2.2  The Immune Status of the Retina  26
2.2.1  The Blood-Retinal Barrier  26
2.2.2  Microglia  29
2.3  Immune Mechanisms in AMD  30
2.3.1  Blood-Retinal Barrier Breakdown  30
2.3.2  Pigmentary Disturbance  30
2.3.3  Drusen  31
2.3.4  Cell-Mediated Immunity and Inflammation  33
2.3.5  Humoral Immunity  35
2.4  Clinical Significance of Drusen  37
2.5  Atrophic ("Dry") Macular Degeneration  37
2.6  Neovascular ("Wet") Macular Degeneration  37
2.7  Involvement of the Retinal Vasculature in AMD  38
2.8  Leucocyte Common Antigen (CD45) Expression in AMD: A Measure of Inflammation  40
2.9  Conclusion  41
References  41

## 2.1
### Introduction: Why the Macula?

Age-related macular degeneration (AMD) specifically affects the macular region of the central retina, where both ganglion cells and cones are present at very high densities. At the approximate centre of the macula is the *fovea centralis* – a small depression in the retinal surface overlying the peak concentration of cone photoreceptors. It is the very high density of cones in the macular region, along with their synaptic relationships with a population of midget ganglion cells, which provide the anatomical bases for our capacity to resolve fine detail (see Chap. 1). Lesions resulting from AMD are often small, but have a high impact on vision because of the number of cells, particularly ganglion cells, affected. Lesions of the dimensions common in AMD but occurring in more peripheral parts of the retina have little, if any, impact on visual acuity or quality of life, because only a relatively small number of cells is affected.

Paradoxically, while the macula has the highest concentration of receptors and neuronal elements in the retina, it has a limited vascular supply. During development, blood vessels are inhibited from entering the incipient fovea, even though initially no anatomical specialization is apparent (Provis et al. 2000). The foveal depression forms within a specified avascular region that blood vessels and astrocytes do not enter at any stage of development. Reduction in vascular density is a specialization associated with high-acuity areas in other species, including the cat (Chan-Ling et al. 1990) and prosimian primates (Woollard 1927; Rohen and Castenholtz 1967; Wolin and Massopust 1970), which have an *area centralis* rather than the *fovea centralis* present in most simian primates including humans (Provis et al. 1998). Thus it appears that a reduction in vascularity of the high-acuity area (*area* or *fovea centralis*) is a specialization associated with improved visual acuity in a variety of species (Provis et al. 2000; Sandercoe et al. 2003). Such a specialization ensures that light reaching the foveal cone mosaic is not diffracted by vascular structures lying in the light path, ensuring optimal quality of the visual information being relayed to the brain. The high density of cones in the foveal cone mosaic is the fundamental substrate of high visual acuity. Amongst primates high cone density has evolved hand-in-hand with a *fovea centralis*, and human *foveae* have amongst the highest densities of cones (Packer et al. 1989; Curcio and Allen 1990). However, photoreceptors are oxygen-hungry, and their very high density within a region where oxygen supply-lines are reduced is an adaptation with evident physiological limitations. While such limitations may not be evident in young, healthy individuals, "…insurance and evolution must each cope with the fact that risk goes up with time" (Jones 1999). The price of high-acuity vision during the early years of life, therefore, may be the risk of photoreceptor loss later in life or vulnerability to degenerative disease.

Macular degeneration occurs in two forms, an atrophic lesion (dry), originally described by Nettleship (1884) and Haab (1885), and a neovascular lesion (wet) originally described by Pagenstecher and Genth (1875). A century later, Gass (1973) suggested that the atrophic and neovascular lesions may be manifestations of the same disease process, although the relationship between their etiologies remains to be established. The term "age-related macular degeneration", has been adopted to include both forms. The histological appearance of the normal macula, incipent forms and end-stage lesions are illustrated in Fig. 2.1.

It would appear, therefore, that in the adult macula there is a critical relationship between limited blood supply and high metabolic demand (Penfold et al. 2001a, 2001b). Even minor perturbations of circulation, for example in incipient vascular disease, may lead to metabolic stress in foveal neurons and/or glia. Such perturbations and the resultant stress may generate signals that induce the neovascular changes associated with *wet* AMD (Figs. 2.1E and 2.5). Chronic failure of choroidal vascular supply at the macula may result in atrophy or *dry* AMD (Figs. 2.1F and 2.7) usually considered to be the natural end-stage of the disease.

## 2.2
## The Immune Status of the Retina

### 2.2.1
### The Blood-Retinal Barrier

The blood-retinal barrier (BRB) is dependent upon the integrity of the retinal pigment epithelium (RPE) (Fig. 2.1), the retinal vasculature and associated *glia limitans*, a sheath of cellular processes which restrict direct access of blood vessels to the neuronal environment (Fig. 2.2A,B,D and E). The BRB exists at two principal sites: an inner barrier consisting of retinal vascular endothelial cells, joined by tight junctions (*zonulae occludens*), and an outer barrier consisting of the RPE in which adjacent cells are similarly joined by *zonulae occludens*. Present evidence indicates that the *glia limitans* is a compound structure in the human retina comprising contributions from at least five cell types: astrocytes, microglia and the terminals of nitric oxide synthase (NOS) and substance P (SP)-immunoreactive amacrine cells in the inner retina, and Müller cells in the outer retina. Constituent SP-immunoreactive neuronal terminals may be involved in autoregulation of the retinal blood supply (Greenwood et al. 2000) and also have the potential to influence retinal immunity.

The outer BRB, the RPE, and the retinal vascular endothelium utilise the same receptor-ligand pairing to control lymphocyte traffic into the retina. The outer BRB is a common site for inflammatory attack, often resulting in breakdown of barrier functions and choroidal neovascularization (Vinores et al. 1994; Devine et al. 1996). Intercellular adhesion molecule-1 (ICAM-1) is constitutively expressed on RPE and vascular endothelial cells surfaces and is an important component of cell-cell interactions during inflammatory responses, mediating leucocyte adhesion, extravasation and diapedesis. Immunohistochemistry of excised AMD neovascular membranes has demonstrated the presence of cell adhesion molecules, including ICAM-1 (Heidenkummer and Kampik 1995). Investigation of the role of ICAM-1 in the development of choroidal neovascularization suggests a role for leucocyte adhesion to vascular

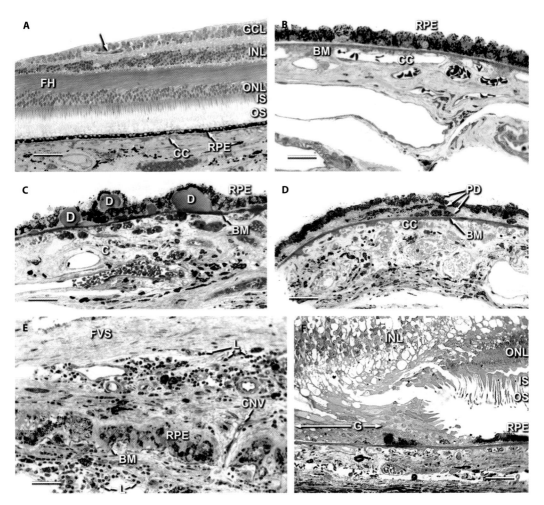

Fig. 2.1. A Photomicrograph of a toluidine blue-stained semithin section at the edge of the fovea in normal adult retina. *Arrow* indicates inner retinal vessel; the ganglion cell layer (*GCL*); the inner nuclear layer (*INL*) and the outer nuclear layer (*ONL*) are separated by a band of foveal cone axons or fibres of Henle (*FH*); inner and outer cone segments (*IS, OS*); retinal pigment epithelium (*RPE*) and choriocapillaris (*CC*) are shown. Bar 50 µm. B At higher magnification the RPE, Bruch's membrane (*BM*) and the CC are shown. Bar 20 µm. C A number of drusen (*D*) are deposited between the RPE and BM, associated with reduced patency of the choroid (*C*). Bar 20 µm. D Section through a region of pigmentary disturbance (*PD*), thickened BM and abnormal CC. Bar 30 µm. E A fibrovascular scar (*FVS*) is shown overlying a mass of leucocytes (*L*) and choroidally derived new vessels (*CNV*). The RPE and BM are disorganised. Bar 40 µm. F Section through an atrophic (*dry*) lesion illustrating normal retinal structures associated with residual RPE. The choroid and CC are significantly reduced and the overlying retina is atrophied, a band of gliotic processes (*G*) has been substituted for the RPE. Bar 30 µm.

endothelium in the development of laser-induced choroidal neovascularization (Sakurai et al. 2003a, 2003b). Increased serum levels of soluble ICAM-1 are associated with retinal vasculitis and compromise BRB integrity (Palmer et al. 1996). Recently, we have employed an epithelial cell line to further investigate the role of ICAM-1 in maintaining the integrity of the BRB. It was shown that triamcinolone acetonide (TA), an anti-inflammatory glucocorticoid, has

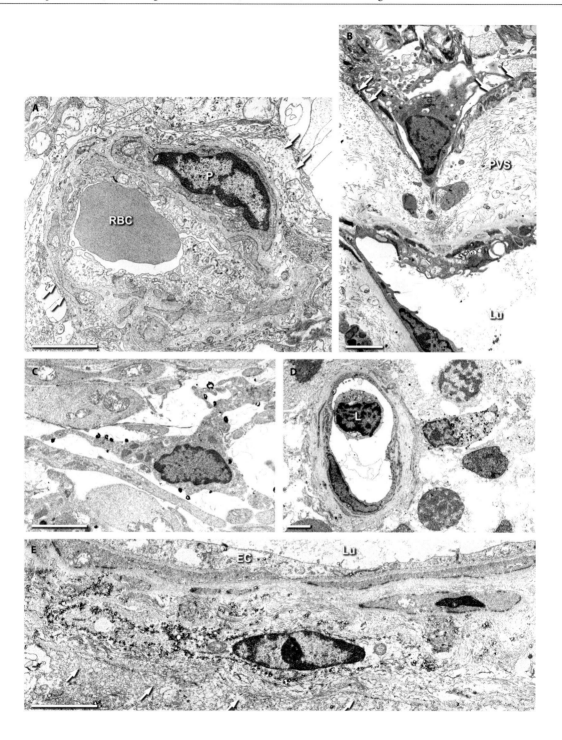

Fig. 2.2. A Normal adult retinal blood vessel: *arrows* indicate the *glia limitans*; a red blood cell (*RBC*) occupies the lumen; P indicates the nucleus of a pericyte within the perivascular space. *Bar* 5 µm. B A microglial cell traversing the *glia limitans* (*arrows*) of a retinal vessel; lumen (*Lu*), perivascular space (*PVS*). *Bar* 15 µm. C A retinal parenchymal microglial cell with MHC-class II immunogold labelling of the plasma membrane. *Bar* 10 µm. D A retinal paravascular microglial cell with MHC-class II immunogold labelling of the cytoplasm; intravascular lymphocyte (*L*). *Bar* 5 µm. E A retinal perivascular macrophage with MHC-class II immunogold labelling of the plasma membrane. *Arrows* indicate the *glia limitans*. Lumen (*Lu*), endothelial cell (*EC*). *Bar* 5 µm

the potential to influence cellular permeability, including the barrier function of the RPE in AMD-affected retinae (Penfold et al. 2000).

## 2.2.2
## Microglia

Retinal microglia are derived from haemopoietic cells which invade from the retinal margin and optic disc *via* the ciliary body and retinal vasculature (Diaz-Araya et al. 1995). In adult human retina, microglia are resident leucocytes associated with the vasculature, with the exception of the fovea (Fig. 2.3A). They comprise both dendritiform cells and macrophages and express leucocyte common antigen (CD45) (Fig. 2.3) (Penfold et al. 1991, 1993; Provis et al. 1995). They occur in three forms: parenchymal microglia (Fig. 2.2C), paravascular macrophages associated with vessels (Fig. 2.2D) and perivascular macrophages found within the perivascular space (Fig. 2.2E) (Penfold et al. 1991; Provis et al. 1995). In the CNS paravascular macrophages are referred to as juxtavascular microglia (Graeber 1993) and perivascular microglia are re-

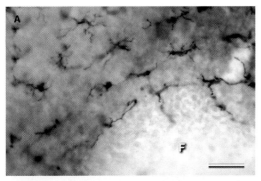

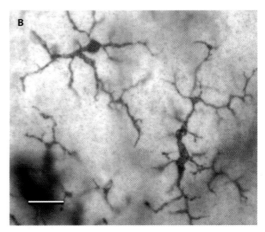

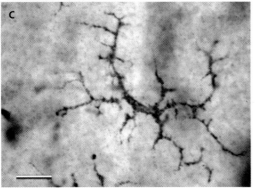

Fig. 2.3. A Retinal flatmount, normal adult, showing parallel chains of CD45-immunolabelled microglia surrounding the fovea (*F*). *Bar* 30 µm. B CD45-immunolabelled parenchymal microglia in retinal flatmount. *Bar* 15 µm. C A further example of a CD45-immunolabelled parenchymal microglial cell in retinal flatmount. *Bar* 15 µm

ferred to as perivascular cells (Graeber 1993; Provis et al. 1995) (Figs. 2.2 and 2.3).

The expression of Major Histocompatability Complex Class II (MHC-II) antigens by human retinal microglia was until recently controversial. Although the functional significance of MHC-II expression remains to be fully determined, both in situ and ex situ (Dick et al. 1997) analyses have shown that constitutive expression of MHC-II antigens is a consistent feature of retinal microglia (Fig. 2.2C–E) (Penfold et al. 1993, 1997, 2001a, 2001b; Provis et al. 1995; Dick et al. 1997; Zhang et al. 1997). In porcine retina the distribution of immunocompetent cells largely resembles that observed in the human retina (Yang et al. 2002). Observations should not be arbitrarily dismissed as artefact, particularly in the context of appropriately controlled studies, merely on the basis that they are derived from postmortem material, whether animal or human biopsies.

Expression of MHC antigens is conventionally associated with antigen presentation and promotion of immune responses, where MHC-II antigens are responsible for the presentation of exogenously derived peptides and MHC-I antigens for the presentation of endogenously derived peptides. In most tissues dendritic cells and macrophages have been found to co-exist, macrophages apparently acting to suppress the antigen-presenting functions of the resident dendritic cells (Holt et al. 1985; Pavli et al. 1993; Liew et al. 1994).

Constitutive expression of MHC-I and -II antigens by human retinal microglia, along with other phenotypic characteristics, illustrates their potential role as dendritic antigen-presenting cells (Figs. 2.2C–E; 2.3B, C) (Streilein et al. 1992; Penfold et al. 1993). Human retinal microglia in vitro also express the accessory molecule CD86 or B7-2 (Diaz et al. 1998); similarly cultured rat retinal microglia constitutively express B7-2, macrophage antigens and ICAM-1 (Matsubara et al. 1999). Expression of co-accessory molecules and the functional changes in antigen expression, after pro-inflammatory stimulation, support an antigen-presenting capability of freshly isolated microglia. However, pro-inflammatory stimulation induces an IL-10-mediated down-regulation of cell surface antigen expression and loss of migratory and phagocytic activity. Therefore, although equipped to act as antigen-presenting cells, microglia are able to modulate their own function and may have the potential to limit inflammation (Broderick et al. 2000). Recent work has demonstrated that microglia are kept in a quiescent state in the intact CNS by local interactions between the microglia receptor CD200 and its ligand, which is expressed on neurons (Neumann 2001). In the retina there is extensive expression of CD200 on neurons and retinal vascular endothelium (Broderick et al. 2002).

## 2.3
## Immune Mechanisms in AMD

### 2.3.1
### Blood-Retinal Barrier Breakdown

The BRB primarily functions to preserve the physiological environment of the neural retina; it limits but does not exclude cell-mediated and humoral immunity, consequentially influencing inflammatory responses. Subretinal neovascularisation disrupts the outer BRB, the most immediate physiological consequence of which is exudation into the subretinal space and subsequent loss of photoreceptor function. However, BRB compromise also exposes sequestered "self" antigens, which, in the presence of microglia, may generate anti-retinal auto-immunity. Consistent with this suggestion, both anti-retinal autoantibody formation and cell-mediated inflammation are associated with RPE disturbance and subretinal neovascularisation (Penfold et al. 1990).

### 2.3.2
### Pigmentary Disturbance

Clinically, retinal pigmentary abnormalities, including hyperpigmentation and hypopigmentation, are significant independent risk factors for the development of wet AMD (Bressler et al. 1990). Pathophysiologically, pigmentary disturbance is an indication that the RPE, the outer aspect of the BRB, is compromised.

Histopathological examination of pigmentary disturbance reveals (Figs. 2.1, 2.4 and 2.5) the presence of sub-RPE macrophages and increased numbers of choroidal leucocytes. Additionally, incipient choroidal neovascular events, including degradation of Bruch's membrane (Fig. 2.5B) and sub-RPE infiltrates (Fig. 2.4B) are apparent. Pigmentary disturbance is also associated with increased expression of the leucocyte common antigen (CD45) (see Sect. 2.8, Leucocyte Common Antigen (CD45) Expression in AMD ) and the formation of anti-astrocyte autoantibodies (Penfold et al. 1990). Taken together these observations indicate that, both clinically and histologically, pigmentary disturbance represents an early manifestation of wet AMD and loss of BRB integrity.

### 2.3.3
### Drusen

It has been suggested (Anderson et al. 1999) that locally derived vitronectin, complement and immunoglobulin components of drusen constitute a chronic inflammatory stimulus

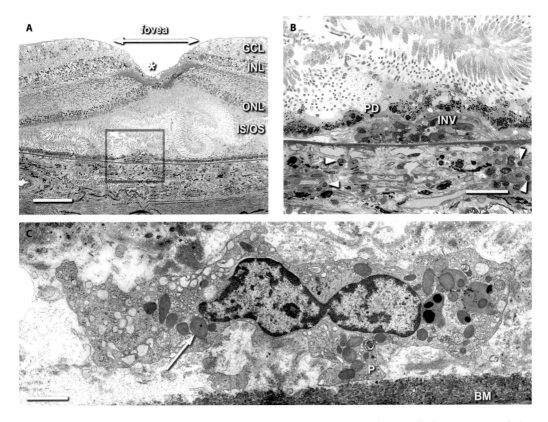

Fig. 2.4. A A light micrograph of a toluidine blue-stained semithin section of adult retina at the fovea. The ganglion cell layer (*GCL*), the inner nuclear layer (*INL*) and the outer nuclear layer (*ONL*) appear well preserved, the arrangement of the inner and outer cone segments (*IS, OS*) is disrupted; a region of gliosis (*asterisk*) is present at the centre of the foveal depression, overlying pigmentary disturbance and incipient new vessels, shown at higher magnification in B. Bar 100 µm. B Both hypo- and hyperpigmented RPE cells form a region of pigmentary disturbance (*PD*). Incipient new vessels (*INV*) are associated with increased numbers of stromal leucocytes (*arrowheads*) in the choroid. Bar 50 µm. C This electron micrograph taken in the region of incipient neovascularisation (A and B above) illustrates the classic morphology of a macrophage; the *arrow* indicates a secondary lysosome, a pseudopodium (*P*) is closely apposed to Bruch's membrane (*BM*). Bar 2.5 µm

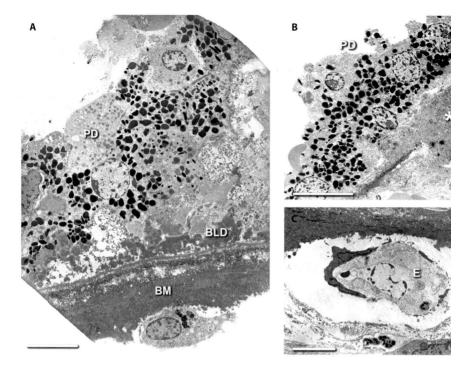

Fig. 2.5. A An electron micrograph illustrating basal laminar deposits (*BLD*) between Bruch's membrane (*BM*) and pigmentary disturbance (*PD*) of the retinal pigment epithelium. *Bar* 10 μm. B Loss of the elastic layer and disorganisation of BM (*asterisk*), associated with PD. *Bar* 10 μm. C Apoptosis of a choriocapillaris endothelial cell (*E*). *Bar* 5 μm

which leads to the binding of choroidally extravasated proteins into drusen. A number of studies have revealed that proteins associated with inflammation and immune-mediated processes are prevalent among drusen-associated constituents. Transcripts that encode a number of these molecules have been detected in retinal, RPE, and choroidal cells. The observation that dendritic cells are intimately associated with drusen development and that complement activation occurs both within drusen and along the RPE-choroid interface supports a direct role of cell- and immune-mediated processes in drusen biogenesis (Hageman et al. 2001).

Various plasma proteins have been identified as molecular components of drusen, including acute-phase reactant proteins. Concentrations of immunoglobulin G and terminal C5b-9 complement complexes are also present in drusen, RPE cells overlying or directly adjacent to drusen, as well as some within apparently normal epithelia. Taken together, these results implicate immune complex deposition as an initiating event in drusen formation. (Johnson et al. 2000). The finding of a role for local inflammation in drusen biogenesis suggests analogous processes to those occuring in other age-related diseases such as Alzheimer's disease and atherosclerosis, where accumulation of extracellular plaques and deposits elicit a local chronic inflammatory response that exacerbates the effects of primary pathogenic stimuli (Anderson et al. 2002).

Data supporting the hypothesis that oxidative injury contributes to the pathogenesis of AMD and that oxidative protein modifications and complement components may have a critical role in drusen formation have recently been published (Crabb et al. 2002); consistent with these data is the suggestion that oxidative stress and inflammation may be interactive pathways. Figure 2.6 illustrates a variety of histological features associated with drusen formation, including the frequent association with choroidal leucocytes.

## 2.3.4
## Cell-Mediated Immunity and Inflammation

### 2.3.4.1
### Choroidal Macrophages

Macrophages have the capability to influence each phase of the angiogenic process, by releasing growth factors and monokines (Sunderkotter et al. 1994). Excised neovascular membranes frequently include chronic inflammatory cells, macrophages, lymphocytes, and plasma cells (Figs. 2.8 and 2.9) (Lopez et al. 1991). Ultrastructural features of subretinal neovascularisation associated with AMD have confirmed the choroidal origin of new vessels and a relationship between macrophages and neovascular structures (Penfold et al. 1987). Wide-banded collagen, phagocytosed by macrophages, has been described associated with AMD lesions both above and below Bruch's membrane (Penfold et al. 1985), and it has been suggested that long-spacing collagen in AMD eyes may be selectively internalised by macrophages (van der Schaft et al. 1993). Recently it has been shown that macrophage depletion reduces size, cellularity and vascularity of choroidal new vessels, supporting the hypothesis that macrophages contribute to the severity of choroidal neovascular lesions (Espinosa-Heidmann et al. 2003). Moreover, mice deficient either in monocyte chemoattractant protein-1 or its cognate C-C chemokine receptor-2 develop cardinal features of AMD, implicating macrophage dysfunction in AMD pathogenesis (Ambati et al. 2003).

Generalized macrophage depletion has been shown to reduce the size and leakage of laser-induced choroidal new vessels (CNV) and is associated with decreased macrophage infiltration and VEGF protein. Angiogenesis is a complex process involving a variety of growth factors, most notably VEGF, and it is significant that macrophage depletion alone almost abolishes CNV (Sakurai et al. 2003a, 2003b). This suggests, at least in this model, that macrophages and cytokines derived from them are requisite in this process and adds to the growing body of evidence implicating leucocytes in the initiation of angiogenesis; it also emphasises the role of the macrophage as a critical component in initiating CNV (Figs. 2.8 and 2.9).

### 2.3.4.2
### Microglia

The sources of the signals which provoke choroidal neovascularisation remain to be defined, although both choroidally derived leucocytes and retinal microglia are increasingly found to be involved (Penfold et al. 2001a, 2001b). Immunological responses in neural retinal microglia are related to early pathogenic changes in RPE pigmentation and drusen formation. An increase in MHC-II immunoreactivity and morphological changes in microglia, are associated with incipient AMD (Penfold et al. 1997; Wong et al. 2001). Activated microglia may also be involved in rod cell death in AMD, retinitis pigmentosa and late-onset retinal degeneration. A recent study has proposed that microglia, activated by primary rod cell death, migrate to the outer nuclear layer, remove rod cell debris and may kill adjacent cone photoreceptors (Gupta et al. 2003). Figures 2.2 and 2.3 further illustrate the morphological and phenotypic characteristics of microglia; quantitative measurements of the role of microglia in the pathogenesis of AMD are presented in Sect. 2.8, below.

### 2.3.4.3
### Chronic Inflammatory Cells

Giant cells and epithelioid cells are typical features of granulomatous inflammation, and a number of studies suggest that they play a role in the pathogenesis of AMD. The involvement of giant cells in the atrophic/dry form of AMD has been described at the light and electron microscope level (Fig. 2.7D) (Penfold et al. 1986). Multinucleated giant cells appear to participate in the breakdown of Bruch's membrane and may provide an angiogenic stimulus for choroidal neovascularisation in AMD (Fig. 2.8C) (Killingsworth and Sarks 1982; Penfold et al. 2001a, 2001b). Multinucleate giant cells have al-

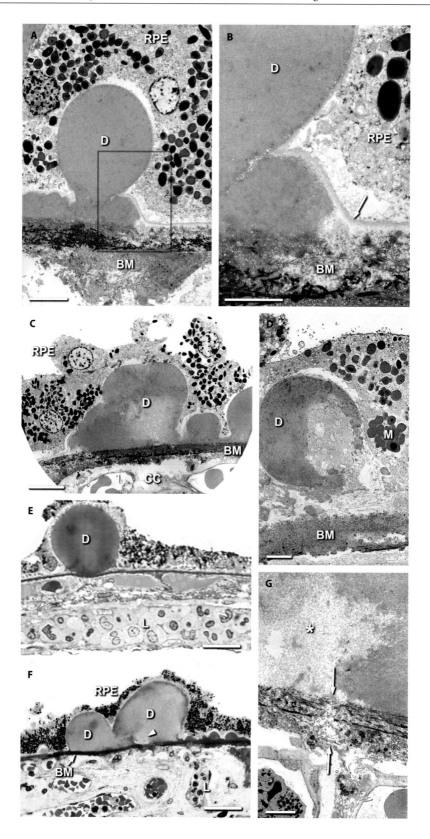

Fig. 2.6. **A** A druse (*D*) is deposited *in continuum* with Bruch's membrane (*BM*), shown at higher power in B. Note increased density of the elastic layer of BM. *Bar* 6 µm. **B** The *arrow* indicates the basement membrane of the retinal pigment epithelium (*RPE*), separating the RPE from the underlying druse (*D*) and BM. *Bar* 2 µm. **C** Drusen (*D*) associated with disruption of the RPE and densification of the elastic layer of BM are situated between chorio-capillaries (*CC*). *Bar* 10 µm. **D** Melanin (*N*) and druse (*D*) detached from BM. *Bar* 4 µm. **E** Toluidine blue semi-thin section illustrating the relationship between a druse (*D*) and an accumulation of choroidal intravascular leucocytes (*L*). *Bar* 15 µm. **F** Two large drusen (*D*) and disturbed RPE showing (*arrowhead*) regions of reduced density. Note accumulations of intravascular leucocytes (*L*) in the choroid. *Bar* 20 µm. **G** Electron micrograph of the region of reduced density (*asterisk*) shown in F. *Arrows* indicate an apparent channel passing through BM and CC. *Bar* 4 µm

so been demonstrated in surgically excised neovascular AMD membranes (Hutchinson et al. 1993). The constituent leucocytes of multinucleated giant cells may be derived from both macrophages and microglia (Dickson 1986).

## 2.3.5
## Humoral Immunity

### 2.3.5.1
### Autoantibodies

Anti-retinal autoantibodies have been reported in association with a number of ocular pathologies, including AMD (Tso 1989). The formation of anti-astrocyte autoantibodies is an early feature of the pathogenesis of AMD, particularly in patients with pigmentary disturbance (Penfold et al. 1990). Additionally, antibodies against retinal proteins of various molecular weights have been detected by Western immunonblot analysis in AMD patients (Chen et al. 1993). Similarly, antibodies immunoreactive with normal human retinal proteins have been detected by Western immunoblot analysis in sera of patients with AMD (Gurne et al. 1991). However, it is not clear whether antiretinal autoantibodies play a primary causative role in the aetiology of AMD or represent a secondary response to retinal damage.

### 2.3.5.2
### Serological factors

An early study suggested relationship between serum ceruloplasmin and the tissue alterations associated with macular degeneration (Newsome et al. 1986). Subsequently profiles of a variety of serum constituents, including immunoglobulins, and alpha and beta globulins, have been examined in AMD patients. The results indicate a higher incidence of serum abnormalities, particularly involving alpha-2 globulin, in patients with disturbance of pigmentation of the RPE (Penfold et al. 1990). Additionally an association between plasma fibrinogen levels and late AMD has been reported (Smith et al. 1996). Acute-phase proteins, including ceruloplasmin, fibrinogen and C-reactive protein, are a heterogeneous group of plasma proteins which alter their blood concentration in association with acute and chronic inflammation. An association with C-reactive protein, implicating inflammation in the pathogenesis of AMD, has also recently been reported (Seddon et al. 2004).

Interleukin-6 concentration in plasma can be a predictor of macular oedema in patients with diabetic retinopathy (Shimizu et al. 2002). Peptide and lipid mediators of the acute inflammatory response appear to enhance adherence of circulating neutrophils to the microvascular endothelium (Tonnesen et al. 1989). Elimination of high molecular weight proteins and lipoproteins from the blood of AMD patients has been shown, in a randomised trial, to improve visual function (Widder et al. 1999). The deposition of phospholipids also may induce an inflammatory reaction resulting in choroidal neovascularisation (Pauleikhoff 1992).

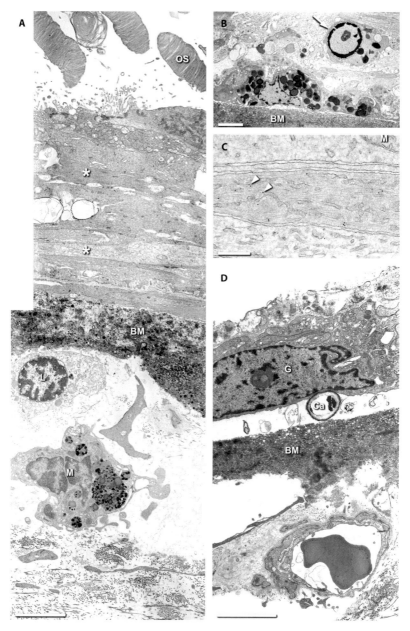

**Fig. 2.7. A** Electron micrograph of an atrophic lesion showing photoreceptor outer segments (*OS*) and Müller glial processes (*asterisk*) substituting for the RPE above Bruch's membrane (*BM*). A lymphocyte (*L*) and macrophage (*M*) are apparent below BM. *Bar* 4 μm. **B** Apoptotic retinal cell (*arrow*) in a region of RPE atrophy. *Bar* 4 μm. **C** High-power electron micrograph of Müller cell processes within a gliotic atrophic lesion; *arrowheads* indicate immunogold labelled glial fibrillary acidic protein positive filaments. Mitochondrion (*M*). *Bar* 300 nm. **D** A giant cell (*G*) within a region of atrophy associated with BM and calcium deposits (*Ca*). *Bar* 4 μm

## 2.4 Clinical Significance of Drusen

Two types of drusen are associated with the pathogenesis of AMD: hard drusen (Figs. 2.1C and 2.6) appear clinically as small, yellow, punctate deposits; soft drusen as paler, larger deposits. The presence of large drusen, bilateral drusen and numerous drusen are significant risk factors for developing late-stage AMD. Hard drusen are more closely associated with the dry form, while patients with soft drusen and pigment clumping have an increased risk of choroidal neovascularisation (Fine et al. 2000). A recent study has defined distinct age-dependent differences in the fluorescence pattern of drusen. This heterogeneity is suggested to be due to differential binding of extravasated fluorescence to the constituents and histological site of the drusen deposits (Chang et al. 2003).

Pulse-labelling or tracer experiments in animal models with macular drusen offer a unique opportunity to establish the origin of drusen constituents. The binding of extravasated indocyanine green dye to drusen material in correlation with early, middle and late phases of the clinical angiogram in a monkey model has been demonstrated histologically (Chang et al. 1998). These observations provide an insight into the dynamics of transport of lipid and protein material from the choroid vessels into drusen, and are consistent with a choroidal derivation of drusen constituents (Fig. 2.6). The conclusion that drusen constituents may be derived from the choroid has important implications for the pathogenesis of AMD and may suggest that drusen formation exacerbates RPE degeneration (Penfold et al. 2001a, 2001b).

## 2.5 Atrophic ("Dry") Macular Degeneration

Dry AMD occurs independently of the neovascular lesion and is associated with choroidal atrophy without the occurrence of breaks in Bruch's membrane and sub-retinal new vessels (Figs. 2.1F and 2.7). Clinical studies have demonstrated reduced blood flow in dry AMD and a decrease in choroidal volume (Grunwald et al. 1998). Reduced choriocapillaris patency and subsequent degeneration of the RPE is associated with involution of the adjacent photoreceptors and outer retinal layers (Fig. 2.1F) (Penfold and Provis 1986; Curcio et al. 2000). The dry form of AMD involves atrophy of the RPE-barrier in the absence of exudation. It appears that barrier function is preserved in regions of RPE atrophy so that the macula remains dry (Fig. 2.7A, C). Consistent with the preservation of BRB function, levels of CD45 expression are lower in dry lesions (see Sect. 2.8, below), indicating reduced involvement of cell-mediated inflammation; although chronic inflammatory/giant cells and occasional choroidal leucocytes are observed (Fig. 2.7A, D).

## 2.6 Neovascular ("Wet") Macular Degeneration

Neovascular (wet) AMD involves invasion of the retina by new blood vessels derived from the underlying choroidal vasculature; recurrent haemorrhage and proliferation of fibrovascular tissue ultimately lead to the formation of a "disciform" scar. Wet AMD occurs in two clinical forms, "classic" and "occult". Approximately 12% of cases present with classic neovascularisation; however, it is estimated that about 50% of all wet AMD cases will develop classic lesions as the disease progresses. Histopathological studies have shown that in the classical form new vessels and exudation directly *penetrate* Bruch's membrane (Figs. 2.8A, B and 2.9), the RPE and neural retina; while in the occult form neovascularisation occurs *between* the RPE and Bruch's membrane, where loss of integrity of the RPE barrier results in subretinal oedema (Penfold et al. 2001a, 2001b).

Classic choroidal neovascularisation (CNV) occurs as a discrete elevation of the RPE, commonly associated with subretinal exudation, blood and lipid deposition. Fluorescein angiography reveals a well-defined area of early vascular hyperfluorescence with progressive leakage at later phases. The majority, more than 85%, of recently diagnosed cases present, as occult CNV, with and without serous pigment epi-

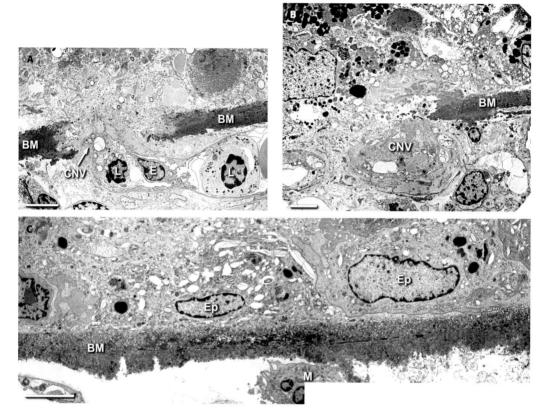

Fig. 2.8. A A choroidal new vessel (*CNV*), containing intravascular leucocytes (*L*), penetrating a break in Bruch's membrane (*BM*); endothelial cell nucleus (*E*). *Bar* 5 µm. B A further section from the same specimen shown in A, showing a larger break in BM. *Bar* 5 µm. C Epithelioid cells (*Ep*) – precursors of giant cell formation – adherent to the retinal aspect of BM; choroidal macrophage (*M*). *Bar* 5 µm

thelial detachment (PED). The clinical appearance of occult CNV reflects the vascular growth pattern, the effect of the overlying RPE (LaFaut et al. 2000) and exudation when present. It is estimated that between 40% and 60% of all wet AMD cases will develop predominantly classic lesions as the disease progresses (Freund et al. 1993; Yannuzzi et al. 2001).

Eyes with occult CNV secondary to AMD can be classified by the presence or absence of an associated serous PED. Patients with unilateral occult CNV have a significant risk of occult CNV developing in the second eye, and the type of occult disease in the first eye is highly predictive of the type of neovascularized disease in the second eye (Chang et al. 1995).

## 2.7
## Involvement of the Retinal Vasculature in AMD

The retinal vasculature, along with the choroid, is increasingly recognised to be involved in pathological neovascularisation associated with AMD. Morphological perturbations in neural vascular elements, including astrocytes, related to early are pathogenic changes in RPE pigmentation and drusen formation (Penfold et al. 1997; Wong et al. 2001).

Why do new vessels arise predominantly from the choroid rather than the retinal vasculature in AMD? It would appear that a number of anatomical and functional considerations

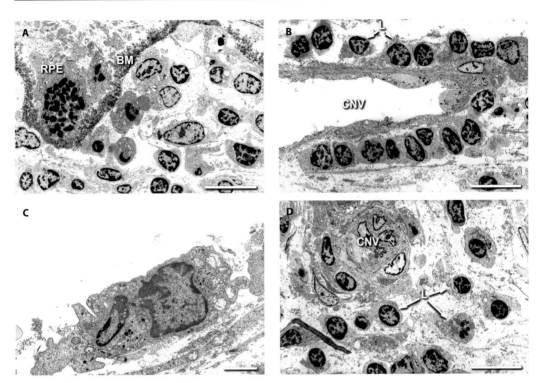

Fig. 2.9. A This electron micrograph, from the same specimen shown in Fig. 2.1E, shows accumulations of leucocytes in the choroid associated with degraded Bruch's membrane (*BM*) and retinal pigment epithelium. *Bar* 10 μm. B An accumulation of extravascular lymphocytes(*L*) is juxtaposed to a choroidal new vessel (*CNV*). *Bar* 10 μm. C A classic macrophage within a fibrovascular scar (same specimen as Fig. 2.1E). *Bar* 2 μm. D Leucocytes (*L*) surrounding a CNV (same specimen as Fig. 2.1E). *Bar* 10 μm

are involved. Developmentally, the outer retina, particularly at the fovea, is primarily dependent on the choroid for supply of nutrients; the demand arising in the outer retinal elements and cones. Physiological stress at the macula may generate neovascular signals, leading to new vessels from the most immediate source of vascular supply, the choroid. In some cases, these new vessels erode the RPE, infiltrate the neural retina, and communicate with the retinal circulation in what has been referred to as a retinal-choroidal anastomosis. However, the reverse also occurs in some cases, when new vessels originating from the retina extend posteriorly into the subretinal space, eventually communicating with choroidal new vessels. This form of neovascularisation can be confused with choroidally derived vessels and appears to be a distinct subgroup of wet AMD (Slakter et al. 2000; Yannuzzi et al. 2001). In a foveal photocoagulation study, a pattern was described as "loculated fluid", consisting of a well-demarcated area of late hyperfluorescence that appeared to represent pooling of fluorescein in a compartmentalized space, within the retina, anterior to the choroidal neovascular leakage. This pattern was unrelated to the extent of choroidal neovascularization and serous detachment or tear of the RPE. One third of baseline angiograms show this unusual pattern of hyperfluorescence (Bressler et al. 1991).

## 2.8
### Leucocyte Common Antigen (CD45) Expression in AMD: A Measure of Inflammation

CD45 is one of the most abundant leucocyte cell surface glycoproteins, established to be a critical component of the signal transduction machinery of lymphocytes; it is expressed *exclusively* by cells of the hematopoietic system. Evidence from genetic experiments indicates that CD45 plays a pivotal role in antigen-stimulated proliferation of T lymphocytes and in thymic development (Trowbridge and Thomas 1994). We have previously reported that microglia can be specifically labelled using monoclonal antisera against the CD45 antigen, establishing their leucocyte lineage (Fig. 2.3A–C) (Penfold et al. 1991). The exclusivity and specificity of CD45 to leucocyte lineage cells, including microglia, together with image-analysis technology, provides a practical and theoretical basis for the measurement of the inflammatory cell content of retinal tissue.

This approach has been used to quantify expression of CD45 on retinal microglia and choreoidal leucocytes in retinas associated with AMD compared with age-matched normal and young adult retinas. Adult eyes ($n=45$) were classified histopathologically into normal and AMD-associated groups. Indirect fluorescence immunohistochemical examination of retinal flatmounts and full-thickness frozen sections was used to estimate immunoreactivity of CD45 antigen. The intensity and distribution of labelling was assessed by scanning laser confocal microscopy and quantified by digital image analysis and masked manual counts.

Image analysis results were calibrated against manual counts to produce a correlation ratio. Increased CD45 microglial immunoreactivity was observed in age-matched retina compared with that seen in normal young retina (Fig. 2.10A). An increase in microglial CD45 was also observed in retinal flatmounts with the exudative form of AMD compared with the age-matched group (Fig. 2.10A). A significant increase ($P<0.05$) in counts of CD45-labeled choroidal leucocytes was also observed in frozen sections of exudative AMD specimens, especially the incipient form (Fig. 2.10B). Hypertrophy of retinal microglia and other morphological signs of activation were also observed in AMD retina compared with young and age-matched specimens. CD45 expression is significantly modulated in AMD-affected tissue. The methodology of fluorescence confocal microscopy and quantitative image analysis consistently correlated with manual counts (Wong et al. 2001).

The significant increase in CD45-labeled leucocytes found in both pigmentary disturbance and disciform AMD specimens is consistent with the earlier suggestion that pigmentary disturbance often occurs as a prelude to subretinal neovascularisation (Bressler et al. 1990). It further correlates with a body of evidence indicating that the pathogenesis of wet AMD in particular involves immunity.

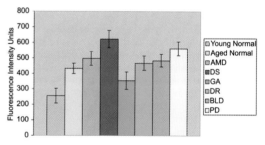

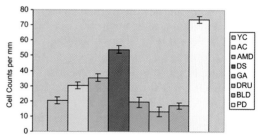

Fig. 2.10A,B. Histograms showing fluorescence intensity of CD45 immunoreactivity in retinal flatmounts, and leucocyte counts in choroidal sections, from control and AMD-affected eyes (Sakurai et al. 2003a, 2003b). (*YC* young controls; *AC* age-matched controls; *AMD* all AMD groups; *DS* disciform scar – neovascular AMD; *GA* geographic atrophy – dry AMD; *DRU* drusen; *BLD* basal laminar deposit; *PD* pigmentary disturbance)

## 2.9 Conclusion

This chapter has reviewed recent advances in understanding the role of immunity in the pathogenesis of AMD. Previously, considerable circumstantial evidence derived from histological and clinical observations aroused conjecture as to the possible "primary or secondary" role of immunity in the aetiology of AMD. Evidence for the involvement of the complement system in drusen formation and the primary role of macrophages in choroidal neovascularisation has now emerged with the availability of new proteome analysis techniques and animal models. Taken together, with the measurement of inflammation in AMD described in this chapter, the evidence that AMD involves immunity in both an exacerbatory and primary causative capacity becomes compelling.

## References

Ambati J, Anand A, Sakurai E, Fernandez S, Lynn B, Kuziel W, Rollins B, Ambati B (2003) An animal model of age-related macular degeneration in senescent Ccl-2 or Ccr-2 deficient mice. Nat Med 9:1390–1397

Anderson DH, Johnson LV, Schneider BL, Nealson M, Mullins RF, Hageman GS (1999) Age-related maculopathy: a model of drusen biogenesis. Invest Ophthalmol Vis Sci 40:922 (Abstract 4863)

Anderson DH, Mullins RF, Hageman GS, Johnson LV (2002) A role for local inflammation in the formation of drusen in the aging eye. Am J Ophthalmol 134:411–431

Bressler SB, Maguire MG, Bressler NM, Fine SL (1990) Relationship of drusen and abnormalities of the retinal pigment epithelium to the prognosis of neovascular macular degeneration. The Macular Photocoagulation Study Group. Arch Ophthalmol 108:1442–1447

Bressler N, Bressler S, Alexander J, Javornik N, Fine S, Murphy R (1991) Loculated fluid. A previously undescribed fluorescein angiographic finding in choroidal neovascularization associated with macular degeneration. Macular Photocoagulation Study Reading Center. Arch Ophthalmol 109:211–215

Broderick C, Duncan L, Taylor N, Dick AD (2000) IFN-gamma and LPS-mediated IL-10-dependent suppression of retinal microglial activation. Invest Ophthalmol Vis Sci 41:2613–2622

Broderick C, Hoek RM, Forrester JV, Liversidge J, Sedgwick JD, Dick AD (2002) Constitutive retinal CD200 expression regulates resident microglia and activation state of inflammatory cells during experimental autoimmune uveoretinitis. Am J Pathol 161:1669–1677

Chang B, Yannuzzi LA, Ladas ID, Guyer DR, Slakter JS, Sorenson JA (1995) Choroidal neovascularization in second eyes of patients with unilateral exudative age-related macular degeneration. Ophthalmology 102:1380–1386

Chang AA, Morse LS, Handa JT, Morales RB, Tucker R, Hjelmeland L, Yannuzzi LA (1998) Histologic localization of indocyanine green dye in aging primate and human ocular tissues with clinical angiographic correlation. Ophthalmology 105:1060–1068

Chang AA, Guyer DR, Orlock DR, Yannuzzi LA (2003) Age-dependent variations in the drusen fluorescence on indocyanine green angiography. Clin Exp Ophthalmol 31:300–304

Chan-Ling T, Halasz P, Stone J (1990) Development of retinal vasculature in the cat: processes and mechanisms. Curr Eye Res 9:459–478

Chen H, Wu L, Pan S, Wu DZ (1993) An immunologic study on age-related macular degeneration. Yan Ke Xue Bao 9:113–120

Crabb JW, Miyagi M, Gu X, Shadrach K, West KA, Sakaguchi H, Kamei M, Hasan A, Yan L, Rayborn ME, Salomon RG, Hollyfield JG (2002) Drusen proteome analysis: an approach to the etiology of age-related macular degeneration. Proc Natl Acad Sci USA 99:14682–14687

Curcio CA, Allen KA (1990) Topography of ganglion cells in human retina. J Comp Neurol 300:5–25

Curcio CA, Saunders PL, Younger PW, Malek G (2000) Peripapillary chorioretinal atrophy: Bruch's membrane changes and photoreceptor loss. Ophthalmology 107:334–343

Devine I, Lightman SI, Greenwood J (1996) Role of LFA-1, ICAM-1, VLA-4 and VCAM-1 in lymphocyte migration across retinal pigment epithelial monolayers in vitro. Immunology 88:456–462

Diaz CM, Penfold PL, Provis JM (1998) Modulation of the resistance of a human endothelial cell line by human retinal glia. Aust NZ J Ophthalmol 26 [Suppl 1]:62–64

Diaz-Araya CM, Provis JM, Penfold PL, Billson FA (1995) Development of microglial topography in human retina. J Comp Neurol 363:53–68

Dick AD, Pell M, Brew BJ, Foulcher E, Sedgwick JD (1997) Direct ex vivo flow cytometric analysis of human microglial cell CD4 expression: examination of central nervous system biopsy specimens from HIV-seropositive patients and patients with other neurological disease. Aids 11:1699–1708

Dickson DW (1986) Multinucleated giant cells in aquired immunodeficiency syndrome encephalopathy. Origin from endogenous microglia? Arch Pathol Lab Med 110:967–968

Espinosa-Heidmann DG, Suner IJ, Hernandez EP, Monroy D, Csaky KG, Cousins SW (2003) Macrophage depletion diminishes lesion size and severity in experimental choroidal neovascularization. Invest Ophthalmol Vis Sci 44:3586–3592

Fine SL, Berger JW, Maguire MG, Ho AC (2000) Age-related macular degeneration. N Engl J Med 342:483-492

Freund KB, Yannuzzi LA, Sorenson JA (1993) Age-related macular degeneration and choroidal neovascularization. Am J Ophthalmol 115:788-791

Gass J (1973) Drusen and disciform macular detachment and degeneration. Arch Ophthalmol 90:206-217

Graeber MB (1993) Microglia, macrophages and the blood-brain barrier. Clin Neuropathol 12:296-297

Greenwood J, Penfold P, Provis J (2000) Evidence for the intrinsic innervation of retinal vessels: anatomical substrate of autoregulation in the retina? In: Burnstock G, Sillito A (eds) Innervation of the Eye. Nervous Control of the Eye, vol 13, chap 5. Harwood Academic, London, pp 155-170

Grunwald JE, Hariprasad SM, DuPont J, Maguire MG, Fine SL, Brucker AJ, Maguire AM, Ho AC (1998) Foveolar choroidal blood flow in age-related macular degeneration. Invest Ophthalmol Vis Sci 39:385-390

Gupta N, Brown KE, Milam AH (2003) Activated microglia in human retinitis pigmentosa, late-onset retinal degeneration, and age-related macular degeneration. Exp Eye Res 76:463-471

Gurne DH, Tso MO, Edward DP, Ripps H (1991) Antiretinal antibodies in serum of patients with age-related macular degeneration. Ophthalmology 98:602-607

Haab O (1885) Erkrankungen der macula lutea. Centralblat Augenheilkd 9:384-391

Hageman GS, Luthert PJ, Victor Chong NH, Johnson LV, Anderson DH, Mullins RF (2001) An integrated hypothesis that considers drusen as biomarkers of immune-mediated processes at the RPE-Bruch's membrane interface in aging and age-related macular degeneration. Prog Ret Eye Res 20:705-732

Heidenkummer HP, Kampik A (1995) Surgical extraction of subretinal pseudotumors in age-related macular degeneration. Clinical, morphologic and immunohistochemical results. Ophthalmologe 92:631-639

Holt PG, Degebrodt A, O'Leary C, Krska K, Plozza T (1985) T cell activation by antigen-presenting cells from lung tissue digests: suppression by endogenous macrophages. Clin Exp Immunol 62:586-593

Hutchinson AK, Grossniklaus HE, Capone A (1993) Giant-cell reaction in surgically excised subretinal neovascular membrane. Arch Ophthalmol 111:734-735

Johnson LV, Ozaki S, Staples MK, Erickson PA, Anderson DH (2000) A potential role for immune complex pathogenesis in drusen formation. Exp Eye Res 70:441-449

Jones S (1999) Almost like a whale. Doubleday, London

Killingsworth MC, Sarks SH (1982) Giant cells in disciform macular degeneration of the human eye. Micron 13:359-360

LaFaut BA, Bartz-Schmidt KU, Vanden Broecke C, Aisenbrey S, De Laey JJ, Heimann K (2000) Clinicopatholgical correlation in exudative age related macular degeneration: histological differentiation between classic and occult choroidal neovascularisation. Br J Ophthalmol 84:239-243

Liew SC, Penfold PL, Provis JM, Madigan MC, Billson FA (1994) Modulation of MHC class II expression in the absence of lymphocytic infiltrates in Alzheimer's retinae. J Neuropathol Exp Neurol 53:150-157

Lopez PF, Grossniklaus HE, Lambert HM, Aaberg TM, Capone AJ, Sternberg PJ, L'Hernault N (1991) Pathologic features of surgically excised subretinal neovascular membranes in age-related macular degeneration. Am J Ophthalmol 112:647-656

Matsubara T, Pararajasegaram G, Wu GS, Rao NA (1999) Retinal microglia differentially express phenotypic markers of antigen-presenting cells in vitro. Invest Ophthalmol Vis Sci 40:3186-3193

Nettleship E (1884) Central senile areolar choroidal atrophy. Trans Ophthalmol Soc 4:165-166

Neumann H (2001) Control of glial immune function by neurons. Glia 36:191-199

Newsome DA, Swartz M, Leone NC, Hewitt AT, Wolford F, Miller ED (1986) Macular degeneration and elevated serum ceruloplasmin. Invest Ophthalmol Vis Sci 27:1675-1680

Packer O, Hendrickson AE, Curcio CA (1989) Photoreceptor topography of the retina in the adult pigtail macaque (*Macaca nemestrina*). J Comp Neurol 288:165-183

Pagenstecher H, Genth C (1875) Atlas der Pathologischen Anatomie des Augapfels. Kreidel, Wiesbaden

Palmer HE, Zaman AG, Ellis BA, Stanford MR, Graham EM, Wallace GR (1996) Longitudinal analysis of soluble intercellular adhesion molecule 1 in retinal vasculitis. Eur J Clin Invest 26:686-691

Pauleikhoff D (1992) Drusen in Bruch's membrane. Their significance for the pathogenesis and therapy of age-associated macular degeneration. Ophthalmologie 89:363-386

Pavli P, Hume DA, Van De Pol E, Doe WF (1993) Dendritic cells, the major antigen-presenting cells of the human colonic lamina propria. Immunology 78:132-141

Penfold PL, Provis JM (1986) Cell death in human retinal development: Phagocytosis of pyknotic and apoptotic bodies by retinal cells. Graefes Arch Clin Exp Ophthalmol 224:549-553

Penfold PL, Killingsworth MC, Sarks SH (1985) Senile macular degeneration: the involvement of immunocompetent cells. Graefes Arch Clin Exp Ophthalmol 223:69-76

Penfold PL, Killingsworth MC, Sarks SH (1986) Senile macular degeneration. The involvement of giant cells in atrophy of the retinal pigment epithelium. Invest Ophthalmol Vis Sci 27:364-371

Penfold PL, Provis JM, Billson FA (1987) Age-related macular degeneration: ultrastructural studies of the relationship of leucocytes to angiogenesis. Graefes Arch Clin Exp Ophthalmol 225:70-76

Penfold PL, Provis JM, Furby JH, Gatenby PA, Billson FA (1990) Autoantibodies to retinal astrocytes associated with age-related macular degeneration. Graefes Arch Clin Exp Ophthalmol 228:270-274

Penfold PL, Madigan MC, Provis JM (1991) Antibodies to human leucocyte antigens indicate subpopulations of microglia in human retina. Vis Neurosci 7:383–388

Penfold PL, Provis JM, Liew SC (1993) Human retinal microglia express phenotypic characteristics in common with dendritic antigen-presenting cells. J Neuroimmunol 45:183–191

Penfold PL, Liew SC, Madigan MC, Provis JM (1997) Modulation of major histocompatibility complex class II expression in retinas with age-related macular degeneration. Invest Ophthalmol Vis Sci 38:2125–2133

Penfold PL, Wen L, Madigan MC, Gillies MC, King NJ, Provis JM (2000) Triamcinolone acetonide modulates permeability and intercellular adhesion molecule-1 (ICAM-1) expression of the ECV304 cell line: implications for macular degeneration. Clin Exp Immunol 121:458–465

Penfold PL, Madigan MC, Gillies MC, Provis JM (2001a) Immunological and aetiological aspects of macular degeneration. Prog Ret Eye Res 20:385–414

Penfold PL, Wong JG, Gyory J, Billson FA (2001b) Effects of triamcinolone acetonide on microglial morphology and quantitative expression of MHC-II in exudative age-related macular degeneration. Clin Exp Ophthalmol 29:188–192

Provis JM, Penfold PL, Edwards AJ, van Driel D (1995) Human retinal microglia: expression of immune markers and relationship to the glia limitans. Glia 14:243–56

Provis JM, Diaz CM, Dreher B (1998) Ontogeny of the primate fovea: a central issue in retinal development. Prog Neurobiol 54:549–580

Provis JM, Sandercoe T, Hendrickson AE (2000) Astrocytes and blood vessels define the foveal rim during primate retinal development. Invest Ophthalmol Vis Sci 41:2827–2836

Rohen JW, Castenholtz A (1967) Über die Zentralisation der Retina bei Primaten. Folia Primatol 5:92–147

Sakurai E, Anand A, Ambati BK, van Rooijen N, Ambati J (2003a) Macrophage depletion inhibits experimental choroidal neovascularization. Invest Ophthalmol Vis Sci 44:3578–3585

Sakurai E, Taguchi H, Anand A, Ambati BK, Gragoudas ES, Miller JW, Adamis AP, Ambati J (2003b) Targeted disruption of the CD18 or ICAM-1 gene inhibits choroidal neovascularization. Invest Ophthalmol Vis Sci 44:2743–2749

Sandercoe TM, Geller SF, Hendrickson AE, Stone J, Provis JM (2003) VEGF expression by ganglion cells in central retina before formation of the foveal depression in monkey retina: evidence of developmental hypoxia. J Comp Neurol 462:42–54

Seddon JM, Gensler G, Milton RC, Klein ML, Rifai N (2004) Association between c-reactive protein and age-related macular degeneration. J Am Med Assoc 291:704–710

Shimizu E, Funatsu H, Yamashita H, Yamashita T, Hori S (2002) Plasma level of interleukin-6 is an indicator for predicting diabetic macular edema. Jpn J Ophthalmol 46(1):78–83

Slakter J, Yannuzzi LA, Schneider U, Sorenson JA, Ciardella A, Guyer DR, Spaide RF, Freund KB, Orlock DA (2000) Retinal choroidal anastomoses and occult choroidal neovascularization in age-related macular degeneration. Ophthalmology 107:742–753

Smith W, Mitchell P, Leeder SR (1996) Smoking and age-related maculopathy. The Blue Mountains Eye Study. Arch Ophthalmol 114:1518–1523

Streilein JW, Wilbanks GA, Cousins SW (1992) Immunoregulatory mechanisms of the eye. J Neuroimmunol 39:185–200

Sunderkotter C, Steinbrink K, Goebeler M, Bhardwaj R, Sorg C (1994) Macrophages and angiogenesis. J Leuk Biol 55:410–422

Tonnesen MG, Anderson DC, Springer TA, Knedler A, Avdi N, Henson PM (1989) Adherence of neutrophils to cultured human microvascular endothelial cells. Stimulation by chemotactic peptides and lipid mediators and dependence upon the Mac-1, LFA-1, p150,95 glycoprotein family. J Clin Invest 83:637–646

Trowbridge I, Thomas M (1994) CD45: an emerging role as a protein tyrosine phosphatase required for lymphocyte activation and development. Ann Rev Immunol 12:85–116

Tso M (1989) Experiments on visual cells by nature and man: In search of treatment for photoreceptor degeneration. Invest Ophthalmol Vis Sci 30:2430–2454

van der Schaft TL, Mooy CM, de Bruijn WC, de Jong PT (1993) Early stages of age-related macular degeneration: an immunofluorescence and electron microscopy study. Br J Ophthalmol 77:657–661

Vinores SA, Amin A, Derevjanik NL, Green WR, Campochiaro PA (1994) Immunohistochemical localization of blood-retinal barrier breakdown sites associated with post-surgical macular oedema. Histochem J 26:655–665

Widder RA, Brunner R, Walter P, Bartz-Schmidt KU, Godehardt E, Heimann K, Borberg H (1999) Modification of vision by change in rheologic parameters in senile macular degeneration through membrane differential filtration – initial results of a randomized study. Klin Monatsbl Augenheilkd 215:43–49

Wolin LR, Massupust LC (1970) Morphology of the primate retina. In: Noback CR, Montagna W (eds) The primate brain, vol 1. Appleton-Century-Crofts, New York, pp 1–27

Wong J, Madigan M, Billson F, Penfold P (2001) Quantification of leucocyte common antigen (CD45) expression in macular degeneration. Invest Ophthalmol Vis Sci 42:S227

Woollard HH (1927) The differentiation of the retina in primates. Proc Zool Soc Lond 1:1–17

Yang P, Chen L, Zwart R, Kijlstra A (2002) Immune cells in the porcine retina: distribution, characterization and morphological features. Invest Ophthalmol Vis Sci 43(5):1488–1492

Yannuzzi L, Negrao S, Iida T, Carvalho C, Rodriguez-Coleman H, Slakter J, Freund K, Sorenson J, Orlock D, Borodoker N, LuEsther T (2001) Retinal angiomatous proliferation in age-related macular degeneration. Retina 21: 416–434

Zhang J, Wu G-S, Ishimoto S-I, Pararajasegaram G, Rao NA (1997) Expression of major histocompatibility complex molecules in rodent retina. Immunohistochemical study. Invest Ophthalmol Vis Sci 38: 1848–1857

Chapter 3

# Photoreceptor Degeneration in Aging and Age-Related Maculopathy

Gregory R. Jackson, Christine A. Curcio, Kenneth R. Sloan, Cynthia Owsley

## Contents

3.1 Introduction to Age-Related Maculopathy   45
3.2 Photoreceptor Loss   47
3.3 Photoreceptor Dysfunction   51
3.3.1 Topography of Loss and Dysfunction   52
3.4 Photoreceptor Function as a Bioassay of RPE and Bruch's Membrane Health   54
3.5 Impairment of Transport Between RPE and Photoreceptors   56
3.6 Summary   57
References   58

## 3.1 Introduction to Age-Related Maculopathy

Age-related maculopathy (ARM) is the major cause of new, untreatable vision loss in the elderly of the industrialized world. In the USA late ARM accounts for 22% of monocular blindness and 75% of legal blindness in adults over age 50 (Klein et al. 1995). As the population ages, ARM will become the largest cause of vision loss among adults (Council 1998).

ARM is a heterogeneous disorder affecting the retinal pigment epithelium (RPE), Bruch's membrane, and choriocapillaris (the RPE/Bruch's membrane complex; Sarks 1976; Green and Enger 1993; Fig. 3.1A) and secondarily the photoreceptors. Early ARM is characterized by minor to moderate acuity loss associated with characteristic extracellular lesions and changes in RPE pigmentation. Lesions between the RPE basal lamina and Bruch's membrane (Fig. 3.1B) can be either focal (drusen) or diffuse (basal linear deposits). A diffuse lesion between the RPE and its basal lamina is basal laminar deposit (Fig. 3.1C). Together, basal laminar and basal linear deposits constitute basal deposits. Late ARM is characterized by severe vision loss associated with extensive RPE atrophy, with or without the sequelae of choroidal neovascularization, that is, in-growth of choroidal vessels through Bruch's membrane and under the RPE in the plane of drusen and basal linear deposits (see Curcio and Millican 1999 for references).

ARM is a multifactorial process, involving a complex interplay of genetic and environmental factors. As described in Chapter 5 recent progress has been made in understanding demographics and natural history of ARM (Klein et al. 1997, 2002), identifying smoking and hypertension as major preventable risk factors (Hyman et al. 2000; Smith et al. 2001), determining the biochemical composition of drusen (Crabb et al. 2002; Malek et al. 2003), and excluding genetic mutations causing some early-onset macular degenerations as risk factors (Stone et al. 2001). Recent studies suggest that statin use (McGwin et al. 2003) and maintaining a healthy body mass index (AREDS 2000; Klein et al. 2001; Seddon et al. 2003) may reduce the risk of the incidence or progression of ARM. Substantial progress has been made in developing mechanisms, animal models, and treatments for choroidal neovascularization (Bressler and Bressler 2000; Campochiaro 2000; Ambati et al. 2003).

The current standards of care include laser photocoagulation of the aberrant vessels or photodynamic therapy, treatments for which only a subset of patients with existing neovas-

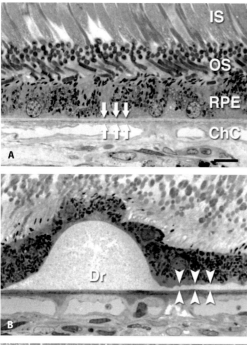

ization, a successful treatment approach may be to arrest the progression of the disease before the onset of late ARM and maintain visual function rather than manage the neovascularization and rescue visual function. The Age-related Eye Disease Study (AREDS) indicated that intake of several antioxidant compounds was beneficial in preventing neovascularization in ARM patients with bilateral drusen (AREDS 2001). However, evaluation of the nutritional intervention was limited for the normal elderly adults and early ARM groups, because very few of these patients progressed to advanced (late) ARM during the course of the trial. Because of the low rate of progression for patients with mild disease, the effectiveness of the treatment for these groups was inconclusive. To increase the feasibility of clinical trials, better methods are needed to select early ARM patients with high risk of progressing to late ARM. A central idea in this review is that development of better diagnostic tests and treatments should be facilitated by and based upon improved understanding of the pathobiology of the earliest disease stages (Ciulla et al. 1998).

Although the most prominent clinical and histopathologic lesions of ARM involve the RPE and Bruch's membrane, it is the degeneration, dysfunction, and death of photoreceptors, through an atrophic process or a neovascular event and its consequences, that accounts for the vision loss associated with ARM. Because the most common clinical endpoint is a loss of acuity, impairment of visual functions mediated by foveal cones has been well characterized in ARM (Brown et al. 1986; Eisner et al. 1987; Sunness et al. 1989; Birch et al. 1992; Mayer et al. 1994; Tolentino et al. 1994; Holopigian et al. 1997; Midena et al. 1997; Huang et al. 2000; Remky et al. 2001; Jurklies et al. 2002; Phipps et al. 2003). Many of the observed visual function changes in ARM could be explained by photoreceptor loss associated with visible lesions. However, alterations in photoreceptor function during the earliest stages of the disease may provide valuable information on clinically invisible changes. Important constituents in the RPE/ Bruch's membrane complex such as basal linear deposit may not be revealed by standard imaging techniques such as fundus photogra-

Fig. 3.1A–C. RPE/Bruch's membrane complex and ARM-associated lesions (from Curcio and Millican 1999). A Normal eye, 63-year-old donor. *Arrows* delimit Bruch's membrane. (*ChC* Choriocapillaris, *OS* outer segments of photoreceptors, *IS* inner segments of photoreceptors, *RPE* retinal pigment epithelium.) B Drusen (*Dr*) and basal linear deposit (*between arrowheads*); 60-year-old donor. C Basal laminar deposits (between *arrows*); 69-year-old donor

cularization qualify (Macular Photocoagulation Study Group 1991; Bressler and Bressler 2000). Another potential treatment approach is to prevent or delay late ARM, which causes the vast majority of legal blindness. Because lesions in Bruch's membrane associated with early ARM (Spraul et al. 1996) precede neovascular-

phy and fluorescein angiography until late in ARM or not at all (Bressler et al. 1994; Curcio and Millican 1999). Therefore, the functional status of photoreceptors may serve as a bioassay of the significance of these changes.

It is important to determine which photoreceptors are most affected by aging and ARM, not only to target potential interventions to the most affected cells, but also to target mechanistic studies toward investigating the earliest disease-related changes. The rate of rod and cone degeneration is a fundamental characteristic of any disorder affecting photoreceptors (LaVail 1981; Jacobson et al. 1986). Because rods and cones have distinctly different biology, the rates at which they die provide important clues to the events initiating their demise. In order to determine the relative rate of degeneration, however, one must obtain comparable information for both rods and cones at matched locations in the same well-characterized study of eyes rather than try to calibrate findings between studies using fundus appearance. Ideally, outcome measurements should be quantifiable and attributable to a specific cell class. Examples of suitable endpoints are cell numbers and cone-mediated and rod-mediated visual sensitivity.

## 3.2
## Photoreceptor Loss

Based on this approach, we characterized the topography of macular photoreceptors in retinal aging and ARM. The macula, defined anatomically as the area with one or more layers of retinal ganglion cells (Polyak 1941) and epidemiologically as the area within the Wisconsin Age-related Maculopathy Grading System grid (Klein et al. 1991), is approximately 6 mm in diameter and centered on the fovea (see Chapter 1). The macula contains two subregions with distinctly different photoreceptor content: a small cone-dominated fovea, 0.8 mm (2.75° of visual angle) in diameter, and a surrounding rod-dominated parafovea. Figure 3.2A–D shows small-cone inner segments in the fovea and large cones with numerous small rods in the parafovea. Figure 3.2E shows that the peak density of cones in the foveal center is high (mean 200,000/mm$^2$) and declines precipitously (tenfold) within 1 mm (3.5°). Rods are absent in the foveal center and rise sharply to a maximum at 2–4 mm eccentricity. In young adults, rods outnumber cones in the macula by 9:1, so the macula is cone-enriched, compared with the eye as a whole (20:1), but it is not cone-dominated (-Østerberg 1935; Curcio et al. 1990).

In flat-mounted retinas from eyes with maculas lacking grossly visible drusen and pigmentary change, the *total* number of cones within the 0.8-mm-diameter cone-dominated fovea was remarkably stable throughout adulthood at about 32,000 (Curcio et al. 1993). Other studies did not detect an age-related change in *peak* foveal cone density (Gao et al. 1990), but the possibility of foveal cone loss at very advanced ages remains open (Feeney-Burns et al. 1990). In contrast, the number of rods in the parafovea of the same eyes decreased by 30% (Curcio et al. 1993), consistent with other results (Panda-Jonas et al. 1995). Within the macula, age-related rod loss was not spatially uniform. Comparison of rod topography in younger and older eyes using difference maps suggests that rod loss is deepest near the fovea and widens toward the periphery over time (Fig. 3.3D–F). In contrast, cone topography changed little with age (Fig. 3.3C–E). The location of age-related rod loss differs from the site of maximal rod density (2–4 mm from the fovea; Østerberg 1935; Curcio et al. 1990) and from the site of cell loss associated with typical retinitis pigmentosa (8–10 mm from the fovea; Heckenlively 1988). The relative rate of rod and cone loss in extramacular retina is uncertain (Gao et al. 1990; Curcio et al. 1993; Panda-Jonas et al. 1995).

Analysis of photoreceptor topography in eyes from 12 ARM donors provided evidence that rods are preferentially affected in ARM as well (Curcio et al. 1996; see Curcio 2001, for review; Medeiros and Curcio 2001). Despite the presence of drusen and thick deposits, the foveal cone mosaic of nonexudative ARM eyes appeared remarkably normal, and the total number of foveal cones fell within the normal range. In contrast, the parafovea was distinctly abnormal, with few rods, broadened cone inner segments, and gaps in the mosaic of inner segments. Figure 3.4 shows premortem fundus ap-

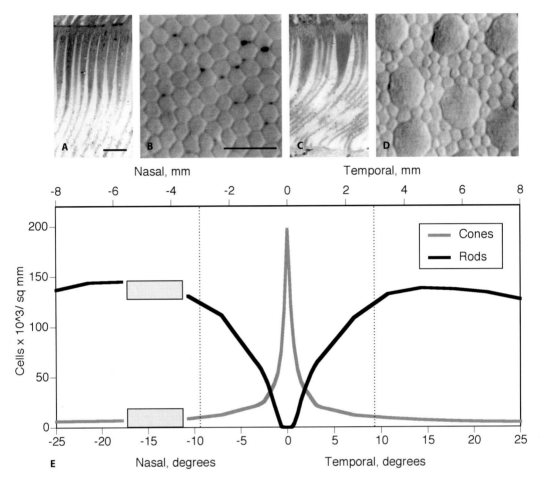

Fig. 3.2A-E. Photoreceptor mosaic and distribution. A Foveal cone inner and outer segments, longitudinal section. B Foveal cone inner segments in a flat-mounted retina of a 34-year-old donor; Nomarski differential interference contrast optics and video. C Nonfoveal cone and rod inner segments, longitudinal section. D Cone inner segments (large) and rod inner segments (small) in the same eye. E Number of cones and rods per square millimeter of retinal surface in nasal and temporal retina, as a function of distance from the foveal center in degrees of visual angle (*bottom*) and millimeters (*top*). *Hatched bar*, optic disc. *Dashed lines* show macular boundaries. (From Curcio et al. 1990)

pearance, fluorescein angiography, and topography of cone and rod loss in an eye with pigment clumping superior to the fovea (case 2 from Curcio et al. 1996). In this eye, photoreceptor loss occurs in relation to funduscopically visible pigment change, the area of loss extends beyond what is visible in the fundus, loss affects both cones and rods, and rod loss is deeper and more extensive in area than cone loss. The other eye of this patient had progressed to neovascular ARM, so the eye in Fig. 3.4 would be considered high risk (Macular Photocoagulation Study Group 1997). In eyes with disciform degeneration and geographic RPE atrophy (see Curcio 2001 for review), many histochemically verified cones survived in pockets of subretinal space enclosed externally by fibrovascular scar. Furthermore, peripheral to the geographic RPE atrophy associated with disciform scars was a transitional zone of thick deposits, degenerating RPE, and a marked decrease in the number of rods. Rod loss was greater than cone loss at

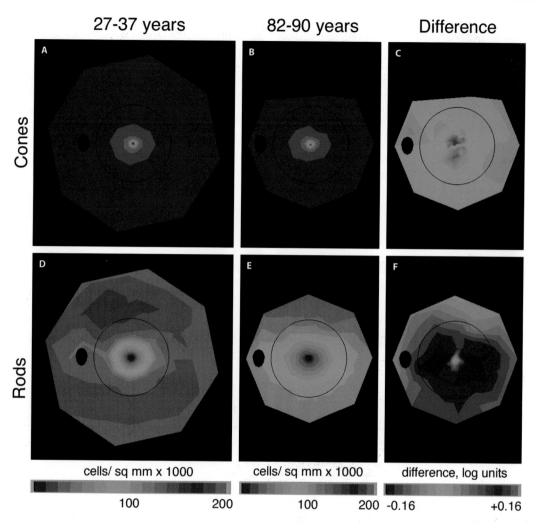

Fig. 3.3A–F. Topography of cones and rods in aging human retina, shown as a fundus of a left eye. *Black oval* is the optic disc, and the *ring* delimits the 6-mm-diameter macula. In C and F, *warm colors* mean that the older group has higher mean density than the young group, and *cool colors* mean that the older group has lower mean density than the young group. A *yellow-green map* means that differences between groups are small. A Cones, 27- to 36-year-old donors. B Cones, 82 to 90-year-old donors. C Log mean difference in cone density between younger adults and older adults is small and inconsistent. D Rods, 27 to 36-year-old donors. E Rods, 82 to 90-year-old donors. F Log mean difference in rod density between younger adults and older adults is greatest at 0.5–3 mm from fovea. *Purple* signifies that the log mean difference (aged–young) was less than –0.16 log units, i.e., that aged eyes had 31% fewer cells than young eyes. (From Curcio et al. 1993)

comparable locations in three-quarters of ARM eyes examined. In summary, although the macula is cone-enriched, rods show the earliest signs of degeneration in most ARM eyes, and the last surviving photoreceptors, appear to be cones. Photoreceptor loss in aging and ARM occurs by apoptosis, and most apoptotic photoreceptors in eyes with geographic atrophy are rods (Xu et al. 1996; Lambooij et al. 2000; Dunaief et al. 2002). In addition to cell loss, photoreceptors suffer sublethal metabolic insults. Outer segments overlying drusen are short-

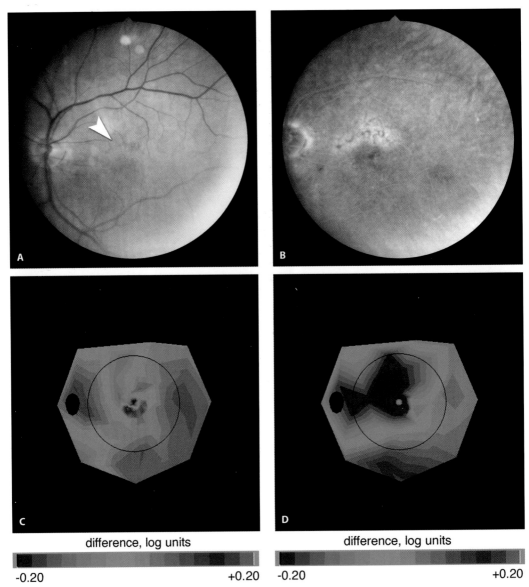

Fig. 3.4A–D. Clinical imaging and topography of cone and rod loss in an eye at risk for late ARM. Eighty-one-year-old donor. Clinical images and topographic maps are not at the same spatial scale. In the difference maps (C, D), *warm colors* mean that ARM eye has higher mean density than age-matched controls, and *cool colors* mean that the ARM eye has lower mean density than age-matched controls (see Fig. 3B, E). *Purple* signifies that the log mean difference (ARM-control) was less than −0.20 log units, i.e., that the ARM eye had 37% fewer cells than age-matched control eyes. A Red-free fundus showing pigment hypertrophy with surrounding pigment atrophy in an arc superior to the fovea. B Fluorescein angiogram showing late hyperfluorescence at 775 s. C Difference in number of cones per square millimeter, relative to age-matched controls. D Difference in the number of rods per square millimeter, relative to age-matched controls. (Data from Curcio et al. 1996)

ened and misoriented, and opsin translocates from the outer segment to the plasma membrane in rod photoreceptors, typical of other degenerations (Johnson et al. 2003).

## 3.3 Photoreceptor Dysfunction

Functional studies have supported the histological evidence for preferential vulnerability of rods in aging and ARM (Jackson and Owsley 2000; Owsley et al. 2000). These functional studies measured photopic and scotopic sensitivity at matched retinal locations in the same cohort of well-characterized eyes. The studies had large samples, involving 106 normal subjects from seven decades of adulthood and 80 early-onset ARM patients. Significantly, macular health was ascertained objectively in all subjects by grading fundus photographs, and the effect of lens density, which reduces retinal illuminance in older persons, was accounted for on an individual basis in interpreting thresholds. These studies demonstrated reduced rod-mediated light sensitivity in older adults in good retinal health, the magnitude of which was similar throughout the parafovea (Jackson et al. 1998; Jackson and Owsley 2000). Scotopic impairment was greater than photopic impairment in 80% of older adults evaluated, and, furthermore, scotopic sensitivity declined throughout adulthood faster than photopic sensitivity declined (Fig. 3.5). With respect to ARM patients, mean scotopic sensitivity within 18° of fixation was significantly lower in early ARM patients as a group than in age-matched controls without ARM (Fig. 3.6). The pattern of scotopic versus photopic sensitivity loss in the central 36° of the visual field varied consider-

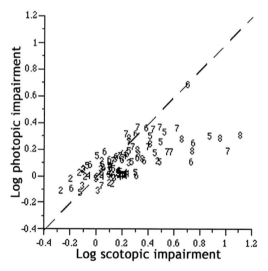

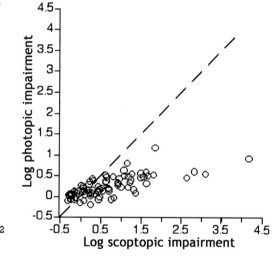

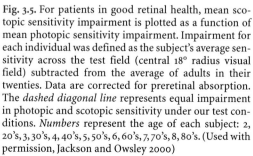

Fig. 3.5. For patients in good retinal health, mean scotopic sensitivity impairment is plotted as a function of mean photopic sensitivity impairment. Impairment for each individual was defined as the subject's average sensitivity across the test field (central 18° radius visual field) subtracted from the average of adults in their twenties. Data are corrected for preretinal absorption. The *dashed diagonal line* represents equal impairment in photopic and scotopic sensitivity under our test conditions. *Numbers* represent the age of each subject: 2, 20's, 3, 30's, 4, 40's, 5, 50's, 6, 60's, 7, 70's, 8, 80's. (Used with permission, Jackson and Owsley 2000)

Fig. 3.6. For ARM patients, mean scotopic sensitivity impairment is plotted as a function of mean photopic sensitivity impairment. Impairment for each individual was defined as the subject's average sensitivity across the test field (central 18 radius visual field) subtracted from the average of elderly normal adults. Data are corrected for preretinal absorption. The *dashed diagonal line* represents equal impairment in photopic and scotopic sensitivity under our test conditions

ably among individual patients with early ARM. Of the patients with reduced light sensitivity in this region, 59% showed reduced scotopic sensitivity, 27% showed both reduced scotopic and photopic sensitivity, and 14% had reduced photopic sensitivity only. In almost all (87%) of these patients, the magnitude of mean scotopic sensitivity loss exceeded the magnitude of mean photopic sensitivity loss.

It is important to note that age-related changes in night vision are not confined to laboratory settings but also impact on daily activities of older adults. One of the most pervasive visual observations reported by older adults, even those free of retinal disease, is a problem seeing under dim illumination (Kosnik et al. 1988), especially problems with night driving (Ball et al. 1998). It is interesting that patients in the earliest phases of ARM report night driving problems, which are associated with their impaired scotopic sensitivity as measured in the laboratory (Scilley et al. 2002).

### 3.3.1
### Topography of Loss and Dysfunction

The topography of photoreceptor loss and dysfunction, which is spatially heterogeneous across the macula, may provide important information about the signals leading to photoreceptor demise. Figure 3.7A, B shows that the scotopic sensitivity loss and the loss of rod photoreceptors in early ARM patients is greatest near the foveal center and declines markedly to the edge of the macula, reminiscent of the deepening and widening of photoreceptor loss around the fovea shown in Fig. 3.2 (Curcio et al. 1996; Owsley et al. 2000; Medeiros and Curcio 2001). In contrast, photopic sensitivity loss and the loss of cone photoreceptors are roughly constant over the same distance (Fig. 3.7A). This heterogeneity can be used to assess the plausibility of potential mechanisms underlying photoreceptor loss and dysfunction, under the assumption that causally related events should exhibit a similar topography. This approach is valuable, because the close proximity and physiological interdependence of photoreceptors, RPE, and Bruch's membrane make it difficult to disentangle the relative contributions of these distinctive layers to the pathogenesis of ARM. In the absence of multiparametric data from many individual eyes, we addressed this question by comparing the best topographic data available from patient-oriented and laboratory studies.

In Fig. 3.7C, D we examine the topography of two features of the normal macula that are frequently mentioned in the context of ARM pathobiology. Lipofuscin is a prominent autofluorescent age-pigment in the RPE that is thought to represent irreducible end-products of outer-segment breakdown. The striking prominence of lipofuscin in adult human macula (Feeney-Burns et al. 1984), its enhanced autofluorescence in areas of incipient RPE atrophy in ARM patients (Holz et al. 2001), its decreased autofluorescence associated with drusen (Delori et al. 2000), and the identification and isolation of a major constituent (A2E; Parish et al. 1998) has lead to several proposed roles in the pathogenesis of ARM. These include inhibition of lysosomal function (Holz et al. 1999), promotion of apoptosis (Suter et al. 2000), enhancement of photo-oxidative injury (Schutt et al. 2000), and detergent-like disruption of the plasma membrane (Eldred 1998). Figure 3.6B shows that funduscopically visible autofluorescence due to lipofuscin follows closely the normal distribution of rods (Delori et al. 1995; von Rückmann et al. 1997), that is, low in the foveal center and high at 2–4 mm.

Macular pigment, comprised of lutein and zeaxanthin, is thought to protect the retina against ARM because it can act as an antioxidant (Beatty et al. 1999). Modulation of macular pigment has attracted attention as a therapeutic route, because factors associated with low macular pigment are also associated with increased risk for ARM, retinal content can be enhanced by dietary supplementation, and pigment is detectable noninvasively in patients (Snodderly 1995; Landrum et al. 1997; Gellermann et al. 2002). Figure 3.7D shows that the macular pigment is highly concentrated at the foveal center, sharply declining within 2 mm. Figure 3.7A–C indicates that the distinctive spatial profiles of ARM-related rod dysfunction/loss, lipofuscin accumulation, and macular pig-

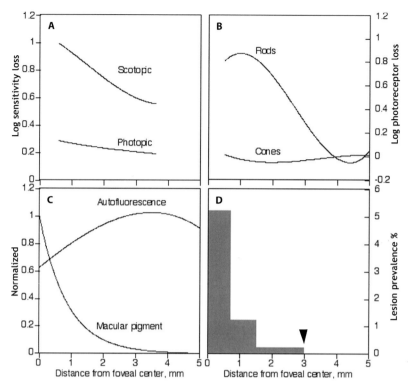

**Fig. 3.7A–D.** Sensitivity loss in ARM eyes (A), photoreceptor loss in ARM eyes (B), funduscopically visible autofluorescence and macular pigment in normal eyes (C), and lesions in ARM eyes (D), plotted as a function of distance from the foveal center. A Dark-adapted (scotopic) and light-adapted (photopic) sensitivity loss for patients with early ARM, with adjustment for preretinal absorption (Owsley et al. 2000); averaged across test loci on the horizontal, nasal and oblique meridians. Sensitivity loss is referenced against old adults in good retinal health with no signs of ARM. B Loss of rods and cones is averaged across four meridians. ARM-related loss is the log of the mean pair-wise differences between early ARM eyes and age-matched controls (Medeiros and Curcio 2001). C Mean autofluorescence due to lipofuscin along the horizontal temporal meridian is normalized to the maximum in three observers (Delori et al. 1995; von Rückmann et al. 1997). The optical density of macular pigment was measured in a normal adult, 23 years old (Werner et al. 1987). D The prevalence of right eyes with soft indistinct drusen and/or RPE hypo- and hyperpigmentation in specific macular regions. Bars indicate the area-weighted prevalence for 247 participants with early ARM in the Beaver Dam Eye Study (Wang et al. 1996) for lesions in regions 0–0.5 mm, 0.5–1.5 mm, and 1.5–3.0 mm, respectively, from the foveal center. These regions correspond to the central subfield, the ring of four inner subfields, and the ring of four outer subfields, respectively, of the Wisconsin Age-related Maculopathy Grading System grading grid (Klein et al. 1991). Arrowhead indicates outer limit of macula

ment are not related in a straightforward manner. A reasonable expectation is that dysfunction and loss should be greater where deleterious factors (lipofuscin) are abundant and/or beneficial factors (macular pigment) are scarce. In fact, the opposite is the case. If lipofuscin and macular pigment are theorized to have a primary role in ARM pathogenesis, the theory must account for the unexpected distribution of these factors in relation to other markers of the disease. The roles of lipofuscin and macular pigment in specific aspects of ARM pathobiology (e.g., RPE cell death) are not excluded by this analysis. However, it should be noted that net RPE cell loss in the aging macula has been difficult to establish (Tso and Friedman 1968; Streeten 1969; Watzke et al. 1993; Harman et al. 1997; Del Priore et al. 2002), possibly due to methodological issues, but also possibly because the RPE cell population is conserved in vivo despite prominent age-related lipofuscin accumulation.

Next we examined the topography of Bruch's membrane pathology. Characteristic debris accumulates within Bruch's membrane throughout adulthood (Feeney-Burns and Ellersieck 1985), accompanied by reduced collagen solubility (Karwatowski et al. 1995), accumulation of advanced glycation end-products (Handa et al. 1999), and deposition of neutral lipids including cholesterol (Pauleikhoff et al. 1990; Curcio et al. 2001; Haimovici et al. 2001). A layer just external to the RPE basal lamina is almost completely occupied by esterified cholesterol-rich droplets (Curcio et al. 2001; Ruberti et al. 2003). Additional material (basal laminar deposit) accumulates between the RPE and Bruch's membrane in older adults and in ARM patients (Sarks 1976; Green and Enger 1993; Spraul and Grossniklaus 1997; Curcio and Millican 1999). Because Bruch's membrane and basal deposits are not directly visible in the fundus (Bressler et al. 1994; Curcio and Millican 1999), we used the distribution of soft drusen and RPE changes visible in the fundus of ARM patients as a surrogate for the invisible pathology, with the caveat that the lateral extent of pathology is likely to be underestimated. Figure 3.7D shows the location of soft drusen and RPE in participants with early ARM in the population-based Beaver Dam Eye Study (Wang et al. 1996). Lesions were localized to defined macular subfields using a validated, semiquantitative grading system of clinical fundus photographs. Soft drusen and RPE changes cluster within the central 1 mm of the macula. Although rod function over focal deposits (drusen) remains to be directly demonstrated, the overall topographic correspondence of RPE/Bruch's membrane pathology and rod dysfunction is striking.

## 3.4
### Photoreceptor Function as a Bioassay of RPE and Bruch's Membrane Health

The photoreceptor layer consists of two intermixed cell types that share a common support system (RPE, Bruch's membrane, and choroid) and environmental exposure to light. Here we consider mechanisms pertaining to the support system, because its age- and disease-related changes are remarkable, and because the role of chronic light exposure in aging and ARM has been difficult to establish. In addition to investigating rod- or cone-specific mechanisms of cell death (Remé et al. 2003), we propose that developing a hypothesis to account for the differential effect on rod and cone survival in the setting of an altered RPE/Bruch's membrane environment may be instructive. A major function of the choroidal vascular system, of which Bruch's membrane is the inner border, is supply of essential nutrients and oxygen to the photoreceptors. The idea that age- and ARM-related changes in the RPE/Bruch's membrane complex ultimately impact the integrity of the photoreceptor resupply route, resulting in degeneration and death, therefore has intuitive appeal. Changes at multiple locations along this route – poor vascular perfusion secondary to choriocapillaris atrophy (Chen et al. 1992; Ramrattan et al. 1994), impaired translocation of plasma nutrients due to Bruch's membrane thickening (Moore et al. 1995), or reduced uptake from plasma or delivery to photoreceptors due to RPE senescence – could impact photoreceptor health. Because photoreceptor function is understood in exquisite detail (Ridge et al. 2003), it may be eventually possible to deduce the specific essentials that photoreceptors lack by careful attention to functional deficits.

Dark adaptation is a good candidate for a test of visual function, because dark adaptation relies on retinoid cycle components contained within the same layers where ARM-associated lesions are located (Lamb et al. 1998; Leibrock et al. 1998; McBee et al. 2001). The classic dark adaptation function describes the recovery of sensitivity following a bright flash of light and consists of an early portion exclusively mediated by cones, a transition to rod function (rod-cone break), and a later portion exclusively mediated by rods (Barlow 1972). The retinoid cycle provides 11-*cis*-retinal, a metabolite of vitamin A, to the photoreceptors for photopigment regeneration (Wald 1935; Hecht and Mandelbaum 1939; Jones et al. 1989). Aging and ARM-related changes may retard dark adaptation by a variety of mechanisms working either independently or in concert to reduce the pool of 11-*cis*-retinal available to the photoreceptors. Debris in Bruch's membrane may slow the passage of vitamin A from the choroid to the RPE, the RPE may process retinoids less efficiently due to age- or disease-related change, transfer of 11-*cis*-retinal from the RPE to the photoreceptor's outer segment could be due to slowed diffusion or impaired interphotoreceptor retinoid binding protein function, and reuptake of all-*trans*-retinol to the RPE for recycling into 11-*cis*-retinal could be compromised. Impairment of any or all of these processes could slow dark adaptation.

In older adults with good macular health, as assessed by grading of fundus appearance, the rod-mediated portion of dark adaptation is significantly slower than in younger adults (Jackson et al. 1999a; Fig. 3.8). During adulthood, the time constant of the rod-mediated component of dark adaptation increases by about 8 s per decade (Jackson et al. 1999a). Rod-mediated dark adaptation is not correlated with scotopic sensitivity in these patients, indicating that the mechanisms underlying these two aspects of rod vision are not identical (Jackson and Owsley 2000). In early ARM patients, even in those with normal acuity, rod-mediated dark adaptation is much slower (13 min on average) than in normal age-matched controls (Owsley et al. 2001; Fig. 3.9). Consistent with the pattern of scotopic sensitivity loss described above, delays in rod-mediated dark adaptation are greater than those for cone-mediated dark adaptation in ARM (Jackson et al. 1999a). Delayed rod-mediated dark adaptation occurs in ARM patients with normal scotopic sensitivity, whereas the opposite pattern, normal dark adaptation with poor scotopic sensitivity, is rare. Rod-mediated dark adaptation may be more sensitive to the effects of early ARM than cone-mediated dark adaptation because of differences in the retinoid cycle of the rods and cones (Mata et al. 2002). Cone photopigment regeneration can occur in the absence of the RPE, whereas rhodopsin regeneration (rod photopigment) is re-

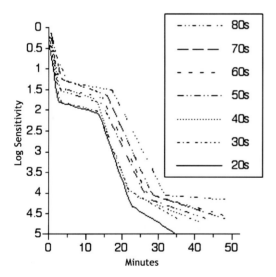

Fig. 3.8. Dark adaptation as a function of decade in persons with normal retinal health. Individual subject's data were grouped by decade and fitted with a four-linear component model. Data are corrected for preretinal absorption. The resulting equations from the nonlinear regression analysis were plotted for illustration purposes. Note that the functions shift to the right with increasing decade, indicating a slowing of the rate of dark adaptation during aging. (Used with permission, Jackson et al. 1999b)

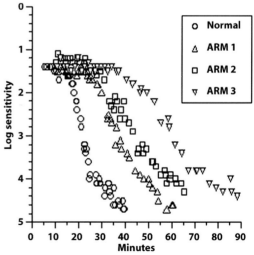

Fig. 3.9. The dark adaptation functions for three patients with ARM and one older adult in good eye health. All patients had 20/25 visual acuity or better. (Used with permission, Owsley et al. 2001)

liant on the RPE/ Bruch's membrane complex (Goldstein and Wolf 1973; Hood and Hock 1973; Jones et al. 1989). Recently, new enzymes for retinoid processing, possibly within Müller cells, have been identified in all-cone retinas (Mata et al. 2002), raising the possibility that fovea, which has a high concentration of Müller cells as well as cones (Yamada 1969), may have additional sources of retinoids. Thus, if cone-mediated dark adaptation is less dependent on the RPE than rod-mediated dark adaptation, impaired rod-mediated dark adaptation may be a better marker for the significant changes in early ARM than measures of cone function.

## 3.5
## Impairment of Transport Between RPE and Photoreceptors

It is reasonable to suggest that the extensive changes found in Bruch's membrane may affect photoreceptor health and function. The accumulation of debris, particularly lipids (see above), are hypothesized to slow the transfer of fluids, essential nutrients, and large proteins across Bruch's membrane. Hydraulic resistivity increases and diffusion of selected compounds decreases with donor age in Bruch's membrane explants (Moore et al. 1995; Marshall et al. 1998; Moore and Clover 2001; Hussain et al. 2002). Thus, lesions in Bruch's membrane may act as a barrier to nutrients moving from the choroidal blood supply to the RPE and ultimately photoreceptors. The hypothesis that rod-mediated dark adaptation is impaired because of slowed translocation of retinoids through Bruch's membrane and ARM-associated lesions is supported by the fact that rod dysfunction and degeneration occur in various late-onset conditions with diffuse subRPE deposits (Jacobson et al. 1995; Kim et al. 1997; Curcio et al. 2000; Hayward et al. 2003). It is possible that subRPE deposits, which differ ultrastructurally and probably biochemically among these disorders, act as nonspecific barriers to the resupply of molecules preferentially essential to rods. The hypothesis is also supported by evidence that dark adaptation improves following dietary vitamin A supplementation in patients with Sorsby's fundus dystrophy, a disorder characterized by thick subRPE deposits (Jacobson et al. 1995). Presumably, the translocation deficit was overcome via mass action in this case. Mutations in genes coding for key visual cycle components also lead to poor night vision (McBee et al. 2001; see review, Thompson and Gal 2003). In some animal models, these deficits can be bypassed by orally administered retinoid compounds (Van Hooser et al. 2000). In mice, vitamin A taken up by RPE is delivered in a complex with retinol-binding protein (Vogel et al. 2002). However, the question of whether perfusion of retinoids through Bruch's membrane is required for the recovery of visual sensitivity remains to be answered. Further, there are currently insufficient data to evaluate the RPE as the site of impaired retinoid translocation, because little is known about age-related changes in retinoid processing. The content of retinyl esters (the storage form) increases with age in monkey macula but not in periphery or in whole human eye (Bridges et al. 1982; Crabtree et al. 1997). If more retinyl esters are stored in aging human macula, it is possible that impaired transport between the RPE and photoreceptor outer segments is responsible for aging-related impairment in dark adaptation.

Although photoreceptor dysfunction and death appear to be related topographically to the lesions in the RPE/Bruch's membrane complex, rod susceptibility to aging and ARM and the mechanism of photoreceptor death are unknown. The results of dark adaptation studies suggest a deficiency of retinoids available to the photoreceptors (Brown et al. 1985; Owsley et al. 2001). Vitamin A deprivation leads to outer segment degeneration and photoreceptor death *in vivo* (Dowling and Wald 1958; Katz et al. 1991, 1993) and accelerated degeneration of photoreceptors with mutant rhodopsins in vitro (Li et al. 1998). Lack of vitamin A affects primarily rods but eventually impacts cones also (Carter-Dawson et al. 1979; Kemp et al. 1988, 1989). It should be noted that vitamin A is necessary not only as the precursor to 11-*cis*-retinal but also as the precursor to other compounds potentially important for RPE and photoreceptor health. Within tissues, retinol is activated to retinoic acid, which binds to nuclear receptors to regulate transcription of more than 300 diverse target genes (Mangelsdorf 1994; Saari 1994) whose exploration in RPE is only beginning (Samuel et al. 2001). Studies to determine the nonvisual effect of retinoids on photoreceptor health would be informative.

It is important to emphasize that the slowing of the retinoid cycle may be simply a marker for another process, such as generalized RPE ill-health, RPE senescence, or Bruch's membrane change, which results in local scarcity of other molecules essential to photoreceptors. For example, lack of oxygen has been implicated in photoreceptor death subsequent to retinal detachment, as mitochondria-rich inner segments are displaced from the high oxygen tension in the choroid (Linsenmeier and Padnick-Silver 2000). Cone inner segments are larger and contain more mitochondria than rods (Hoang et al. 2002) but the relative oxygen requirements of rod and cones are unknown. Determining the impact of thickened deposits and shortened outer segments associated with ARM to oxygen levels at the level of the inner segment, the relative usage of oxygen by cones and rods, and the effects of modulating oxygen availability on cone and rod survival would be useful. Another nutrient important for photoreceptor survival is vitamin E (tocopherol). A constituent of the AREDS anti-oxidant formulation (AREDS 2000), vitamin E, is normally delivered to tissues by plasma low density lipoproteins, (Kayden and Traber 1993) and it protects unsaturated lipids in membranes and in vitamin A. Of the four naturally occurring vitamin E isomers, $\alpha$-tocopherol is abundant in the RPE/choroid, and within neural retina, it is highest in the fovea (Crabtree et al. 1996a, 1996b, 1997). Because macular $\alpha$-tocopherol concentrations are independent of plasma levels, it is thought that the retina regulates its vitamin E content closely. Vitamin E increases with age in RPE-choroid in macaque and humans (Organisciak et al. 1987; Friedrichson et al. 1995). Systemic vitamin E deficiency in rat leads to photoreceptor degeneration and loss but the relative rates of cone and rod loss are not known (Robison et al. 1980). Interestingly, the retina may be capable of generating its own vitamin E. Mutations of the gene coding for $\alpha$-tocopherol transfer protein are associated with pigmentary retinopathy in humans and mice, and this gene is expressed in retina and brain (Gotoda et al. 1995; Yokota et al. 1997, 2001; Copp et al. 1999). Thus, it is possible that the potential advantage afforded to cones via an intraretinal retinoid delivery pathway (see above) may represent a class of local mechanisms that nourish and protect photoreceptors.

## 3.6
## Summary

In summary, anatomical and functional studies have converged to demonstrate that photoreceptor degeneration and loss occurs before disease in the RPE/Bruch's membrane complex progresses to late ARM. Furthermore, macular rods are affected earlier and more severely than cones in aging and ARM. These findings are significant for both clinical and basic research. In many patients tests of rod function may permit detection of ARM at earlier stages than do standard tests of cone function such as visual acuity. The preferential vulnerability of rods in aging and ARM is a phenomenon which should be accounted for by mechanistic theories. These findings provide a standard against which the relevance of emerging animal models (Mata et al. 2001; Rakoczy et al. 2002; Weber et al. 2002) and other potentially pathogenic phenomena in the macula should be assessed. Since rods secrete factors that enhance cone survival (Mohand-Said et al. 1998), early interventions that target rod photoreceptors may have an indirect salutatory effect on cones as well.

The link between photoreceptor dysfunction, assessed by vision function studies, and risk for neovascularization in Bruch's membrane, alluded to at the beginning of this review, is most plausibly attributable to the common cause of poor RPE health, long postulated as central to ARM pathogenesis (Hogan 1972). Because the RPE is polarized, problems pertaining to the resupply of photoreceptors on the apical aspect of the RPE (leading to photoreceptor death) should be conceptually separated from problems pertaining to waste removal on the basal aspect of the RPE (leading to Bruch's membrane damage and neovascularization), at least for the purposes of designing mechanistic experiments. These processes are governed by different proteins and pathways at the cellular level and will be reflected by different risk factors and genetic predispositions at the population level. Rigorous test of a nutrient deficiency hypothesis of ARM-associated photoreceptor death, awaits more information about normal nutrient delivery mechanisms across the RPE/Bruch's membrane complex, intraretinal contributions to photoreceptor nutrition, changes in these mechanisms with age and pathology, and differential effects on rods and cones.

## References

Age-Related Eye Disease Study Research (AREDS) Group (2000) Risk factors associated with age-related macular degeneration. A case-control study in the age-related eye disease study: age-related eye disease study report number 3. Ophthalmology 107:2224–2232

Age-Related Eye Disease Study Research AREDS Group (2001) Report No. 8: a randomized, placebo-controlled, clinical trial of high-dose supplementation with vitamins C and E, beta carotene, and zinc for age-related macular degeneration and vision loss. Arch Ophthalmol 119:1417–1436

Ambati J, Ambati BK, Yoo SH, Ianchulev S, Adamis AP (2003) Age-related macular degeneration: etiology, pathogenesis, and therapeutic strategies. Surv Ophthalmol 48:257–293

Ball K, Owsley C, Stalvey B, Roenker DL, Graves M (1998) Driving avoidance and functional impairment in older drivers. Accident Anal Prevent 30:313–322

Barlow H (1972) Dark and light adaptation: psychophysics. In: Hurvich L (ed) Handbook of sensory physiology, vol VII. Springer, New York, pp 1–28

Beatty S, Boulton M, Henson D, Koh H-H, Murray IJ (1999) Macular pigment and age-related macular degeneration. Br J Ophthalmol 83:867–877

Birch DG, Anderson JL, Fish GE, Jost BF (1992) Pattern-reversal electroretinographic acuity in untreated eyes with subfoveal neovascular membranes. Invest Ophthalmol Vis Sci 33:2097–2104

Bressler NM, Bressler SB (2000) Photodynamic therapy with verteporfin (Visudyne): impact on ophthalmology and visual sciences. Invest Ophthalmol Vis Sci 41:624–628

Bressler NM, Silva JC, Bressler SB, Fine SL, Green WR (1994) Clinicopathological correlation of drusen and retinal pigment epithelial abnormalities in age-related macular degeneration. Retina 14:130–142

Bridges C, Alvarez R, Fong S (1982) Vitamin A in human eyes: amount, distribution, and composition. Invest Ophthalmol Vis Sci 22:706–714

Brown B, Adams AJ, Coletta NJ, Haegerstrom-Portnoy G (1985) Dark adaptation in age-related maculopathy. Ophthalmol Physiol Opt 6:81–84

Brown B, Tobin C, Roche N, Wolanowski A (1986) Cone adaptation in age-related maculopathy. Am J Ophthalmol Physiol Opt 63:450–454

Campochiaro PA (2000) Retinal and choroidal neovascularization. J Cell Physiol 184:301–310

Carter-Dawson L, Kuwabara T, O'Brien P, Bieri J (1979) Structural and biochemical changes in vitamin A-deficient rat retinas. Invest Ophthalmol Vis Sci 18:437–446

Chen JC, Fitzke FW, Pauleikhoff D, Bird AC (1992) Functional loss in age-related Bruch's membrane change with choroidal perfusion defect. Invest Ophthalmol Vis Sci 33:334–340

Ciulla TA, Danis RP, Harris A (1998) Age-related macular degeneration: a review of experimental treatments. Surv Ophthalmol 43:134–146

Copp RP, Wisniewski T, Hentati F, Larnaout A, Ben Hamida M, Kayden HJ (1999) Localization of alpha-tocopherol transfer protein in the brains of patients with ataxia with vitamin E deficiency and other oxidative stress related neurodegenerative disorders. Brain Res 822:80–87

Council NAE (1998) Vision research – a national plan: 1999–2003. Executive summary. National Institutes of Health, Bethesda, MD

Crabb JW, Miyagi M, Gu X, Shadrach K, West KA, Sakaguchi H, Kamei M, Hasan A, Yan L, Rayborn ME, Salomon RG, Hollyfield JG (2002) Drusen proteome analysis: an approach to the etiology of age-related macular degeneration. Proc Natl Acad Sci USA 99:14682–14687

Crabtree DV, Adler AJ, Snodderly DM (1996a) Radial distribution of tocopherols in rhesus monkey retina and retinal pigment epithelium-choroid. Invest Ophthalmol Vis Sci 37:61–76

Crabtree DV, Adler AJ, Snodderly DM (1996b) Vitamin E, retinyl palmitate, and protein in rhesus monkey retina and retinal pigment epithelium-choroid. Invest Ophthalmol Vis Sci 37:47–60

Crabtree D, Snodderly D, Adler A (1997) Retinyl palmitate in macaque retina-retinal pigment epithelium-choroid: distribution and correlation with age and vitamin E. Exp Eye Res 64:455–463

Curcio CA (2001) Photoreceptor topography in ageing and age-related maculopathy. Eye 15:376–383

Curcio CA, Millican CL (1999) Basal linear deposit and large drusen are specific for early age-related maculopathy. Arch Opthalmol 117:329–339

Curcio CA, Sloan KR, Kalina RE, Hendrickson AE (1990) Human photoreceptor topography. J Comp Neurol 292:497–523

Curcio CA, Millican CL, Allen KA, Kalina RE (1993) Aging of the human photoreceptor mosaic: evidence for selective vulnerability of rods in central retina. Invest Ophthalmol Vis Sci 34:3278–3296

Curcio CA, Medeiros NE, Millican CL (1996) Photoreceptor loss in age-related macular degeneration. Invest Ophthalmol Vis Sci 37:1236–1249

Curcio CA, Saunders PL, Younger PW, Malek G (2000) Peripapillary chorioretinal atrophy: Bruch's membrane changes and photoreceptor loss. Ophthalmology 107:334–343

Curcio CA, Millican CL, Bailey T, Kruth HS (2001) Accumulation of cholesterol with age in human Bruch's membrane. Invest Ophthalmol Vis Sci 42:265–274

Delori FC, Dorey CK, Staurenghi G, Arend O, Goger DG, Weiter JJ (1995) In vivo fluorescence of the ocular fundus exhibits retinal pigment epithelium lipofuscin characteristics. Invest Ophthalmol Vis Sci 36:718–729

Delori FC, Fleckner MR, Goger DG, Weiter JJ, Dorey CK (2000) Autofluorescence distribution associated with drusen in age-related macular degeneration. Invest Ophthalmol Vis Sci 41:496–504

Del Priore LV, Kuo Y-H, Tezel TH (2002) Age-related changes in human RPE cell density and apoptosis proportion in situ. Invest Ophthalmol Vis Sci 43:3312–3318

Dowling J, Wald G (1958) Vitamin A deficiency and night blindness. Proc Natl Acad Sci USA 44:648–661

Dunaief JL, Dentchev T, Ying GS, Milam AH (2002) The role of apoptosis in age-related macular degeneration. Arch Opthalmol 120:1435–1442

Eisner A, Fleming SA, Klein ML, Mauldin WM (1987) Sensitivities in older eyes with good acuity: eyes whose fellow eye has exudative AMD. Invest Ophthalmol Vis Sci 28:1832–1837

Eldred GE (1998) Lipofuscin and other lysosomal storage deposits in the retinal pigment epithelium. In: Wolfensberger TJ (ed) The retinal pigment epithelium: function and disease. Oxford University Press, New York, pp 651–668

Feeney-Burns L, Ellersieck MR (1985) Age-related changes in the ultrastructure of Bruch's membrane. Am J Ophthalmol 100:686–697

Feeney-Burns L, Hilderbrand E, Eldridge S (1984) Aging human RPE: morphometric analysis of macular, equatorial, and peripheral cells. Invest Ophthalmol Vis Sci 25:195–200

Feeney-Burns L, Burns RP, Gao C-L (1990) Age-related macular changes in humans over 90 years old. Am J Ophthalmol 109:265–278

Friedrichson T, Kalbach HL, Buck P, Kuijk FJGM van (1995) Vitamin E in macular and peripheral tissues of the human eye. Curr Eye Res 14:693–701

Gao H, Rayborn ME, Meyers KM, Hollyfield JG (1990) Differential aging of neurons during aging of human retina. Invest Ophthalmol Vis Sci 31:357

Gellermann W, Ermakov IV, Ermakova MR, McClane RW, Zhao DY, Bernstein PS (2002) In vivo resonant Raman measurement of macular carotenoid pigments in the young and the aging human retina. J Opt Soc Am A Opt Image Sci Vis 19:1172–1186

Goldstein EB, Wolf BM (1973) Regeneration of the green-rod pigment in the isolated frog retina. Vision Res 13:527–534

Gotoda T, Arita M, Arai H, Inoue K, Yokota T, Fukuo Y, Yazaki Y, Yamada N (1995) Adult-onset spinocerebellar dysfunction caused by a mutation in the gene for the alpha-tocopherol-transfer protein. N Engl J Med 333:1313–1318

Green WR, Enger C (1993) Age-related macular degeneration histopathologic studies: the 1992 Lorenz E. Zimmerman Lecture. Ophthalmology 100:1519–1535

Macular Photocoagulation Study Group (1991) Subfoveal neovascular lesions in age-related macular degeneration. Guidelines for evaluation and treatment in the Macular Photocoagulation Study. Arch Ophthalmol 109:1242–1257

Haimovici R, Gantz DL, Rumelt S, Freddo TF, Small DM (2001) The lipid composition of drusen, Bruch's membrane, and sclera by hot stage polarizing microscopy. Invest Ophthalmol Vis Sci 42:1592–1599

Handa JT, Verzijl N, Matsunaga H, Aotaki-Keen A, Lutty GA, Koppele JM te, Miyata T, Hjelmeland LM (1999) Increase in the advanced glycation end product pentosidine in Bruch's membrane with age. Invest Ophthalmol Vis Sci 40:775–779

Harman AM, Fleming PA, Hoskins RV, Moore SR (1997) Development and aging of cell topography in the human retinal pigment epithelium. Invest Ophthalmol Vis Sci 38:2016–2026

Hayward C, Shu X, Cideciyan AV, Lennon A, Barran P, Zareparsi S, Sawyer L, Hendry G, Dhillon B, Milam AH, Luthert PJ, Swaroop A, Hastie ND, Jacobson SG, Wright AF (2003) Mutation in a short-chain collagen gene, CTRP5, results in extracellular deposit formation in late-onset retinal degeneration: a genetic model for age-related macular degeneration. Hum Mol Genet 12:2657–2667

Hecht S, Mandelbaum J (1939) The relation between vitamin A and dark adaptation. J Am Med Assoc 112:1910–1916

Heckenlively JR (1988) Retinitis pigmentosa. Lippincott, Philadelphia

Hoang Q, Linsenmeier RA, Chung C, Curcio CA (2002) Photoreceptor inner segments in monkey and human retina: mitochondrial density, optics, and regional variation. Vis Neurosci 19:395–407

Hogan MJ (1972) Role of the retinal pigment epithelium in macular disease. Trans Am Acad Ophthalmol Otolaryngol 76:64–80

Holopigian K, Seiple W, Greenstein V, Kim D, Carr RE (1997) Relative effects of aging and age-related macular degeneration on peripheral visual function. Optom Vis Sci 74:152–159

Holz FG, Schütt F, Kopitz J, Eldred GE, Kruse FE, Völcker HE, Cantz M (1999) Inhibition of lysosomal degradative functions in RPE cells by a retinoid component of lipofuscin. Invest Ophthalmol Vis Sci 40:737–743

Holz FG, Bellman C, Staudt S, Schutt F, Volcker H (2001) Fundus autofluorescence and development of geographic atrophy in age-related macular degeneration. Invest Ophthalmol Vis Sci 42:1051–1056

Hood DC, Hock PA (1973) Recovery of cone receptor activity in the frog's isolated retina. Vis Res 13:1943–1951

Huang S, Wu D, Jiang F, Ma J, Wu L, Liang J, Luo G (2000) The multifocal electroretinogram in age-related maculopathies. Doc Ophthalmol 101:115–124

Hussain AA, Rowe L, Marshall J (2002) Age-related alterations in the diffusional transport of amino acids across the human Bruchs-choroid complex. J Opt Soc Am A 19:166–172

Hyman L, Schachat AP, He Q, Leske MC (2000) Hypertension, cardiovascular disease, and age-related macular degeneration. Arch Opthalmol 117:351–358

Jackson GR, Owsley C (2000) Scotopic sensitivity during adulthood. Vis Res 40:2467–2473

Jackson GR, Owsley C, Cordle EP, Finley CD (1998) Aging and scotopic sensitivity. Vis Res 38:3655–3662

Jackson GR, Edwards DJ, McGwin G, Owsley C (1999a) Changes in dark adaptation in early AMD. Invest Ophthalmol Vis Sci (Suppl) 40:739

Jackson GR, Owsley C, McGwin G (1999b) Aging and dark adaptation. Vis Res 39:3975–3982

Jacobson SG, Voigt WJ, Parel J-M, Apathy PP, Nghiem-Phu L, Myers SW, Patella VM (1986) Automated light- and dark-adapted perimetry for evaluating retinitis pigmentosa. Ophthalmology 93:1604–1611

Jacobson SG, Cideciyan AV, Regunath G, Rodriguez FJ, Vandenburgh K, Sheffield VC, Stone EM (1995) Night blindness in Sorsby's fundus dystrophy reversed by vitamin A. Nat Genet 11:27–32

Johnson PT, Lewis GP, Talaga KC, Brown MN, Kappel PJ, Fisher SK, Anderson DH, Johnson LV (2003) Drusen-associated degeneration in the retina. Invest Ophthalmol Vis Sci 44:4481–4488

Jones GJ, Crouch RK, Wiggert B, Cornwall MC, Chader GJ (1989) Retinoid requirements for recovery of sensitivity after visual-pigment bleaching in isolated photoreceptors. Proc Natl Acad Sci USA 86:9606–9610

Jurklies B, Weismann M, Husing J, Sutter EE, Bornfeld N (2002) Monitoring retinal function in neovascular maculopathy using multifocal electroretinography – early and long-term correlation with clinical findings. Graefes Arch Clin Exp Ophthalmol 240:244–264

Karwatowski W, Jeffried T, Duance V, Albon J, Bailey A, Easty D (1995) Preparation of Bruch's membrane and analysis of the age-related changes in the structural collagens. Br J Ophthalmol 79:944–952

Katz M, Kutryb M, Norberg M, Gao C, White R, Stark W (1991) Maintenance of opsin density in photoreceptor outer segments of retinoid-deprived rats. Invest Ophthalmol Vis Sci 32:1968–1980

Katz M, Gao C, Stientjes H (1993) Regulation of the interphotoreceptor retinoid-binding protein content of the retina by vitamin A. Exp Eye Res 57:393–401

Kayden HJ, Traber MG (1993) Absorption, lipoprotein transport, and regulation of plasma concentrations of vitamin E in humans. J Lipid Res 34:343–358

Kemp C, Jacobson S, Faulkner D, Walt R (1988) Visual function and rhodopsin levels in humans with vitamin A deficiency. Exp Eye Res 46:185–197

Kemp C, Jacobson S, Borruat F, Chaitin M (1989) Rhodopsin levels and retinal function in cats during recovery from vitamin A deficiency. Exp Eye Res 49:49–65

Kim RY, Faktorovich EG, Kuo CY, Olson JL (1997) Retinal function abnormalities in membranoproliferative glomerulonephritis type II. Am J Ophthalmol 123:619–628

Klein BE, Klein R, Lee KE, Jensen SC (2001) Measures of obesity and age-related eye diseases. Ophthalmic Epidemiol 8:251–262

Klein R, Davis MD, Magli YL, Segal P, Klein BEK, Hubbard L (1991) The Wisconsin Age-Related Maculopathy Grading System. Ophthalmology 98:1128–1134

Klein R, Wang Q, Klein BEK, Moss SE, Meuer SM (1995) The relationship of age-related maculopathy, cataract, and glaucoma to visual acuity. Invest Ophthalmol Vis Sci 36:182–191

Klein R, Klein BEK, Jensen SC, Meuer SM (1997) The five-year incidence and progression of age-related maculopathy. Ophthalmology 104:7–21

Klein R, Klein BE, Tomany SC, Meuer SM, Huang GH (2002) Ten-year incidence and progression of age-related maculopathy: the Beaver Dam eye study. Ophthalmology 109:1767–1779

Kosnik W, Winslow L, Kline D, Rasinski K, Sekuler R (1988) Visual changes in daily life throughout adulthood. J Gerontol 43:P63–70

Lamb TD, Cideciyan AV, Jacobson SG, Pugh EN (1998) Towards a molecular description of human dark adaptation. J Physiol (Lond) 506:88

Lambooij AC, Kliffen M, Kuijpers RWAM, Houtsmuller AB, Broese JJ, Mooy CM (2000) Apoptosis is present in the primate macula at all ages. Graefes Arch Clin Exp Ophthalmol 238:508–514

Landrum JT, Bone RA, Joa H, Kilburn MD, Moore LL, Sprague KE (1997) A one year study of the macular pigment: the effect of 140 days of a lutein supplement. Exp Eye Res 65:57–62

LaVail MM (1981) Analysis of neurological mutants with inherited retinal degeneration. Invest Ophthalmol Vis Sci 21:638–657

Leibrock CS, Reuter T, Lamb TD (1998) Molecular basis of dark adaptation in rod photoreceptors. Eye 12:511–520

Li T, Sandberg MA, Pawlyk BS, Rosner B, Hayes KC, Dryja TP, Berson EL (1998) Effect of vitamin A supplementation on rhodopsin mutants threonine-17 →methionine and proline-347→serine in transgenic mice and in cell cultures. Proc Natl Acad Sci USA 95:11933–11938

Linsenmeier RA, Padnick-Silver L (2000) Metabolic dependence of photoreceptors on the choroid in the normal and detached retina. Invest Ophthalmol Vis Sci 41:3117–3123

Macular Photocoagulation Study Group (1997) Risk factors for choroidal neovascularization in the second eye of patients with juxtafoveal or subfoveal choroidal neovascularization secondary to age-related macular degeneration. Arch Ophthalmol 115:741–747

Malek G, Li C-M, Guidry C, Medeiros NE, Curcio CA (2003) Apolipoprotein B in cholesterol-containing drusen and basal deposits in eyes with age-related maculopathy. Am J Pathol 162:413–425

Mangelsdorf DJ (1994) The retinoid receptors. In: Goodman D (ed) The retinoids: biology, chemistry, and medicine, 2nd edn. Raven Press, New York, pp 319–350

Marshall J, Hussain AA, Starita C, Moore DJ, Patmore AL (1998) Aging and Bruch's membrane. In: Wolfensberger TJ (ed) The retinal pigment epithelium: function and disease. Oxford University Press, New York, pp 669–692

Mata N, Tzekov R, Liu X, Weng J, Birch D, Travis G (2001) Delayed dark-adaptation and lipofuscin accumulation in abcr+/– mice: implications for involvement of ABCR in age-related macular degeneration. Invest Ophthalmol Vis Sci 42:1685–1690

Mata N, Radu R, Clemmons R, Travis G (2002) Isomerization and oxidation of vitamin a in cone-dominant retinas. A novel pathway for visual-pigment regeneration in daylight. Neuron 36:69

Mayer MJ, Ward B, Klein R, Talcott JB, Dougherty RF, Glucs A (1994) Flicker sensitivity and fundus appearance in pre-exudative age-related maculopathy. Invest Ophthalmol Vis Sci 35:1138–1149

McBee JK, Palczewski K, Baehr W, Pepperberg DR (2001) Confronting complexity: the interlink of phototransduction and retinoid metabolism in the vertebrate retina. Prog Retin Eye Res 20:469–529

McGwin GJ, Owsley C, Curcio CA, Crain RJ (2003) The association between statin use ang age related maculopathy. Br J Ophthalmol 87:1–5

Medeiros NE, Curcio CA (2001) Preservation of ganglion cell layer neurons in age-related macular degeneration. Invest Ophthalmol Vis Sci 42:795–803

Midena E, Degli Angeli C, Blarzino MC, Valenti M, Segato T (1997) Macular function impairment in eyes with early age-related macular degeneration. Invest Ophthalmol Vis Sci 38:469–477

Mohand-Said S, Deudon-Combe A, Hicks D, Simonutti M, Forster V, Fintz AC, Leveillard T, Dreyfus H, Sahel JA (1998) Normal retina releases a diffusible factor stimulating cone survival in the retinal degeneration mouse. Proc Natl Acad Sci USA 95:8357–8362

Moore DJ, Clover GM (2001) The effect of age on the macromolecular permeability of human Bruch's membrane. Invest Ophthalmol Vis Sci 42:2970–2975

Moore DJ, Hussain AA, Marshall J (1995) Age-related variation in the hydraulic conductivity of Bruch's membrane. Invest Ophthalmol Vis Sci 36:1290–1297

Organisciak DT, Berman ER, Wang HM, Feeney-Burns L (1987) Vitamin E in human neural retina and retinal pigment epithelium: effect of age. Curr Eye Res 6:1051–1055

Østerberg GA (1935) Topography of the layer of rods and cones in the human retina. Acta Ophthalmol 13 (Suppl 6):1–103

Owsley C, Jackson GR, Cideciyan AV, Huang Y, Fine SL, Ho AC, Maguire MG, Lolley V, Jacobson SG (2000) Psychophysical evidence for rod vulnerability in age-related macular degeneration. Invest Ophthalmol Vis Sci 41:267–273

Owsley C, Jackson GR, White M, Feist R, Edwards DJ (2001) Delays in rod-mediated dark adaptation in early age-related maculopathy. Ophthalmology 108:1196–1202

Panda-Jonas S, Jonas JB, Jokobczyk-Zmija (1995) Retinal photoreceptor density decreases with age. Ophthalmology 102:1853–1859

Parish CA, Hashimoto M, Nakanishi K, Dillon J, Sparrow J (1998) Isolation and one-step preparation of A2E and iso-A2E, fluorophores from human retinal pigment epithelium. Proc Natl Acad Sci USA 95:14609–14613

Pauleikhoff D, Harper CA, Marshall J, Bird AC (1990) Aging changes in Bruch's membrane: a histochemical and morphological study. Ophthalmology 97:171–178

Phipps JA, Guymer RH, Vingrys AJ (2003) Loss of cone function in age-related maculopathy. Invest Ophthalmol Vis Sci 44:2277–2283

Polyak SL (1941) The retina. University of Chicago, Chicago

Rakoczy PE, Zhang D, Robertson T, Barnett NL, Papadimitriou J, Constable IJ, Lai CM (2002) Progressive age-related changes similar to age-related macular degeneration in a transgenic mouse model. Am J Pathol 161:1515–1524

Ramrattan RS, Schaft TL van der, Mooy CM, Bruijn WC de, Mulder PGH, Jong PTVM de (1994) Morphometric analysis of Bruch's membrane, the choriocapillaris, and the choroid in aging. Invest Ophthalmol Vis Sci 35:2857–2864

Remé CE, Grimm C, Hafezi F, Iseli HP, Wenzel A (2003) Why study rod cell death in retinal degenerations and how? Doc Ophthalmol 106:25–29

Remky A, Lichtenberg K, Elsner AE, Arend O (2001) Short wavelength automated perimetry in age related maculopathy. Br J Ophthalmol 85:1432–1436

Ridge KD, Abdulaev NG, Sousa M, Palczewski K (2003) Phototransduction: crystal clear. Trends Biochem Sci 28:479–487

Robison WG Jr, Kuwabara T, Bieri JG (1980) Deficiencies of vitamins E and A in the rat. Retinal damage and lipofuscin accumulation. Invest Ophthalmol Vis Sci 19:1030–1037

Ruberti JW, Curcio CA, Millican CL, Menco BPM, Huang J-D, Johnson M (2003) Quick-freeze/deep-etch visualization of age-related lipid accumulation in Bruch's membrane. Invest Ophthalmol Vis Sci 44:1753–1759

Rückmann A von, Fitzke FW, Bird AC (1997) Fundus autofluorescence in age-related macular disease imaged with a laser scanning ophthalmoscope. Invest Ophthalmol Vis Sci 38:478–486

Saari J (1994) Retinoids in photosensitive systems. In: Goodman D (ed) The retinoids: biology, chemistry, and medicine, 2nd edn. Raven Press, New York, pp 351–385

Samuel W, Kutty RK, Nagineni S, Gordon JS, Prouty SM, Chandraratna RAS, Wiggert B (2001) Regulation of stearoyl coenzyme A desaturase expression in human retinal pigment epithelial cells by retinoic acid. J Biol Chem 276:28744–28750

Sarks SH (1976) Ageing and degeneration in the macular region: a clinico-pathological study. Br J Ophthalmol 60:324–341

Schutt F, Davies S, Kopitz J, Holz F, Boulton M (2000) Photodamage to human RPE cells by A2-E, a retinoid component of lipofuscin. Invest Ophthalmol Vis Sci 41:2303–2308

Scilley K, Jackson GR, Cideciyan AV, Maguire MG, Jacobson SG, Owsley C (2002) Early age-related maculopathy and self-reported visual difficulty in daily life. Ophthalmology 109(7):1235–1242

Seddon JM, Cote J, Davis N, Rosner B (2003) Progression of age-related macular degeneration: association with body mass index, waist circumference, and waist-hip ratio. Arch Opthalmol 121:785–792

Smith W, Assink J, Klein R, Mitchell P, Klaver CCW, Klein BEK, Hofman A, Jensen S, Wang JJ, Jong PTVM de (2001) Risk factors for age-related macular degeneration. Pooled findings from three continents. Ophthalmology 108:697–704

Snoddery DM (1995) Evidence for protection against age-related macular degeneration (AMD) by carotenoids and antioxidant vitamins. Am J Clin Nutr 62 (Suppl):1448S–1461S

Spraul CW, Grossniklaus HE (1997) Characteristics of drusen and Bruch's membrane in postmortem eyes with age-related macular degeneration. Arch Opthalmol 115:267–273

Spraul CW, Lang GE, Grossniklaus HE (1996) Morphometric analysis of the choroid, Bruch's membrane, and retinal pigment epithelium in eyes with age-related macular degeneration. Invest Ophthalmol Vis Sci 37:2724–2735

Stone E, Sheffield V, Hageman G (2001) Molecular genetics of age-related macular degeneration. Hum Mol Genet 10:2285–2292

Streeten BW (1969) Development of the human retinal pigment epithelium and the posterior segment. Arch Opthalmol 81:383–394

Sunness J, Massof R, Johnson M, Rubin G, Fine SL (1989) Dimished foveal sensitivity may predict the development of advanced age-related macular degeneration. Ophthalmology 96:375–381

Suter M, Reme C, Grimm C, Wenzel A, Jaattela M, Esser P, Kociok N, Leist M, Richter C (2000) Age-related macular degeneration. The lipofusion component N-retinyl-N-retinylidene ethanolamine detaches pro-apoptotic proteins from mitochondria and induces apoptosis in mammalian retinal pigment epithelial cells. J Biol Chem 275:39625–39630

Thompson DA, Gal A (2003) Vitamin A metabolism in the retinal pigment epithelium: genes, mutations, and diseases. Prog Retin Eye Res 22:683–703

Tolentino MJ, Miller S, Gaudio AR, Sandberg MA (1994) Visual field deficits in early age-related macular degeneration. Vis Res 34:409–413

Tso MOM, Friedman E (1968) The retinal pigment epithelium. III. Growth and development. Arch Opthalmol 80:214–216

Van Hooser JP, Aleman TS, He Y-G, Cideciyan AV, Kuksa V, Pittler SJ, Stone EM, Jacobson SG, Palczewski K (2000) Rapid restoration of visual pigment and function with oral retinoid in a mouse model of childhood blindness. Proc Natl Acad Sci USA 97:8623–8628

Vogel S, Piantedosi R, O'Byrne SM, Kako Y, Quadro L, Gottesman ME, Goldberg IJ, Blaner WS (2002) Retinol-binding protein-deficient mice: biochemical basis for impaired vision. Biochemistry 41:15360–15368

Wald G (1935) Carotenoids and the visual cycle. J Gen Physiol 19:351

Wang Q, Chappell RJ, Klein R, Eisner A, Klein BEK, Jensen SC, Moss SE (1996) Pattern of age-related maculopathy in the macular area. The Beaver Dam eye study. Invest Ophthalmol Vis Sci 37:2234–2242

Watzke RC, Soldevilla JD, Trune DR (1993) Morphometric analysis of human retinal pigment epithelium: correlation with age and location. Curr Eye Res 12:133–142

Weber BH, Lin B, White K, Kohler K, Soboleva G, Herterich S, Seeliger MW, Jaissle GB, Grimm C, Reme C, Wenzel A, Asan E, Schrewe H (2002) A mouse model for Sorsby fundus dystrophy. Invest Ophthalmol Vis Sci 43:2732–2740

Werner JS, Donnelly SK, Kliegl R (1987) Aging and human macular pigment density. Vis Res 27:257–268

Xu GZ, Li WW, Tso MO (1996) Apoptosis in human retinal degenerations. Trans Am Ophthalmol Soc 94:411–430; 430–431

Yamada E (1969) Some structural features of the fovea centralis of the human retina. Arch Opthalmol 82:151–159

Yokota T, Shiojiri T, Gotoda T, Arita M, Arai H, Ohga T, Kanda T, Suzuki J, Imai T, Matsumoto H, Harino S, Kiyosawa M, Mizusawa H, Inoue K (1997) Friedreich-like ataxia with retinitis pigmentosa caused by the His101Gln mutation of the alpha-tocopherol transfer protein gene. Ann Neurol 41:826–832

Yokota T, Igarashi K, Uchihara T, Jishage K-i, Tomita H, Inaba A, Li Y, Arita M, Suzuki H, Mizusawa H, Arai H (2001) Delayed-onset ataxia in mice lacking alpha-tocopherol transfer protein: model for neuronal degeneration caused by chronic oxidative stress. Proc Natl Acad Sci USA 98:15185–15190

# Genes and Age-Related Macular Degeneration

Robyn H. Guymer, Niro Narendran, Paul N. Baird

## Contents

4.1 Age-Related Macular Degeneration, a Complex Genetic Disease   63

4.2 Genetic Basis of Disease   64
4.2.1 Family Studies   64
4.2.2 Sibling Studies   64
4.2.3 Twin Studies   65

4.3 Approaches to Genetic Investigation of AMD   65
4.3.1 Linkage Analysis   65
4.3.2 Candidate-Gene Screening   67

4.4 Future Directions   73
4.4.1 Single-Nucleotide Polymorphisms   73
4.4.2 Microarrays   73
4.4.3 Proteomics   74

4.5 Conclusions   74

References   75

## 4.1 Age-Related Macular Degeneration, a Complex Genetic Disease

The completion of the Human Genome Project (HGP), which provided the first high-quality, comprehensive sequence of the entire human genome offers a unique and powerful resource for the elucidation of genes involved in human disease. Completion of this huge task has, however, led to more questions than it has answered. It has shown that only 1–2% of the sequence encodes for proteins and, of the estimated 30,000 genes in the human genome, only a small percentage have so far been identified (The International Human Genome Sequencing Consortium 2001). Little is known about the regulatory elements that occur in the noncoding portion of the genome and how they affect the function of these genes. While the completion of the HGP is without doubt a huge step forward towards understanding genetic diseases, genes and their products do not function independently but interact with other genes through complex pathways; and for many common diseases, there is interplay between genetic and nongenetic risk factors (environment) that dictate susceptibility to disease and treatment response. Deciphering the role of genes in human disease, given this complexity of interactions, is a formidable task. However, we are now in an unprecedented positon, with the advances in genetic techniques and the emergence of the study of genetic epidemiology, to finally make discoveries about the genetic associations with complex genetic diseases such as age-related macular degeneration (AMD).

The aging population and increasing life expectancy necessitate urgent answers as to the causes and risk factors for AMD, and present vision researchers with a major challenge for the new millennium. AMD is considered a complex genetic disease, where environmental risk factors impact on a genetic background. Finding the genes that determine susceptibility or modify the disease process offers possibly the greatest chance of understanding the underlying disease processes and the development of preventative strategies and treatments.

This chapter explores our current knowledge about the genetic influences on AMD and indicates possible directions for future study.

## 4.2
## Genetic Basis of Disease

The genetic basis to AMD has really only been established over the last 15–20 years. There is now an abundance of evidence from family, sibling and twin studies to support a genetic basis for AMD.

### 4.2.1
### Family Studies

An early report by Gass found that 38 of 200 patients (19%) with macular drusen reported a family history (Gass 1973). Later, Hyman et al. (1983) documented that AMD patients were 3 times more likely to report a positive family history than age-matched controls, with 21.6% of 228 AMD patients giving a positive family history. A practice-based mail survey found that affected individuals' reporting of a family history of AMD was 28.4% of 391 subjects (Keverline et al. 1998). Studies such as these are subject to the level of knowledge of the individual. Early-stage disease which is asymptomatic is not diagnosed without ophthalmic examination, thus an under-reporting of disease is possible in both controls and cases. Given the general lack of awareness of this disease in the population, individuals diagnosed with AMD may be more likely to recall information about their relatives with the same disease than individuals who do not have a diagnosis of AMD, introducing another possibility of biased reporting (Rosenthal and Thompson 2003).

Studies that perform ophthalmic examination of families are more reliable and several of these have been published over the last 10 years. Examination of 119 AMD patients from ophthalmology clinics has revealed a prevalence of AMD of 23.7% in first-degree relatives compared with 11.6% in first degree relatives of 72 nonaffected cases (Seddon et al. 1997). The difference is significant in relation to exudative features, but not atrophic features. Another study has examined families that were derived from index cases seen in the population-based Rotterdam Study (Klaver et al. 1998a, 1998b, 1998c). The families were chosen regardless of family history and other known risk factors, and all available siblings and children of the index cases have been seen. The occurrence of early- and late-stage AMD is higher in first-degree relatives of late-stage AMD patients than in relatives of controls. This is independent of known risk factors such as smoking. The calculated lifetime absolute risk of developing late-stage AMD in first-degree relatives of late-stage AMD index cases is 50% (95%CI=26–73%), compared with 12% in relatives of controls (95%CI=2–16%; Klaver et al. 1998a, 1998b, 1998c). Their calculated relative risk for a genetic component to AMD is 23%. There is low concordance of similarity of clinical features between members of the same family, except in relation to stage of disease. In addition the relatives of patients develop AMD features at a younger age than those relatives of controls.

### 4.2.2
### Sibling Studies

Sib-pairs are easier to recruit than large family pedigrees of AMD. Segregation analysis performed on sibling data from the Beaver Dam Eye Study (564 families) has shown significant sibling correlations where a younger sibling has an increased likelihood of developing the specific macular lesion possessed by an older sibling (soft drusen, pigmentary changes and exudative changes) after 5 years. (Heiba et al. 1994; Klein et al. 2001)

Piguet et al. have compared the drusen characteristics in 53 siblings pairs with 50 long-term spouse pairs. This type of comparison minimises environmental variation between pairs. In the sibling pairs, but not the spouse pairs, a trend has been found towards concordance of drusen features (number, size and confluence), suggesting that genetic factors are influential in the manifestation of retinal disease (Piguet et al. 1993). In 1994, Silvestri et al. examined 36 index cases with AMD and identified 20 of 81 of their siblings to be affected with AMD, compared with 36 unaffected controls who had only 1 of 78 siblings with AMD. Once again, there appeared to be segregation of a gene(s) with disease (Silvestri et al.1994).

## 4.2.3
## Twin Studies

Monozygotic (MZ) twins share 100% of their DNA and dizygotic (DZ) twins share 50% of their DNA, thus diseases that have a large genetic component should manifest more commonly in MZ than DZ twins and more commonly in both types of twins than in unrelated individuals. Twins are thus a valuable resource for investigating genetic tendencies in disease.

The occurrence of choroidal neovascular membranes (CNVM) in the same eye of a set of MZ twins was reported for the first time in 1985 (Melrose et al. 1985). Following this, Meyer and Zachary documented the first occurrence of bilateral CNVM in a set of zygosity-confirmed monozygotic twins (Meyer and Zachary 1988). A study involving nine MZ twin pairs has observed concordance of AMD features in all sets of twins and found that there is a similarity of features and vision in eight of the nine pairs (Klein et al. 1994).

Meyer et al. again investigated twins, this time looking at both MZ and DZ twin pairs. They have found 100% concordance of clinical features in 25 MZ twin pairs affected with AMD and concordance of 42% in 12 DZ twins studied (Meyer et al. 1995). Another twin study has found the concordance of AMD features in 50 MZ pairs to be 90%, which is significantly higher than a comparison of twins with their spouse pairs (Gottfredsdottir et al. 1999). All the twin studies thus far described have used small numbers, reflecting the difficulty in collecting appropriate subjects.

The most recently published twin study has investigated early AMD features in a large number of MZ (226) and DZ (280) females (Hammond et al. 2002). They confirm a significant influence of genetic factors and conclude that the features that appear to be most heritable are large drusen (more than or equal to 125 µm) and more than or equal to 20 hard drusen, and calculate the heritability of these features as 45% (95%CI=35–53%). The authors suggest that the lower concordance found in this study compared with previous studies could be explained by the use of a population-based approach rather than selection of twin pairs.

Twin, sibling and family studies strongly suggest a genetic tendency in AMD. However, it is not known how many genes are involved, what sort of inheritance pattern the disease has, nor whether the genetic influence has a causative or modifying effort.

## 4.3
## Approaches to Genetic Investigation of AMD

Identifying the gene(s) involved in AMD is a high priority and currently involves a large number of international groups and collaborations. Success in genetic investigation is, to a large degree, dependent upon the sample type and analysis used. Family studies are usually preferred, if available, followed by twins/siblings and then population-based studies. With regard toAMD, the main techniques employed to investigate the genetic influences have thus far been linkage analysis and candidate-gene screening.

## 4.3.1
## Linkage Analysis

Linkage is the study of the co-segregation of markers at different, but close genetic loci, on the basis of phenotypic observations. It relies on the fact that the pedigree relationship and preferably the inheritance of the disease to be studied are known. It also assumes that the gene for the disease being studied lies close to a marker under test and that these are inherited and segregate with the disease phenotype.

Measurement of the LOD score (logarithm of the odds) determines the likelihood that the marker studied segregates with disease. A significant LOD score is considered to be 3.00, representing a $P$-value of 0.001, which means that there is only a 1:1000 chance that the segregation could have occurred as a chance event. Once a disease locus has been identified, the genes that lie within it need to be identified and analysed in order to locate the putative causative gene.

#### 4.3.1.1
#### Difficulties with Linkage Analysis in AMD

Optimal linkage analysis occurs if a disease gene segregates with disease over several generations and multiple individuals are affected. However, in AMD there is a paucity of large families due to the late onset of the disease, the lack of living parents, and the young age of offspring where manifestation of disease may not yet have occurred. Even with a positive history in the parents' generation, the absence of medical documentation, which is often the case, means that the mode of inheritance cannot be reliably ascertained.

Definition of clinical phenotypes is critical for linkage analysis. In the case of AMD, this is difficult due to the great variation in clinical picture and similarity with other macular disease phenotypes. Clinical heterogeneity is common even within the same family. Whether these are all indeed the same disease is still unclear and may have implications for the accuracy of linkage results (De La Paz et al. 1997a, 1997b). Given the well-established rise in prevalence with increasing age, the assignment of an "unaffected status" is difficult, as it is not known whether that individual will develop disease later in life. In addition, the strong evidence for an environmental component and the possibility that AMD is a multigene disease further complicates linkage analysis.

#### 4.3.1.2
#### Results of Linkage Studies in AMD

*4.3.1.2.1*
*Family Studies*

A significant LOD score for AMD has only been demonstrated for one large family in the published literature. Klein et al. describe a large family (ten affected members) spanning 3 generations with predominantly atrophic AMD inherited in an autosomal dominant fashion. Linkage analysis reveals that the likely disease locus maps to chromosome 1q25-q31 with a significant LOD score of 3.00 (Klein et al. 1998). Sequencing of many genes in this region has revealed that the AMD phenotype seen in this family is associated with a Gln5345Arg change in the *Fibulin 6* (*FIBL-6*) gene (Schultz et al. 2003a, 2003b). This change is also found in some affected members of other AMD families and in a sporadic case of AMD. Interestingly it has also been identified in some unaffected members of AMD families and two control subjects (Schultz et al. 2003a, 2003b). Given the fact that the change was found in controls and unaffected family members, and is present in only one family so far, this base change may be a polymorphism rather than a disease-associated mutation. Clearly further studies are required to clarify the involvement of this gene in AMD.

*4.3.1.2.2*
*Genome-Wide Scans*

Genome-wide scans have typically involved the collection of large numbers of affected families, with the majority of collected individuals consisting of affected sib pairs. Several hundred microsatellite markers spread at regular intervals across the genome have been used to genotype individuals and genetic linkage analysis has been performed to identify potential disease-containing regions.

Some studies undertaking genome-wide scans have concentrated on late (end-stage) disease, while others have looked at additional phenotypic features such as drusen and degree of pigmentation (Weeks et al. 2000, 2001; Majewski et al. 2003; Schick et al. 2003). The first genome-wide scan of 222 families, reported by Weeks et al., resulted in regions of interest on chromosomes 5, 9, 10 and 12 being found (LOD score <1.69; Weeks et al. 2000). Increasing the study to include 364 families has resulted in consistent linkage being found for chromosome 10, with a LOD score of 1.42 (Weeks et al. 2000). Using a second cohort of 391 families, chromosomal regions at 1q31, 9p13, 10q26 and 17q25 have been found, with LOD scores of greater than 2.0, region 17q25 having a maximum LOD score of 3.16 (Weeks et al. 2001). Additional loci on chromosomes 2 and 12 were also identified when the LOD score threshold was reduced to 1.5. On the other hand, Majewski et al. have undertaken a genome-wide scan using 70 families that consist of between 3 and 14 affected individuals with late-stage disease. Several regions with LOD scores greater than 2.0 have been identified at chromosomal regions 1q31, 3p13, 9q33 and 10q26, with the latter region presenting with the highest LOD score of 3.06 (Majewski et al. 2003). Stratification of AMD into atrophic or neovascular disease has uncovered an additional locus at chromosome 4q32 in individuals with atrophic disease (Majewski et al. 2003). Genome-wide scan analysis has also been performed by Schick et al. using 102 families collected through the Beaver Dam Eye Study, with initial linkage and subsequent fine mapping giving the strongest indication for linkage on chromosome 12, while 3 other regions on chromosomes 5, 6 and 15 are also linked (Schick et al. 2003).

The results of the genome scans, performed on AMD, have all identified a number of putative chromosomal regions of interest. However, none have so far reached a significance level of a LOD score of 3.6, recommended for allele-sharing methods based on sib pairs (Lander and Kruglyak 1995). The strong candidates are the loci that can be reproduced in new studies. Several regions have now been reproduced, such as the regions reported in chromosomes 1q and 10q (Weeks et al. 2001; Majewski et al. 2003) and chromosome 12 (Weeks et al. 2000; Schick et al. 2003). The studies showing linkage to chromosome 1q overlap with the locus originally described by Klein et al. in a large AMD family (Klein et al. 1998). However, problems relating to clinical heterogeneity and mode of inheritance of AMD make it difficult to "pin down" and replicate regions with convincing LOD scores through the use of genome-wide scans until a better understanding of the genotype-phenotype relationship exists in this disease. The studies conducted so far have provided several potential regions that can be followed up in subsequent studies. In relation to linkage analysis, replication of results in different studies using different populations provides valuable evidence that a chromosomal region is indeed involved in disease. It remains to be seen how many of these regions will stand the test of replication or will achieve statistical significance.

## 4.3.2
## Candidate-Gene Screening

Given the difficulty of finding large AMD families and a wide range of suggestive chromosomal linkages, an alternative means of genetic investigation is to screen genes that have already been identified as playing a role in diseases that are clinically similar to AMD (Fig. 4.1).

### 4.3.2.1
### Genes Associated with Other Early-Onset Maculopathies

In the case of AMD, there are a number of phenotypically similar, yet distinctly different, maculopathies that have an early-disease onset and established Mendelian inheritance (Table 4.1). Screening the genes that cause these diseases, in AMD patients, could yield clues to the genetic background of disease. Several genes identified from the early-onset maculopathies have already been explored in AMD.

#### 4.3.2.1.1
#### RDS/Peripherin Gene
#### and Macular Pattern Dystrophies

The RDS, or 'retinal degeneration slow' gene was initally identified as causing a photoreceptor disease of the same name in mice (Travis et al. 1989). Later, human mutations in the RDS gene were found to be involved in macular pattern dystrophies. Pattern dystrophy describe the phenotypic appearance of bilateral macular pigmentary changes, which can take many forms. Butterfly macular dystrophy was the first to be described, caused by a G-to-A transition in codon 167 of the RDS gene (Nichols et al. 1993). Wells et al. discovered that other mutations in the same gene can cause both macular dystrophy (arginine to tryptophan substitution at codon 172) and retinitis pigmentosa (cysteine deletion at codon 118/119; Wells et al. 1993). In the same year, a report was published describing retinitis pigmentosa, pattern dystrophy and fundus flavimaculatus in the same family, all caused by the same 3-bp deletion of codon 153 or 154 in the RDS gene (Weleber et al. 1993).

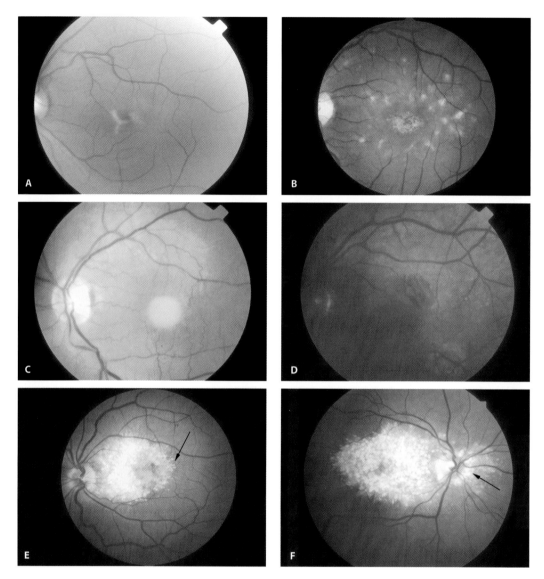

Fig. 4.1A–F. Inherited maculopathies that resemble AMD. The genes causing these diseases provide possible candidates genes for AMD. A Pattern dystrophy; B Stargardt's disease; C Best disease; D Sorsby's fundus dystrophy; E Malattia Leventinese syndrome (*arrow* shows the radial distribution of the drusen); F Doyne's honeycomb retinal dystrophy (*arrow* shows drusen surrounding the nerve head)

Other mutations in the RDS gene are responsible for another retinal phenotype, adult vitelliform macular dystrophy, an autosomal dominant condition, with bilateral yellow foveal deposits similar to those seen in Best disease (Sect. 4.3.2). The disease manifests in the fourth or fifth decade and is thus also known as adult Best disease (Felbor et al. 1997).

Atrophic AMD-like phenotypes, sometimes with drusen, can also result from mutations in the RDS gene, including the Arg172Trp mutation and a proline-to-arginine mutation in co-

Table 4.1. Early-onset macular degenerations and their causative genes

| Phenotype | Inheritance | Genomic locus | Gene |
|---|---|---|---|
| Pattern dystrophy (butterfly macular dystrophy) | AD | 6p21.2 | RDS |
| "Macular" dystrophy with Geographic Atrophy | AD | 6p21.2-cen | RDS |
| Adult vitelliform macular dystrophy | AD | 6p21.2-cen | RDS |
| Stargardt disease | AR | 1p22 | ABCA4 |
| Stargardt-like fundus dystrophy | AD | 6q11-q15 | ELOVL4 |
| Best disease | AD | 11q13 | VMD2 |
| Sorsby fundus dystrophy | AD | 22q13-qter | TIMP3 |
| Doyne honeycomb retinal dystrophy/Malattia Leventinese | AD | 2p16 | EFEMP1 |

don 210 (Wroblewski et al. 1994; Gorin et al. 1995). The end stage results in marked loss of vision and clinically resembles the geographic atrophy of AMD.

RDS represents an interesting example of a single gene, with different mutations, being responsible for a number of different retinal phenotypes ranging from macular-specific disease to more widespread retinal disease. The RDS gene is also an early illustration of digenic inheritance, a phenomenon occurring when mutations in two different and unlinked genes are required to produce a phenotype. Three families have been described with mutations in both the RDS gene on chromosome 6 and in the ROM1 gene on chromosome 11 (both coding for photoreceptor genes) which produce the phenotype of retinitis pigmentosa (Kajiwara et al. 1994).

With regard to AMD, Shastry and Trese have concluded that there is no involvement of RDS mutations in two unrelated families with AMD (Shastry and Trese 1999). Stone et al., in a review of the literature regarding the genetics of AMD, describe their unpublished findings that sequence variations in the RDS gene are not found in 182 AMD patients (Stone et al. 2001). Screening of the RDS gene in 293 AMD patients in our laboratory also does not reveal any involvement (personal communication).

*4.3.2.1.2*
*ABCA4 Gene and Stargardt's Disease*

Stargardt's disease is an autosomal recessive condition characterised by early-onset "pisciform" flecks in the posterior pole at the level of the retinal pigment epithelium (RPE). Later macular atrophy develops. It is the commonest inherited macular dystrophy and occurs in 1 in 10,000 individuals (Duker 1998). Like AMD, it is associated with abnormal accumulation of lipofuscin pigment in the RPE (Petrukhin et al. 1998).

Allikmets et al. identified the *ABCA4* gene as being associated with Stargardt's disease in 1997, when 19 different mutations were identified (Allikmets et al. 1997a). The photoreceptor cell-specific ATP-binding transporter gene (*ABCA4* or, previously, *ABCR*) is located on chromosome 1p and is expressed only in the retina (Kaplan et al. 1993). The *ABCA4* gene is very polymorphic, with 213 different variations in the *ABCA4* gene being reported in 374 index cases with Stargardt's disease (Webster et al. 2001). Not all DNA changes are thought to be disease-causing. Like the RDS gene described above, the *ABCA4* gene has been found to harbour several mutations responsible for a variety of different conditions, including cone-rod dystrophy and retinitis pigmentosa (Martinez-Mir et al. 1998; Cremers et al. 1998).

The role *ABCA4* plays in AMD is controversial. Allikmets et al. initially claimed that mutations in this gene were responsible for 16% of 167 AMD cases examined (Allikmets et al. 1997b). Stone et al., however, did not find any association of the *ABCA4* gene in 182 patients with AMD (Stone et al. 1998). Later, Allikmets and the International ABCR Screening Consortium reported that two of the most frequently

AMD-associated variants (*G1961E* and *D2177N*) were indeed significantly associated with a large number (n=1218) of patients with AMD from Europe and the USA (Allikmets and the International ABCR Screening Consortium 2000). However, these findings have been debated widely in the literature (Klaver et al. 1998c; Stone et al. 1998; De la Paz MA et al. 1999). Allikmets et al. have revised down their estimate of the role of *ABCA4* in AMD to about 3% (Allikmets et al.1997b).

Importantly, Guymer et al. have reported the *G1961E* is in significantly high proportions of individuals of Somalian descent who do not have AMD, showing it to be a normal allelic variant rather than a disease-causing mutation (Guymer et al. 2001). This example highlights the importance of obtaining normal controls from the same population as the cases. Guymer et al. have also reported that there is no significant association of the *G1961E* and *D2177N* variant in individuals with AMD (Guymer et al. 2001).

Thus far, the *ABCA4* gene has not been shown to have a convincing role in AMD. The high frequency of allelic changes makes it difficult to know whether any change is a true disease-causing mutation or a nondisease-associated variant. What role, if any, the *ABCA4* gene plays in AMD needs further clarification.

*4.3.2.1.3*
*ELOVL4 and Autosomal Dominant Stargardt-Like Disease*

Autosomal dominant Stargardt-like disease, as the name suggests, is similar to Stargardt disease but has an autosomal dominant inheritance in contrast to the expected autosomal recessive inheritance.The responsible gene codes for a photoreceptor cell-specific factor that is involved in the elongation of very long fatty acids (*ELOVL4*; Zhang et al. 2001). No significant association of the gene with AMD susceptibility has been found in a study of 778 patients (Ayyagari et al. 2001).

*4.3.2.1.4*
*VMD2/Bestrophin Gene and Best Disease*

Best disease is a rare, bilateral dominantly inherited macular dystrophy that usually appears in childhood. Like Stargardt disease it shares a similarity with AMD because it is associated with an abnormal accumulation of lipofuschin in RPE cells (Petrukhin et al. 1998). Clinically, a yellow lesion appears first in the central macula. This is termed the vitelliform stage owing to the resemblance to an egg yolk. This lesion is subsequently disrupted and results in macular scarring (Duker 1998). The vitelliform macular dystrophy 2 (*VMD2*) gene, lying on chromosome 11q13, is responsible for Best disease and codes for a 585-amino acid protein, bestrophin (Petrukhin et al. 1998).

In a study of 321 AMD patients, the frequency of mutations in the *VMD2* gene is the same as the control population (1.5%), and thus it is concluded that this gene is unlikely to be significantly involved in AMD (Lotery et al. 2000).

*4.3.2.1.5*
*TIMP3 Gene and Sorsby Fundus Dystrophy*

Sorsby fundus dystrophy, another autosomal dominant dystrophy, resembles AMD. Patients present with fine drusen at a young age, progressing to severe exudative disease in the fourth decade (Duker 1998). Mutations in the *TIMP3* gene (which maps to 22q13-qter) are the cause of Sorsby fundus dystrophy (Weber et al. 1994). The protein product belongs to a family of enzymes involved in extracellular matrix modelling. A study of 38 families with AMD has found no linkage or association of the *TIMP3* gene in these families (De La Paz et al. 1997b).

*4.3.2.1.6*
*EFEMP1 Gene and the Fibulin Family*

The presence of drusen as a hallmark feature of AMD has led investigators to aggressively study early-onset dominant drusen phenotypes in order to gain a better understanding of the aetiology and pathogenesis of AMD. In fact Gass has suggested that "familial drusen" and "senile macular degeneration" may be part of the same disease spectrum, differing only in age of onset (Gass 1973).

An early description of a dominant drusen phenotype came from Robert W. Doyne in 1899, when he described the appearance of macular drusen with concurrent drusen around the optic nerve head, particularly nasally. (Doyne 1899). This appearance was termed Doyne honeycomb retinal dystrophy (DHRD), and it was subsequently thought to be due to a dominantly inherited gene (Pearce 1968).

Other phenotypic patterns of dominant drusen have been described and given different names. Examples include Malattia Leventinese (ML), where the classic pattern is radial drusen in the macular region (Klainguti 1932), Hutchinson-Tay choroiditis (Hutchinson and Tay 1875) and "dominant drusen of Bruch's membrane" (Deutman and Jansen 1970). It was anticipated that the genes involved in these dominant drusen phenotypes could play a role in the more complex genetic condition of AMD. Thus the hunt for a causative genes was pursued vigorously and the result eagerly anticipated.

Although dominant drusen can take a number of phenotypic forms and have been given a number of different names, further genetic investigation has revealed that the locus for some classic patterns is the same. Malattia Leventinese and Doyne honeycomb retinal dystrophy are both localised to chromosome 2p16–21 (Gregory et al. 1996, Heon et al. 1996). A detailed phenotypic study of DHRD has found some individuals with a minor degree of radial drusen, a feature more commonly associated with ML (Evans et al. 1997). In 1998 a clinical and genetic study of a large family with ML confirmed the linkage to chromosome 2p16–21 (maximum LOD score of 3.72) and described a variable phenotypic picture in affected members, from numerous radial drusen to a single parapapillary drusen (Edwards et al. 1998). These two studies illustrate some overlap in clinical features between DHRD and ML (Piguet et al. 1995).

Further refinement of the chromosome 2 locus has identified a mutation in the epidermal growth factor (EGF)-containing fibrillin-like extracellular matrix protein-1 (*EFEMP1*) gene in DHRD and ML patients (Stone et al. 1999). The mutation, a single C>T base change in codon 345, results in an arginine-to-tryptophan substitution (Arg345Trp mutation). To date it accounts for all cases of ML and DHRD where the mutation is known. More recently, cases are being reported where the typical phenotype is present without the accompanying genotype (Toto et al. 2002)

*EFEMP1* has been found to accumulate within and beneath RPE cells in areas of drusen in ML and AMD. It does not accumulate in otherwise normal eyes, in areas of drusen formation (RPE, Bruch's membrane, choroid), or in parts of the retina without pathological change (Marmorstein et al. 2002). The *EFEMP1* gene shows homology with the family of extracellular matrix proteins called fibrillins. The protein product of the *EFEMP1* gene was identified in 1995 but its function remains unknown (Lecka-Czernik et al. 1995).

Disappointingly, after the anticipation that a gene causing dominant drusen may be involved in AMD, no mutations in the *EFEMP1* gene have been found to be associated with disease in 494 typical AMD patients (Stone et al. 1999). In addition, the Arg345Trp mutation has not been found to be involved in 54 familial AMD patients (Guymer et al. 2002).

A gene with similarities to *EFEMP1* was discovered in 2000 and given the name *EFEMP2* (Katsanis et al. 2000). This gene occurs on chromosome 11q13, where other retinal disorders have been mapped, and thus the authors suggest that *EFEMP2* may be a candidate gene worth studying.

The Fibulin family is of further interest in the search for an AMD-causing gene. As mentioned previously a familial AMD phenotype has been associated with a Gln5345Arg change in the *Fibulin 6* (*FBLN-6*) gene, suggesting that the mutation may be responsible for a number of cases of AMD (Schultz et al. 2003b). Similarly, results of genome-wide scans also identify the same region as being of interest. (Weeks et al. 2000, Majewski et al. 2003) This gene may be worthy of further investigation in relation to AMD. Recently Stone et al. have reported missense variations in the fibulin 5 gene in a small percent of AMD cases (Stone et al. 2004).

### 4.3.2.2
### Genes Involved in Other Diseases

Investigations have also been carried out on genes that play a role in other degenerative diseases that share similarities with AMD. One example is the investigation of the *APOE* gene, which has been associated with diseases such as cardiovascular diseases and Alzheimer's disease (Strittmatter et al. 1993; Davignon et al. 1999).

*4.3.2.2.1*
*Apolipoprotein E (APOE)*

It is now thought possible that common variations in the DNA, called single-nucleotide polymorphisms (SNPs), may contribute significantly to genetic risk for common diseases. SNPs are DNA sequence variations that occur when a single nucleotide is altered, giving alternatives (alleles) in normal individuals. The APOE4 allele association with common complex diseases is an important example of these polymorphisms. It is thought that the marginal effects of common variations could account for a substantial proportion of the population risk for complex diseases (Amaratunga et al. 1996)

The *APOE* gene, which lies on chromosome 19q13.2, has three co-dominant alleles, ε2, ε3 and ε4, producing 3 isoforms E2, E3 and E4, respectively. The allelic changes are due to point mutations causing amino acid substitutions at positions 112 and 158 on exon 4 of the *APOE* gene (Table 4.2). The ε3 allele is the commonest and is thought to be the ancestral allele, with the ε2 and ε4 alleles being variants.

Possession of the ε4 allele increases the risk and decreases the age of onset of familial Alzheimer's disease, while the ε2 allele decreases the risk and increases the age of onset of familial and sporadic Alzheimer's disease (Corder et al. 1993, 1994). The E4 allele has also been associated with an increased risk of coronary heart disease (Davignon et al. 1999) and reduced longevity (Schachter et al. 1994). *APOE* also plays an important role in the regulation of cholesterol and lipid metabolism (Mahley 1988). *APOE* is an ideal candidate to study in AMD due to epidemiological, anatomical and pathological associations between AMD and lipids and its similarity to these other degenerative diseases.

Studies in AMD have found *APOE* to be the first gene consistently associated with disease. Converse to the situation in cardiovascular disease and Alzheimer's disease, three published studies have indicated that the ε4 allele of the *APOE* gene decreases the risk of AMD, suggesting a protective effect. Klaver et al. have shown a significant decrease in ε4 in 901 controls compared with 88 cases [odds ratio 0.43 (95%CI= 0.21–0.88) (Klaver et al. 1998b]. Souied et al. have analysed 116 patients with exudative disease only and found a significant difference in the ε4 allele frequency (12.1% in the cases compared with 28.6% in the controls; Souied et al. 1998). Schmidt et al. have found the protective effect of ε4 to be significant in only some familial cases of AMD under the age of 70 years (Schmidt et al. 2000). Klaver et al. have also reported that the ε2 allele may increase the risk of AMD [odds ratio 1.5 (95%CI=0.8–2.82); Klaver et al. 1998b]. Schmidt et al. used pooled data from four studies comprising 617 AMD cases and 1260 controls and confirmed the protective effect of the ε4 allele (odds ratio for ε4 carriers 0.54, 95% CI=0.41–0.70; Schmidt et al. 2002). More recently, we have shown a protective effect for AMD in individuals with the ε3 ε4 genotype of APOE in 322 cases and 123 unrelated but ethnically matched control subjects, that was greatest in individuals with atrophic disease (OR 0.35, 95%CI 0.13, 0.92) (Baird et al. 2004). In addition, we were able to show that individuals with late AMD and the ε2 ε3 genotype had a significantly earlier mean age of diagnosis of disease (3.4 years, p=0.015) compared to those with the ε3 v3 genotype and this was most noticeable in females (3.9 years, p=0.011) and in

Table 4.2. Allelic changes responsible for amino acid substitutions in the *APOE* gene

| Allele | Amino Acid 112 | Amino Acid 158 |
|---|---|---|
| ε2 | Cys | Cys |
| ε3 | Cys | Arg |
| ε4 | Arg | Arg |

individuals with neovascular disease (4.7 years, p=0.003) (Baird et al. 2004).

In a Chinese population the ε4 allele has been reported to be unlikely to reduce the risk of AMD (Pang et al. 2000), perhaps highlighting the differences in certain allele frequencies in populations of varying ethnicity. A more recent report, however, has found no association of the ε2 or ε4 allele in 56 families and 104 unrelated cases of AMD (Schultz et al. 2003a).

It is of interest that APOE is associated with the RPE and Bruch's membrane and that APOE immunoreactivity occurs in most types of drusen regardless of phenotype (Anderson et al. 2001). Further analysis of the *APOE* gene is warranted to establish its involvement with the disease, how the alleles interact with the lipid pathway to influence AMD development, and why the influences appear to be opposite to that in cardiovascular disease and Alzheimer's disease.

*4.3.2.2.2*
*Paraoxonase 1 Gene (PON1)*

Another gene involved in the cholesterol pathway has sparked some interest following the potential association of APOE with AMD. The *PON1* gene on chromosome 7 plays a role in lipid metabolism and is involved in cardiovascular disease. A study of Japanese patients has found that certain alleles in the *PON1* gene are associated with an increased likelihood of exudative AMD (Ikeda et al. 2001). Baird et al. have not find any involvement of the *PON1* gene in AMD in an Anglo-Celtic cohort, perhaps reflecting differences in the ethnicity of the two studies (Baird et al. 2003).

### 4.3.2.3
### Other Possible Candidate Genes

Some other genes studied in relation to AMD in single reports without replication to date are: angiotensin converting enzyme (*ACE*; Hamdi et al. 2002); cystatin C (*CST3*; Zurdel et al. 2002); and manganese superoxide dismutase (Kimura et al. 2000). New reports of possible associations with candidate genes clearly need to be reproduced before any credence can be given to them. Undoubtedly as knowledge is gained about pathways and proposed pathways that play a role in AMD pathogenesis, new potential candidate genes will be identified for further study.

## 4.4
## Future Directions

### 4.4.1
### Single-Nucleotide Polymorphisms

Single-nucleotide polymorphisms (SNPs) are DNA sequence variations that occur when a single nucleotide in the DNA sequence is altered, giving alternatives (alleles) in normal individuals. As discussed above (Sect. 4.3.2), the *APOE* gene has three allelic forms as a result of SNPs. These small changes in the DNA sequence of the *APOE* gene are thought to influence an individual's risk of AMD. It is thought that the marginal effects of common variations could account for a substantial proportion of the population risk for complex diseases such as cancer and diabetes (Amaratunga et al. 1996).

SNPs are indeed the most abundant source of genetic variants in the genome, as they occur approximately once every 300 bp (Lander 1996; Chakravartic 1999). There is therefore a potential that they could be used as markers in linkage studies to establish patterns of linkage disequilibrium across the genome, with the eventual aim of performing large-scale genetic association studies. A public collaboration, the International HapMap Project, to capture this type of information, has now been established (http://www.genome.gov/Pages/Resarch/HapMap).

### 4.4.2
### Microarrays

Microarrays consist of the precise positioning of DNA fragments (probes) at a high density on a solid support and have rapidly evolved as a means of generating large amounts of data in many areas of biology, including human disease. Many studies are now looking at the profiles of gene expression or how genes may be clustered into groups of genes with different functions or in different biological pathways. This technology could have a major impact on the study of AMD. Such studies have already been underaken to compare gene expression profiling in young versus elderly retinae to identify candidate genes in aging-associated retinal diseases (Yoshida et al. 2002). In this study, 2400 human genes were arrayed on a slide and probed using RNA from young (13–14 years) and elderly (62–74 years) human retinae. The expression of the majority of these genes was not altered during aging, but 24 genes did show a difference with age. This and other studies demonstrate the utility of gene microarrays in identifying changes in gene expression across many thousands of retinal genes.

### 4.4.3
### Proteomics

The ultimate function of a gene resides with its protein product and it is here that there is a large amount of diversity. This reflects in part the changes that can occur to a protein that range from protein splicing through to structural changes and post-translational modifications. In view of this complexity at the protein level, the study of the sequence, modification and function of all proteins in a biological system has been termed proteomics. This is a young and rapidly changing field; however, studies have already been undertaken to compare the protein composition of drusen from normal versus AMD donor eyes (Crabb et al.2002). Preliminary work from this study has identified 129 proteins that could be isolated from drusen, and evidence of oxidative modification of proteins has also found. It may be possible in the future to use the knowledge acquired through proteomics to ultimately piece together how environmental risk factors (such as smoking) and changes at the genome level (such as apoE) can lead to changes typically associated with AMD.

### 4.5
### Conclusions

It is clear that there is an underlying genetic tendency to disease in AMD; however, its nature and extent remain obscure. AMD is likely to be a complex genetic disease with multiple genetic loci or susceptibility loci that interact with environmental factors. For common diseases with complex inheritance, finding the "guilty" genes poses a great challenge. The lack of strong linkage signals indicates that there will probably not be a single locus that confers a high degree of risk for AMD. There may be one or many mutations for each clinical subtype of disease. Because mutations in any single gene are neither necessary nor sufficient to cause the disease, the correlation between genotype and phenotype is imperfect. Mutations or SNPs contributing to complex traits may cause subtle alterations of gene function or regulation. They may be difficult to recognize, as there may not be major disruptions in the DNA coding sequence.

The elucidation of the genetic basis of AMD is one of the greatest challenges in ophthalmology today. It is imperative that the international community work together to unify the subclassifation of the disease to obtain homogeneous phenotypes so that identification of genes contributing to only part of the phenotype can more easily be achieved. Once genes are known, it will allow us to identify individuals who have an increased risk of AMD before they are symptomatic and to understand the pathogenesis of AMD at the molecular level. This new knowledge will help us develop therapies to minimize the impact of genetic susceptibility.

Over the next few decades there will be genes linked to AMD, but genetic evidence alone will not be sufficient to establish a causative gene. Ultimate proof will need to be established through biological, cellular and animal models of disease. The use of animal and cellular models will provide information not only about the function of a particular gene but also the underlying pathogenesis of AMD and the mechanisms by which alterations in the gene bring about the disease.

# References

Allikmets R, Singh N, Sun H, Shroyer NF, Hutchinson A, Chidambaran A Gerrard B, Baird L, Stauffer D, Peiffer A, Rattner A, Smallwood P, Li Y, Anderson KL, Lewis RA, Nathans J, Leppert M, Dean M, Lupski JR (1997a) A photoreceptor cell-specific ATP-binding transporter gene (ABCR) is mutated in recessive Stargardt macular dystrophy. Nat Genet 15:236–246

Allikmets R, Shroyer NF, Singh N, Seddon JM, Lewis RA, Bernstein PS, Peiffer A, Zabriskie NA, Li Y, Hutchinson A, Dean M, Lupski JR, Leppert M (1997b) Mutation of the Stargardt disease gene (ABCR) in age-related macular degeneration. Science 277:1805–1807

Allikmets R, International ABCR Screening Consortium (2000) Further Evidence for an association of *ABCR* alleles with age-related macular degeneration. Am J Hum Genet 67:487–491

Amaratunga A, Abraham CR, Edwards RB et al. (1996) Apolipoprotein E is synthesized in the retina by Muller glial cells, secreted into the vitreous, and rapidly transported into the optic nerve by retinal ganglion cells. J Biol Chem 271:1805–1807

Anderson DH, Ozaki S, Nealon M, Neitz J, Mullins RF, Hageman GS, Johnson LV (2001) Local cellular sources of apolipoprotein E in the human retina and retinal pigmented epithelium: implications for the process of drusen formation. Am J Ophthalmol 131:767–781

Ayyagari R, Zhang K, Hutchinson A, Yu Z, Swaroop A, Kakuk LE, Seddon JM, Bernstein PS, Lewis RA, Tammur J, Yang Z, Li Y, Zhang H, Yashar BM, Liu J, Petrukhin K, Sieving PA, Allikmets R (2001) Evaluation of the *ELOVL4* gene in patients with age-related macular degeneration. Ophthalmol Genet 22(4):233–239

Baird PN, Guida E, Chu DT, Vu HT, Guymer RH (2004a) The ε2 and ε4 alleles of the apolipoprotein gene are associated with age-related macular degeneration. Invest Ophthalmol Vis Sci 45:1311–1315

Baird PN, Chu D, Guida E, Cain M, Vu HTV, Guymer RH (2004b) The Gln192Arg SNP of the Paraxonase gene (PON1) is associated with age-related macular degeneration. Am J Ophthalmol (in press)

Chakravartic A (1999) Population genetics – making sense out of sequence. Nat Genet 21:56–60

Corder EH, Saunders AM, Strittmatter WJ, Schmechel DE, Gaskell PC, Small GW, Roses AD, Haines JL, Pericak-Vance MA (1993) Gene dose of apolipoprotein E type 4 allele and the risk of Alzheimer's disease in late-onset families. Science 261(5123):921–923

Corder EH, Saunders AM, Risch NJ, Strittmatter WJ, Schmechel DE, Gaskell PC Jr, Rimmler JB, Locke PA, Conneally PM, Schmader KE, Small GW, Roses AD, Haines JL, Pericak-Vance MA (1994) Protective effect of apolipoprotein E type 2 allele for late onset Alzheimer disease. Nat Genet 7(2):180–184

Crabb JW, Miyagi M, Gu X, Shadrach K, West KA, Sakaguchi H, Kamei M, Hasan A, Yan L, Rayborn ME, Salomon RG, Hollyfield JG (2002) Drusen proteome analysis: an approach to the etiology of age-related macular degeneration [comment]. Proc Natl Acad Sci USA 99(23):14682–14687

Cremers FP, Pol DJ van de, Driel M van, Hollander AI den, Haren FJ van, Knoers NV, Tijmes N, Bergen AA, Rohrschneider K, Blankenagel A, Pinckers AJ, Deutman AF, Hoyng CB (1998) Autosomal recessive retinitis pigmentosa and cone-rod dystrophy caused by splice site mutations in the Stargardt's disease gene *ABCR*. Hum Mol Genet 7(3):355–362

Davignon J, Cohn JS, Mabile L, Bernier L (1999) Apolipoprotein E and atherosclerosis: insight from animal and human studies. Clin Chim Acta 286:115–143

De la Paz MA, Pericak-Vance MA, Haines JL, Seddon JM (1997a) Phenotypic heterogeneity in families with age-related macular degeneration. Am J Ophthalmol 124(3):331–343

De La Paz MA, Pericak-Vance MA, Lennon F, Haines JL, Seddon JM (1997b) Exclusion of *TIMP3* as a candidate locus in age-related macular degeneration. Invest Ophthalmol Vis Sci 38(6):1060–1065

De la Paz MA, Abau-Donia S, Heinis R et al. (1999) Analysis of the Stargardt disease gene (*ABCR*) in age-related macular degeneration. Ophthalmology 106:153

Deutman AF, Jansen LMAA (1970) Dominantly inherited drusen of Bruch's membrane. Br J Ophthalmol 34:373–382

Doyne RW (1899) Peculiar condition of choroiditis occuring in several members of the same family. Trans Ophthalmol Soc UK 19:71–71

Duker JS (1998) Retina and vitreous. In: Yanoff M, Duker JS (eds) Ophthalmology. Mosby, pp 8.1.1–8.48.4

Edwards AO, Klein ML, Berselli CB, Hejtmancik JF, Rust K, Wirtz MK, Weleber RG, Acott TS (1998) Malattia Leventinese: refinement of the genetic locus and phenotypic variability in autosomal dominant macular drusen. Am J Ophthalmol 126:417–424

Evans K, Gregory CY, Wijesuriya SD, Kermani S, Jay MR, Plant C, Bird AC (1997) Assessment of the phenotypic range seen in Doyne honeycomb retinal dystrophy. Arch Ophthalmol 115:904–910

Felbor U, Schilling H, Weber BHF (1997) Adult vitelliform macular dystrophy is frequently associated with mutations in the peripherin/RDS gene. Hum Mutat 10:301–309

Gass JD (1973) Drusen and disciform macular detachment and degeneration. Arch Ophthalmol 90:206–217

Gorin MB, Jackson KE, Ferrell RE, Sheffield VC, Jacobson SG, Gass JD, Mitchell E, Stone EM (1995) A peripherin/retinal degeneration slow mutation (Pro-210-Arg) associated with macular and peripheral retinal degeneration. Ophthalmology 102(2):246–255

Gottfredsdottir MS, Sverrisson T, Musch DC, Stefansson E (1999) Age related macular degeneration in monozygotic twins and their spouses in Iceland. Acta Ophthalmol Scand 77:422–425

Gregory CY, Evans K, Wijesuriya SD, Kermani S, Jay MR, Plant C, Cox N, Bird AC, Bhattacharya SS (1996) The gene responsible for autosomal dominant Doyne's honeycomb retinal dystrophy (DHRD) maps to chromosome 2p16. Hum Mol Genet 5(7):1055–1059

Guymer RH, Heon E, Lotery AJ, Munier FL, Schorderet DF, Baird PN, McNeil RJ, Haines H, Sheffield VC, Stone EM (2001) Variation of codons 1961 and 2177 of the Stardardt disease gene is not associated with age-related macular degeneration. Arch Ophthalmol 119: 745–751

Guymer RH, McNeil R, Cain M, Tomlin B, Allen PJ, Dip CL, Baird PN (2002) Analysis of the Arg345Trp disease-associated allele of the *EFEMP1* gene in individuals with early-onset drusen or familial age-related macular degeneration. Clin Exp Ophthalmol 30: 419–423

Hamdi HK, Reznik J, Castellon R, Atilano SR, Ong JM, Udar N, Tavis JH, Aoki AM, Nesburn AB, Boyer DS, Small KW, Brown DJ, Kenney MC (2002) Alu DNA polymorphism in ACE gene is protective for age-related macular degeneration. Biochem Biophys Res Commun 295(3):668–672

Hammond CJ, Webster AR, Snieder H, Bird AC, Gilbert CE, Spector TD (2002) Genetic influence on early age-related maculopathy: a twin study. Ophthalmology 109(4):730–736

Heiba IM, Elston RC, Klein BE, Klein R (1994) Sibling correlations and segregation analysis of age-related maculopathy: the Beaver Dam Eye Study. Genet Epidemiol 11:51–67

Heon E, Piguet B, Munier F, Sneed SR, Morgan CM, Forni S, Pescia G, Schorderet D, Taylor CM, Streb LM, Wiles CD, Nishimura DY, Sheffield VC, Stone EM (1996) Linkage of autosomal dominant radial drusen (malattia leventinese) to chromosome 2p16–21. Arch Ophthalmol 114(2):193–198

Hutchinson J, Tay W (1875) Symmetrical central chorioretinal disease occurring in senile persons. R Lond Ophthalmol Hosp Rep 8:231–44

Hyman LG, Lilienfeld AM, Ferris FL 3rd, Fine SL (1983) Senile macular degeneration: a case-control study. Am J Epidemiol 118(2):213–227

Ikeda T, Obayashi H, Hasegawa G, Nakamura N, Yoshikawa T, Imamura Y, Koizumi K, Kinoshita S (2001) Paraoxonase gene polymorphisms and plasma oxidized low-density lipoprotein level as possible risk factors for exudative age-related macular degeneration. Am J Ophthalmol 132(2):191–195

Kajiwara K, Berson EL, Dryja TP (1994) Digenic retinitis pigmentosa due to mutations at the unlinked peripherin/RDS and ROM1 level. Science 264(5165): 1604–1607

Kaplan J, Gerber S, Larget-Piet D, Rozet JM, Dollfus H, Duffer JL, Odent S, Postel-Vinay A, Janin N, Briard ML, et al. (1993) A gene for Stargardt's disease (fundus flavimaculatus) maps to the short arm of chromosome 1. Nat Genet 5:308–311 (Erratum in Nat Genet 6:214)

Katsanis N, Venable S, Smith JR, Lupski JR (2000) Isolation of a paralog of the Doyne honeycomb retinal dystrophy gene from the multiple retinopathy critical region on 11q13. Hum Genet 106(1):66–72

Keverline MR, Mah TS, Keverline PO, Gorin MB (1998) A practice-based survey of familial age-related maculopathy. Ophthalmol Genet 19(1):19–26

Kimura K, Isashiki Y, Sonoda S, Kakiuchi-Matsumoto T, Ohba N (2000) Genetic association of manganese superoxide dismutase with exudative age-related macular degeneration. Am J Ophthalmol 130(6): 769–773

Klainguti R (1932) Die Tapeto-retinal degeneration im Kanton Tessin. Klin Monatsbl Augenheilkd 89: 253–254

Klaver CC, Wolfs RC, Assink JJ, Duijn CM van, Hofman A, Jong PT de (1998a) Genetic risk of age-related maculopathy. Population-based familial aggregation study. Arch Ophthalmol 116(12):1646–1651

Klaver CC, Kliffen M, Duijn CM van, Hofman A, Cruts M, Grobbee DE, Broeckhoven C van, Jong PT de (1998b) Genetic association of apolipoprotein E with age-related macular degeneration. Am J Hum Genet 63(1):200–206

Klaver CCW, Assink JJM, Bergen AAB, et al. (1998c) *ABCR* gene and age-related macular degeneration. Sci Online 279:1107

Klein BEK, Klein R, Lee KE, Moore EL, Danforth L (2001) Risk of incident age-related disease in people with an affected sibling: The Beaver Dam Eye Study. Am J Epidemiol 154:207–211

Klein ML, Mauldin WM, Stoumbos VD (1994) Heredity and age-related macular degeneration. Observations in monozygotic twins. Arch Ophthalmol 112(7): 932–937

Klein ML, Schultz DW, Edwards A, Matise TC, Rust K, Berselli CB, Trzupek K, Weleber RG, Ott J, Wirtz MK, Acott TS (1998) Age-related macular degeneration. Clinical features in a large family and linkage to chromosome 1q. Arch Ophthalmol 116(8):1082–1088

Lander E (1996) The new genomics: global views of biology. Science 274:536–539

Lander E, Kruglyak L (1995) Genetic dissection of complex traits: guidelines for interpreting and reporting linkage results. Nat Genet 11:241–247

Lecka-Czernik B, Lumpkin CK Jr, Goldstein S (1995) An overexpressed gene transcript in senescent and quiescent human fibroblasts encoding a novel protein in the epidermal growth factor-like repeat family stimulates DNA synthesis. Mol Cell Biol 15(1): 120–128

Lotery AJ, Munier FL, Fishman GA, Weleber RG, Jacobson SG, Affatigato LM, Nichols BE, Schorderet DF, Sheffield VC, StoneEM (2000) Allelic variation in the *VMD2* gene in best disease and age-related macular degeneration. Invest Ophthalmol Vis Sci 41(6): 1291–1296

Mahley RW (1988) Apolipoprotein E: cholesterol transport protein with expanding role in cell biology. Science 240 (4852):622–630

Majewski J, Schultz DW, Weleber RG, Schain MB, Edwards AO, Matise TC, Acott TS, Ott J, Klein ML (2003) Age-related macular degeneration-a genome scan in extended families. Am J Hum Genet 73(3):540–550

Marmorstein LY, Munier FL, Arsenijevic Y, Schorderet DF, McLaughlin PJ, Chung D et al. (2002) Aberrant accumulation of *EFEMP1* underlies drusen formation in Malattia Leventinese and age-related macular degeneration. Proc Natl Acad Sci USA 99:13067–13072

Martinez-Mir A, Paloma E, Allikmets R, Ayuso C, Rio T del, Dean M, Vilageliu L, Gonzalez-Duarte R, Balcells S (1998) Retinitis pigmentosa caused by a homozygous mutation in the Stargardt disease gene *ABCR*. Nat Genet 18:11–12

Melrose MA, Magargal LE, Lucier AC (1985) Identical twins with subretinal neovascularization complicating senile macular degeneration. Ophthalmol Surg 16:648–651

Meyer SM, Zachary AA (1988) Monozygotic twins with age-related macular degeneration. Arch Ophthalmol 106:651–653

Meyer SM, Greene T, Gutman FA (1995) A twin study of age-related macular degeneration. Am J Ophthalmol 120(6):757–766

Nichols BE, Sheffield VC, Vandenburgh K, Drack AV, Kimura AE, Stone EM (1993) Butterfly-shaped pigment dystrophy of the fovea caused by a point mutation in codon 167 of the RDS gene. Nat Genet 3:202–207

Pang CP, Baum L, Chan WM, Lau TC, Poon PM, Lam DS (2000) The apolipoprotein E epsilon-4 allele is unlikely to be a major risk factor of age-related macular degeneration in Chinese. Ophthalmologica 214(4):289–291

Pearce WG (1968) Doyne's honeycomb retinal degeneration. Br J Ophthalmol 52:73–78

Petrukhin K, Koisti MJ, Bakall B, Li W, Xie G, Marknell T, Sandgren O, Forsman K, Holmgren G, Andreasson S, Vujic M. Bergen AAB, McCarty-Dugan V, Figueroa D, Austin CP, Metzker ML, Caskey CT, Wadelius C (1998) Identification of the gene responsible for Best macular dystrophy. Nat Genet 19:241–247

Piguet B, Wells JA, Palmvang IB, Wormald R, Chisholm IH, Bird AC (1993) Age-related Bruch's membrane change: a clinical study of the relative role of heredity and environment. Br J Ophthalmol 77(7):400–403

Piguet B, Haimovici R, Bird AC (1995) Dominantly inherited drusen represent more than one disorder: a historical review. Eye 9:34–41

Rosenthal B, Thompson B (2003) Awareness of age-related macular degeneration in adults: the results of a large scale international survey. Optometry 74(1):16–24

Schachter F, Faure-Delanef L, Guenot F, Rouger H, Froguel P, Lesueur-Ginot L, Cohen D (1994) Genetic associations with human longevity at the *APOE* and *ACE* loci. Nat Genet 6:29–32

Schick JH, Iyengar SK, Klein BE, Klein R, Reading K, Liptak R, Millard C, Lee KE, Tomany SC, Moore EL, Fijal BA, Elston RC (2003) A whole-genome screen of a quantitative trait of age-related maculopathy in sibships from the Beaver Dam Eye Study. Am J Hum Genet 72(6):1412–1424

Schmidt S, Saunders AM, De La Paz MA, Postel EA, Heinis RM, Agarwal A, Scott WK, Gilbert JR, McDowell JG, Bazyk A, Gass DM, Haines JL, Pericak-Vance MA (2000) Association of the apolipoprotein E gene with age-related macular degeneration: possible effect modification by family history, age, and gender. Mol Vis 6:287–293

Schmidt S, Klaver C, Saunders A, Postel E, De La Paz M, Agarwal A, Small K, Udar N, Ong J, Chalukya M, Nesburn A, Kenney C, Domurath R, Hogan M, Mah T, Conley Y, Ferrell R, Weeks D, Jong PT de, Duijn C van, Haines J, Pericak-Vance M, Gorin M (2002) A pooled case-control study of the apolipoprotein E (*APOE*) gene in age-related maculopathy. Ophthalmol Genet 23(4):209–223

Schultz DW, Klein ML, Humpert A, Majewski J, Schain M, Weleber RG, Ott J, Acott TS (2003a) Lack of an association of the apolipoprotein E gene with familial age-related macular degeneration. Arch Ophthalmol 121(5):679–683

Schultz DW, Humpert AJ, Luzier CW, Persun V, Schain M, Weleber RG, Acott TS, Klein ML (2003b) Evidence that FIBL-6 is the *ARMD1* Gene. Invest Ophthalmol Vis Sci (ARVO abstract 2017)

Seddon JM, Ajani UA, Mitchell BD (1997) Familial aggregation of age-related maculopathy. Am J Ophthalmol 123(2):199–120

Shastry BS, Trese MT (1999) Evaluation of the peripherin/RDS gene as a candidate-gene in families with age-related macular degeneration. Ophthalmologica 213(3):165–170

Silvestri G, Johnston PB, Hughes AE (1994) Is genetic predisposition an important risk factor in age-related macular degeneration? Eye 8(5):564–568

Souied EH, Benlian P, Amouyel P, Feingold J, Lagarde JP, Munnich A, Kaplan J, Coscas G, Soubrane G (1998) The epsilon-4 allele of the apolipoprotein E gene as a potential protective factor with age-related macular degeneration. Am J Ophthalmol 125:353–359

Stone EM, Webster AR, Vandenburgh K, Streb LM, Hockey RR, Lotery AJ, Sheffield VC (1998) Allelic variation in ABCR associated with Stargardt disease but not age-related macular degeneration. Nat Genet 20:328–329

Stone EM, Lotery AJ, Munier FL, Heon E, Piguet B, Guymer RH, Vandernburgh K, Cousin P, Nishimura D, Swiderski RE, Silvestri G, Mackey DA, Hageman GS, Bird AC, Sheffield VC, Schorderet DF (1999) A single *EFEMP1* mutation associated with both Malattia Leventinese and Doyne honeycomb retinal dystrophy. Nat Genet 22:199–202

Stone EM, Sheffield VC, Hageman GS (2001) Molecular genetics of age-related macular degeneration. Hum Mol Genet 10(20):2285–2292

Stone EM, Russell TA, Kuehn MH, Lotery AJ, Moore PA, Eastman CG, Casavant TL, Sheffield VC (2004) Missense Variations in the Fibulin 5 Gene and Age-Related Macular Degeneration. The New England Journal of Medicine 351:346–353

Strittmatter WJ, Saunders AM, Schmechel D, Pericak-Vance M, Enghild J, Salvesen GS, Roses AD (1993) Apolipoprotein E: high-avidity binding to beta-amyloid and increased frequency of type 4 allele in late-onset familial Alzheimer disease. Proc Natl Acad Sci USA 90:1977–1981

The International Human Genome Sequencing Consortium (2001) Initial sequencing and analysis of the human genome. Nature 409:860–921

Toto L, Parodi MB, Baralle F et al. (2002) Genetic heterogeneity in Malattia Leventinese. Clin Genet 62:399–403

Travis GH, Brennan MB, Danielson PE, Kozak CA, Sutcliffe JG (1989) Identification of a photoreceptor-specific mRNA encoded by the gene responsible for retinal degeneration slow (RDS). Nature 338:70–73

Weber BHF, Vogt G, Pruett RC, Stthr H, Felbor U (1994) Mutations in the tissue inhibitor metalloproteinases-3 (*TIMP3*) in patients with Sorsby's fundus dystrophy. Nat Genet 8:352–356

Webster AR, Heon E, Lotery AJ, Vandenburgh K, Casavant TL, Oh KT, Beck G, Fishman GA, Lam BL, Levin A, Heckenlively JR, Jacobson SG, Weleber RG, Sheffield VC, Stone EM (2001) An analysis of allelic variation in the *ABCA4* gene. Invest Ophthalmol Vis Sci 42(6):1179–1189

Weeks DE, Conley YP, Mah TS, Paul TO, Morse L, Ngo-Chang J, Dailey JP, Ferrell RE, Gorin MB (2000) A full genome scan for age-related maculopathy. Hum Mol Genet 9(9):1329–1349

Weeks DE, Conley YP, Tsai HJ, Mah TS, Rosenfeld PJ, Paul TO, Eller AW, Morse LS, Dailey JP, Ferrell RE, Gorin MB (2001) Age-related maculopathy: an expanded genome-wide scan with evidence of susceptibility loci within the 1q31 and 17q25 regions. Am J Ophthalmol 132(5):682–692

Weleber RG, Carr RE, Murphey WH, Sheffield VC, Stone EM (1993) Phenotypic variation including retinitis pigmentosa, pattern dystrophy, and fundus flavimaculatus in a single family with a deletion of codon 153 or 154 of the peripherin/RDS gene. Arch Ophthalmol 111:1531–1542

Wells J, Wroblewski J, Keen J, Inglehearn C, Jubb C, Eckstein A, Jay M, Arden G, Bhattacharya S, Fitzke F, et al. (1993) Mutations in the human retinal degeneration slow (RDS) gene can cause either retinitis pigmentosa or macular dystrophy. Nat Genet 3:213–218

Wroblewski JJ, Wells JA 3rd, Eckstein A, et al. (1994) Macular dystrophy associated with mutations at codon 172 in the human retinal degeneration slow gene. Ophthalmology 101(1):12–22

Yoshida S, Yashar BM, Hiriyanna S, Swaroop A (2002) Microarray analysis of gene expression in the aging human retina. Invest Ophthalmol Vis Sci 43:2554–2560

Zhang K, Kniazeva M, Han M, et al. (2001) A 5-bp deletion in *ELOVL4* is associated with two related forms of autosomal dominant macular dystrophy. Nat Genet 27:89–93

Zurdel J, Finckh U, Menzer G, Nitsch RM, Richard G (2002) CST3 genotype associated with exudative age related macular degeneration. Br J Ophthalmol 86(2):214–219

# Chapter 5

# Epidemiology of Age-Related Macular Degeneration

Ronald Klein

## Contents

5.1 Prevalence and Incidence of Age-Related Macular Degeneration  79
5.1.1 Introduction  79
5.1.2 Prevalence of Age-Related Macular Degeneration  80
5.1.3 Prevalence: Race/Ethnicity  81
5.1.4 Incidence of Age-Related Macular Degeneration  82

5.2 Risk Factors for Age-Related Macular Degeneration  82
5.2.1 Introduction  82
5.2.2 Familial Factors  83
5.2.3 Systemic Factors  83
5.2.4 Lifestyle Behavior  87
5.2.5 Environmental Factors  89
5.2.6 Ocular Factors  90
5.2.7 Socioeconomic Factors and Work Exposures  92

5.3 Age-Related Macular Degeneration and Survival  93

5.4 Public Health Issues  93

5.5 Conclusions  93

References  94

## 5.1 Prevalence and Incidence of Age-Related Macular Degeneration

### 5.1.1 Introduction

While recognition and description of lesions characteristic of age-related macular degeneration (AMD) have occurred since the invention of the ophthalmoscope, an important recent advance has been the development of systematic grading schemes (Klein et al. 1991a; Bird et al. 1995). These are based on fundus photography and grading protocols using plastic grids consisting of concentric circles, 500 µm, 1,500 µm, and 3,000 µm in radius, to define the macular area, and standard-sized circles on plastic grids to measure drusen size and area, and extent and distribution in the fundus of other AMD lesions (Figs. 5.1 and 5.2; Klein et al. 1991a). Use of

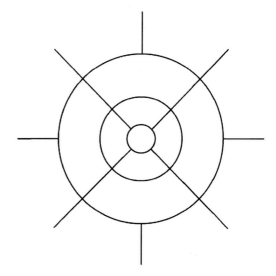

Fig. 5.1. Grid used to define subfields in the macula in the Wisconsin Age-Related Maculopathy Grading Scheme. (Reprinted from Klein R, Davis MD, Magli YL, Segal P, Klein BE, Hubbard L 1991, The Wisconsin Age-Related Maculopathy Grading System. Ophthalmology 98:1128–1134, with permission from The American Academy of Ophthalmology)

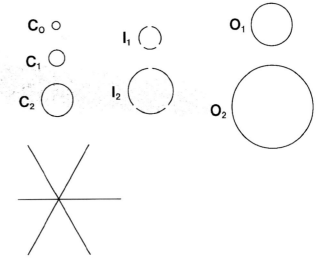

Fig. 5.2.
Grid used to estimate size of drusen, area involved by drusen and area involved by increased retinal pigment in the Wisconsin Age-Related Maculopathy Grading Scheme. (Reprinted from Klein R, Davis MD, Magli YL, Segal P, Klein BE, Hubbard L, 1991, The Wisconsin Age-Related Maculopathy Grading System. Ophthalmology 98:1128–1134, with permission from The American Academy of Ophthalmology)

these grading systems has permitted comparisons of prevalence and incidence of AMD and its risk factors among different racial/ethnic groups living in different geographic locations (Klein et al. 1995b; Smith et al. 2001; Tomany et al. 2003).

### 5.1.2
### Prevalence of Age-Related Macular Degeneration

While the presence of signs of neovascular AMD (e.g., retinal pigment epithelial, RPE, and sensory retinal detachment, disciform scar, subretinal and retinal hemorrhage, hard exudate) and geographic atrophy define advanced or late stages of AMD, there have been few internationally agreed-upon definitions of early AMD, making comparisons among studies difficult (Bird et al. 1995). For this reason, many studies that have arbitrarily defined early AMD as the presence of soft drusen or hard drusen with pigmentary abnormalities have also reported the prevalence of drusen size and type, and pigmentary abnormalities that characterize early AMD.

The Beaver Dam Eye Study has been described in detail elsewhere (Klein R et al. 1996, 1991b, 2001a). In brief, it is a population-based study of 4,926 Whites 43–86 years of age identified by a private census in 1987–1988 and examined in 1988–1990, 1993–1995, 1998–2000, and 2003–2005. The overall prevalence of early, late, or any AMD in Beaver Dam is presented in Table 5.1. The prevalence of early AMD at baseline, defined as the presence of hard drusen with pigmentary abnormalities or soft, indistinct drusen, was 15.6% and it varied from 8.4% to 29.7% in persons 43–54 years old and ≥75 years old, respectively (Table 5.1; Klein et al. 1992). The prevalence of late AMD, defined as the presence of signs of neovascular AMD or geographic atrophy, was 1.6% and varied from 0.1% in persons 43–54 years old to 7.1% in persons ≥75 years old (Table 5.1). Based on the Beaver Dam population, it is estimated that 640,000 Americans of 75–86 years of age have signs of AMD. Prevalences of AMD and its component lesions are comparable in the Blue Mountains Eye Study but higher than found in the Rotterdam Eye Study. Both of these population-based studies used protocols similar to those used in Beaver Dam for detection of AMD. Variations in prevalence of AMD among populations may be due to differences in genetic and environmental factors among the studies (see below; Klein et al. 1992; Mitchell et al. 1995; Vingerling et al. 1995b; Smith et al. 2001).

Data from population-based cohort studies (Beaver Dam Eye Study, Blue Mountains Eye Study, the Rotterdam Eye Study, the Salisbury

Table 5.1. Relation of age to the prevalence and 10-year incidence of AMD in the Beaver Dam Eye Study

| Age (year) | Prevalence AMD | | | 10-year incidence | | | |
|---|---|---|---|---|---|---|---|
| | No. at risk | Early (%) | Late (%) | No. at risk at baseline | Early (%) | No. at risk at baseline | Late (%) |
| 43–54 | 1504 | 8.4 | 0.1 | 1136 | 4.7 | 1250 | 0.1 |
| 55–64 | 1301 | 13.8 | 0.6 | 856 | 10.6 | 1032 | 1.0 |
| 65–74 | 1249 | 18.0 | 1.4 | 654 | 21.7 | 899 | 4.4 |
| 75+ | 717 | 29.7 | 7.1 | 185 | 28.7 | 315 | 9.5 |
| Total | 4771 | 15.6 | 1.6 | 2831 | 10.9 | 3496 | 2.1 |

Eye Evaluation project, and the Melbourne Vision Impairment Project for Whites; and the Baltimore Eye Survey, the Barbados Eye Study and the Salisbury Eye Evaluation project for Blacks) have recently been pooled to provide estimates of AMD in the US population (Bressler et al. 1989, Klein et al. 1992; Mitchell et al. 1995; Schachat et al. 1995; Vingerling et al. 1995b; VanNewKirk et al. 2000; Smith et al. 2001).

## 5.1.3
**Prevalence: Race/Ethnicity**

Racial/ethnic differences in the prevalence of AMD have been found in studies using similar methodology to grade AMD. In the National Health and Nutrition Examination Survey (NHANES) III, a survey from 1988 to 1994 that involved non-Hispanic Whites, Mexican Americans, and non-Hispanic Blacks, rates of large soft drusen, a lesion defining early AMD, were similar among the three groups 40–65 years of age (Klein et al. 1995b, 1999c). Late AMD was rare in persons this age. However, after 65 years of age, there was significantly higher age-specific prevalence of signs of late AMD among non-Hispanic Whites compared with Blacks and Mexican Americans. Higher prevalence of late AMD has been found in Whites compared with Blacks in some population-based studies (Atherosclerosis Risk in Communities, ARIC, Klein et al. 1999a; Cardiovascular Health Study, CHS, Klein et al. 2003a) but not in all (Salisbury Eye Evaluation, SEE, Bressler et al. 1998).

Population-based data on the prevalence of AMD in Mexican-Americans in Los Angeles (the Latino Eye Study, LALES) and in Tucson and Nogales, Arizona (the Proyecto VER Study) have shown higher frequency of large soft drusen (>125 µm in diameter) and a lower frequency of late AMD in Mexican Americans, especially in persons 40–59 years of age, compared with Whites (Muñoz et al. 2003; Tong et al. 2003). In the LALES, the overall prevalence of large soft drusen was 14.5% (varying from 8.6% in those 40–49 years of age to 45.5% in those aged 80 years and older). For late AMD in the LALES, it was 0.4%. The estimates of early and late AMD in the Proyecto VER cohort of Mexican Americans in Arizona were similar.

Racial/ethnic differences in AMD prevalence have been attributed to variations in pigmentation. Eyes with more pigmentation are thought to be protected against development of neovascular AMD (Weiter et al. 1985). One proposed physiologic explanation is that melanin in the retinal pigment epithelium and choroid protects the retina from free radicals associated with photo-oxidation and light absorption (Jampol and Tielsch 1992). Further support for a possible role of pigmentation in the pathogenesis of neovascular macular degeneration comes from observations in transgenic mice (Rohan et al. 2000). A higher rate of iris neovascularization and hyphema after implantation of corneal pellets of bFGF in albino mice (C57BL/6J-Tyrc-2J and 129/SV mice) compared with their pigmented relatives (C57BL/6J and 129/SylmJ mice, respectively) has been reported. However, caution is needed when evaluating data from animal models whether genetically modified or not. There may be limited analogy to diseases in humans. Also, it is possible that RPE and choroidal pigmentation may not have a direct protective effect, but are markers for those genes associated with AMD.

### 5.1.4
### Incidence of Age-Related Macular Degeneration

Less is known about the long-term incidence and progression of AMD. Data from Beaver Dam and other population-based studies show that the disease is slowly progressive in its earliest stages (Bressler et al. 1995; Klein et al. 1997, 2002a; Sparrow et al. 1997; Klaver et al. 2001; Mitchell et al. 2002; van Leeuwen et al. 2003; Wang et al. 2003a). In Beaver Dam, the 10-year incidence of early AMD was 10.9% and it varied from 4.7% to 28.7% in persons 43–54 years old and ≥75+ years old, respectively, at baseline (Table 5.1; Klein et al. 2002a). The incidence of late AMD was 2.1% and varied from 0.1% in persons 43–54 years old to 9.5% in persons ≥75 years old (Table 5.1). Based on the data from the Beaver Dam Eye Study, it is estimated that 1.9 million people 65–84 years of age will develop signs of late AMD over a 10-year period.

Epidemiological data from Beaver Dam have also provided some insights regarding when the disease begins (Klein et al. 2002a). In Beaver Dam, one or two small hard drusen were found in 94% of the population and eyes with this number of small drusen had almost no risk of progression to late AMD over 10 years of follow-up. Because of this, eyes with these lesions are not considered to have AMD or to be at risk of developing the disease (Bressler et al. 1995; Klein et al. 2002a). However, the 10-year results from the Beaver Dam Eye Study show that when large areas of small hard drusen (157,686 $\mu m^2$) are present at baseline, there is an approximate 2.5 times increase in the risk of developing soft, indistinct drusen, a 3.3 times increase in the risk of developing pigmentary abnormalities, and a 2.7 times increase in the risk of developing large soft drusen (125 µm in diameter) compared with eyes with smaller areas of small hard drusen. In the Chesapeake Bay Waterman Study, larger numbers of small hard drusen were associated with a twofold increase in the 5-year incidence of larger soft drusen (Bressler et al. 1995).

Large-sized drusen (≥125 µm in diameter), soft, indistinct drusen, a large area of drusen, and pigmentary abnormalities have been found to be predictive of late AMD (Bressler et al. 1995; Klein et al. 1997, 2002a; Age-Related Eye Disease Study Research Group 2001; Klaver et al. 2001; Mitchell et al. 2002; van Leeuwen et al. 2003; Wang et al. 2003a). For example, the presence of large drusen involving an area of 157,683–393,743 $\mu m^2$ (equivalent to 13–32 drusen, 125 µm in diameter) is associated with a 10-year cumulative incidence of 14.0% for late AMD. This is consistent with previous observations that these lesions increase the risk of end-stage disease (Gass 1973; Gregor et al. 1977; Strahlman et al. 1983; Smiddy and Fine 1984; Bressler et al. 1990; Roy and Kaiser-Kupfer 1990; Macular Photocoagulation Study Group 1993, 1997; Holz et al. 1994b). These observations, as well as results of clinicopathologic studies, suggest that these lesions indicate the presence of early AMD (Sarks SH 1976; Sarks SH et al. 1980; Sarks JP et al. 1988; Feeney-Burns and Ellersieck 1985; Green and Enger 1993). These data, along with those from other population-based studies (Klein BEK et al. 1994; Klaver et al. 1998; AREDS 1999), will facilitate the development of a severity scale for AMD similar to the Early Treatment Diabetic Retinopathy Study severity scale, which will be useful in epidemiological studies.

## 5.2
## Risk Factors for Age-Related Macular Degeneration

### 5.2.1
### Introduction

The only factors that have been consistently found to be associated with AMD are age and a family history. The inability to find consistent associations with other risk factors may be due to differences in study design (case-control vs cohort studies), methods used to measure and define risk factors, sample-size issues, and other biases such as selective ascertainment that are associated with epidemiological studies. The risk factors for AMD that have been examined in epidemiological studies include systemic diseases such as hypertension, cardiovascular disease, inflammatory diseases and their markers, lifestyle behaviors such as smoking, drinking, and diet, environmental exposures such as light, and ocular disease such as cataract and cataract surgery. In this section associations between AMD and these risk factors will be reviewed, and detailed references will be provided. The genetics of AMD are presented in Chap. 4.

## 5.2.2
## Familial Factors

After age, the strongest and most consistent risk factor for developing late AMD is having a sibling or parent with this condition. For example, in Beaver Dam, the odds of a younger sibling having the same lesion as an older sibling (e.g., neovascular AMD) over a 5-year period was 10.3 (Klein R et al. 2001c).

## 5.2.3
## Systemic Factors

### 5.2.3.1
### Age

Every study of AMD has shown a strong relationship of increasing prevalence and incidence of AMD with increasing age (Leibowitz et al. 1980; Klein and Klein 1982; Martinez et al. 1982; Gibson et al. 1985; Jonasson and Thordarson 1987; Bressler et al. 1989, 1995, 1998; Vinding 1989; Klein et al. 1992, 1995b; 1997, 1999a, 1999c, 2002a, 2003a; Mitchell et al. 1995, 2002; Schachat et al. 1995; Vingerling et al. 1995b; Hirvela et al. 1996; Cruickshanks et al. 1997; Sparrow et al. 1997; Delcourt et al. 1999; VanNewKirk et al. 2000; Klaver et al. 2001; Muñoz et al. 2003; Tong et al. 2003; van Leeuwen et al. 2003; Wang et al. 2003a). An issue with AMD has been whether it is due to biologicalal aging in addition to a specific disease process. One measure of biological (rather than chronological) aging is handgrip strength (Dubina et al. 1984). In two studies, the Framingham Eye Study (Kahn et al. 1977) and the case-control study of Hyman et al. (1983), weaker handgrip strength was associated with AMD, independent of age. In Beaver Dam, handgrip strength was marginally related to AMD (Klein et al., unpublished data). While controlling for age and sex, there was an 18% lower prevalence of geographic atrophy and a 12% lower prevalence of neovascular AMD per 5-unit increase in dominant handgrip strength. A limitation in interpreting these results is the difficulty of controlling adequately for unmeasured concurrent illness or other confounding factors that may be associated with AMD.

### 5.2.3.2
### Blood Pressure

Increased blood pressure by virtue of its effect on the choroidal circulation has been hypothesized to increase the risk of developing AMD (Kornzweig 1977; Bischoff and Flower 1983; Pauleikhoff et al. 1990). However, epidemiological data have been inconsistent, with some studies showing a relation (MPS Group 1997; AREDS 2000; Eye Disease Case-Control Study Group 1992; Hyman et al. 2000; Klein et al. 2003b) while others have not (Vinding et al. 1992; Vingerling et al. 1996; Klein et al. 2003a). Persons in Beaver Dam with treated and controlled hypertension at baseline were approximately twice as likely and persons with treated uncontrolled hypertension were about thrice as likely to develop exudative macular degeneration after 10 years of follow-up than persons who were normotensive (Klein et al. 2003b). Furthermore, in Beaver Dam an increase in systolic blood pressure between baseline and the 5-year follow-up was associated with an increased incidence of late AMD at the 10-year follow-up compared with persons in whom the systolic blood pressure decreased.

In a case-control study, Hyman et al. (2000) have reported that persons with neovascular AMD are more likely than controls to have a diastolic blood pressure greater than 95 mmHg (OR 4.4; 95% CI 1.4, 14.2). In the Eye Disease Case-Control Study (EDCCS), a significant association has been found between systolic blood pressure and neovascular AMD (OR 1.7; 95% CI 1.0, 2.8 for systolic blood pressure of 155 mmHg or more compared with systolic blood pressure of 118 mmHg or less; ESCCS Group 1992). In the AREDS, persons with hypertension were 1.5 times as likely (95% CI 1.2, 1.8) to have neovascular macular degeneration compared with persons without hypertension (AREDS 2000). In a follow-up study of 670 patients enrolled in the Macular Photocoagulation Study, the presence of definite systemic hypertension at baseline was associated with the development of subretinal new vessels in unaffected fellow eyes (OR 1.7, 95% CI 1.2, 2.4; MPS Group 1997).

Epidemiological data have not shown a protective association of antihypertensive medications with the incidence of late AMD (van Leeuwen et al. 2004). There have been no randomized controlled clinical trials to show that lowering blood pressure is associated with a reduction in the incidence and progression of AMD.

In Beaver Dam, while controlling for other factors including either systolic or diastolic blood pressure, higher pulse pressure at baseline was associated with an increase in the 10-year incidence of RPE depigmentation (RR per 10 mmHg 1.17; 95% CI 1.07, 1.28; $P<0.001$), increased retinal pigment (RR 1.10; 95% CI 1.01, 1.19; $P=.03$), neovascular AMD (RR 1.34; 95% CI 1.14, 1.60; $P<0.001$), and the progression of AMD (RR 1.08; 95% CI 1.01, 1.17; $P=.03$; Klein et al. 2003b). Higher pulse pressure is presumably related to decreased arterial distensibility, resulting in a greater difference between systolic and diastolic blood pressures (O'Rourke 1982). The decrease in distensibility of the large blood vessels is thought to be due to age-related degenerative changes in the collagen and elastin. Similar age-related changes have been found in Bruchs membrane (Newsome et al. 1987). It is possible that higher pulse pressure may be a marker for such degenerative changes in eyes developing AMD.

### 5.2.3.3 Atherosclerosis

Atherosclerosis of the choroidal circulation and lipid deposition in Bruch's membrane have been hypothesized to increase the risk of AMD (Pauleikoff et al. 1990). However, atherosclerosis as manifest by clinical cardiovascular disease is inconsistently associated with AMD (Kahn et al. 1977; Klein and Klein 1982; Ferris 1983; Hyman et al. 1983, 2000; ESCCS Group 1992; Vingerling et al. 1995a; MPS Group 1997; Smith et al. 1998, 2001; Klein et al. 1999a, 1993b, 1999c, 2003a; Snow and Seddon 1999; Friedman 2000; Delcourt et al. 2001a, 2001b; Evans 2001; Tomany et al. 2004a). In Beaver Dam, a history of neither heart attack nor stroke was related to the 10-year incidence of AMD (Klein et al. 2003b). On the other hand, data from the Rotterdam Study suggests a possible relationship between subclinical atherosclerotic disease and AMD. Vingerling et al. (1995a) have reported that, in participants in that study, those with plaques at the carotid bifurcation, as measured by ultrasonography, are 4.7 times as likely to have late AMD (95% CI 1.8 to 12.2) as those without plaques. In the same study, greater intima-media wall thickness of the carotid artery, another measure of subclinical atherosclerosis, is associated with an increase in incident AMD. Similarly, while controlling for age and sex, persons in the Atherosclerosis Risk in Communities Study with carotid artery plaque are 1.8 times as likely to have RPE depigmentation present compared with those without (Klein et al. 1999a).

High serum LDL-cholesterol and total cholesterol, risk factors for atherosclerosis, have not been found to be related to AMD in most studies (EDCCS Group 1992; Evans 2001). In Beaver Dam, high serum HDL-cholesterol was associated with a higher 10-year incidence of geographic atrophy (Klein et al. 2003b). In a case-control study, Hyman et al. (2000) have also reported that eyes with neovascular AMD have higher serum HDL-cholesterol (OR highest quintile vs lowest quintile 2.3; 95% CI 1.1, 4.7). In the Rotterdam Study, higher serum HDL-cholesterol is directly associated with geographic atrophy (OR per mmol/L 2.39; 95% CI 1.02, 5.59; Vingerling et al. 1995a). However, in the Blue Mountains Eye Study, serum HDL-cholesterol is not associated with either geographic atrophy (OR 0.82 per milligrams per deciliter; 95% CI 0.19, 3.46) or neovascular macular degeneration (OR 1.73; 95% CI 0.83, 3.62; Smith et al. 1998). Consistent with these epidemiological data, Dithmar et al. have reported no association of cholesterol in Bruch's membrane with plasma cholesterol in animal studies (Dithmar et al. 2001).

The reason for the HDL-cholesterol association with AMD in Beaver Dam and Rotterdam is not known. It is contrary to what one would expect if the cholesterol found in normal Bruch's membrane and small drusen (Curcio et al. 2001) arose by direct infusion of plasma LDL into the extracellular matrix. It is known that the retinal pigment epithelium has LDL receptor activity in vitro (Hayes et al. 1989), and the expression of LDL receptors is downregulated by high levels of plasma cholesterol (Goldstein et al. 2001). Based on these observations, Malek et al. have speculated that if the retinal pigment epithelium contributes LDL-derived cholesterol to Bruch's membrane and drusen, then it is possible that downregulation of LDL receptors in the retinal pigment epithelium, subsequent to high plasma cholesterol, may contribute to the reduction of risk for AMD (Malek et al. 2003).

In the Rotterdam Study, the frequency of apolipoprotein E allele E4, associated with Alzheimer's disease and neuronal repair, is lower in subjects with geographic atrophy and neovascular AMD (Poirier et al. 1995; Klaver et al. 1998). This has also been found in other studies (Souied et al. 1998; Gorin et al. 1999; Schmidt et al. 2000). Klaver et al. have speculated that this might be due to linkage disequilibrium with AMD or it might be a susceptibility factor (Klaver et al. 1998). However, this association is not found in other studies (Pang et al. 2000; Schultz et al. 2003).

Recent data from two-case control studies showed an inverse association of the lipid-lowering statins and age-related maculopathy (Hall et al. 2001; McCarty et al. 2001). The cholesterol lowering, antioxidant, and antiendothelial dysfunction effect of statins are hypothesized to explain why statins might reduce the incidence and progression of AMD (Beatty et al. 2000; Curcio et al. 2001; Hall et al. 2001; Lip et al. 2001). In Beaver Dam, statin use has not been found to be associated with the 5-year incidence of early AMD (age-sex adjusted OR=1.12, 95% CI 0.47, 2.67), progression of age-related maculopathy (OR=1.22, 95% CI 0.54, 2.76), or incidence of late AMD (OR=0.41, 95% CI 0.12, 1.45; Klein et al. 2003d). Data from the Pathologies Oculaires Liees a l'Age (POLA) and Rotterdam studies also do not show a relationship of statins with AMD (Delcourt et al. 2001b; van Leeuwen et al. 2004). High serum total cholesterol has been associated with reduced prevalence of advanced AMD (Klein 1999; Klein et al. 2003a). It is possible that the protective relationship found in the case-control studies may be due to unmeasured confounding or to a protective effect of high cholesterol levels (as noted before) which was incidentally treated with statins.

#### 5.2.3.4
**Pulmonary Disease**

There are few epidemiological data regarding pulmonary disease and AMD (Kahn et al. 1977; Hymen et al. 1983). In Beaver Dam a history of emphysema at baseline, independent of smoking history, was associated with the incidence of retinal pigment epithelial depigmentation (RR 2.84, 95% CI 1.40, 5.78), increased retinal pigment (RR 2.20, 95% CI 1.11, 4.35), and neovascular AMD (RR 5.12, 95% CI 1.63, 16.06; Klein et al. 2003c). It has been speculated that this may be due, in part, to the association of emphysema with inflammatory disease (see below). However, it is possible that the associations may be due to noninflammatory processes such as low oxygen tensions in persons with emphysema rather than an inflammatory process. However, peak flow volume, usually reduced in persons with emphysema, was not found to be associated with AMD in the Beaver Dam Eye Study (Klein R, unpublished data).

#### 5.2.3.5
#### Gout

In Beaver Dam, a history of gout was associated with a higher 10-year incidence of pure geographic atrophy (RR 3.48, 95% CI 1.27, 9.53; Klein et al. 2003c). Medications taken for gout at baseline in that study were also associated with the incidence of pure geographic atrophy. It has been speculated that this association may be due, in part, to inflammation associated with this disease (see below).

#### 5.2.3.6
#### Arthritis

The AREDS data has shown a cross-sectional association of arthritis with large drusen (OR 1.20 95% CI 1.04,1.39) but not with geographic atrophy or neovascular AMD. This association was not found prospectively (AREDS 2000). In Beaver Dam, arthritis at baseline was not significantly associated with the incidence of AMD (R. Klein, unpublished data).

#### 5.2.3.7
#### Diabetes

Diabetes, through its effect on the normal structure and functioning of the RPE, Bruch's membrane, and the choroidal circulation, has been hypothesized to increase the risk of AMD (Hidayat and Fine 1985; Fryczkowski et al. 1988; Vinores et al. 1988; Marano and Matchinsky 1989; Chakrabarti et al. 1990; Miceli and Newsome 1991). However, most epidemiological studies have not shown an association of diabetes with AMD.

#### 5.2.3.8
#### Obesity

Obesity has not been consistently associated with AMD (AREDS 2000). In a prospective cohort study in a hospital-based retinal practice, those who were obese (body mass index of at least 30 kg/m$^2$) were at increased risk of developing late AMD (RR 2.35, 95% CI 1.27, 4.34) compared with those with a body mass index of <25 kg/m$^2$ (Seddon et al. 2003). Higher waist-hip circumference was also found to be associated with AMD in that study. Others have speculated that excessive caloric intake might increase the risk of AMD because of an increased risk of oxidative damage, a hypothesized pathogenetic factor, in obese persons (Hirvela et al. 1996). It is also possible that obesity and increased waist circumference are associated with increased inflammation, another possible pathogenetic factor for AMD (see below). In Beaver Dam, greater body mass was associated with the prevalence of early but not late AMD; there was no association with the incidence of AMD (Klein et al. 1993b, 2003b).

Lean body mass has also been reported to be associated with the incidence of "dry" AMD (OR lean vs normal body mass 1.43, 95% CI 1.01, 2.04; Schaumberg et al. 2001). In Beaver Dam, those who were lean at baseline were more likely to develop geographic atrophy (Klein et al. 2003b). The reason for this finding in these studies is not known.

#### 5.2.3.9
#### Inflammation and Its Markers

Inflammation has been hypothesized to have a role in the pathogenesis of AMD (Sarks 1976; Penfold et al. 1986, 2001). Hageman et al. (2001) have shown that drusen contain proteins associated with immune-mediated processes and inflammation. Chronic inflammatory cells have also been observed on the outer surface of Bruch's membrane in eyes with neovascular macular degeneration (Penfold et al. 2001). The presence of these cells may cause microvascular injury by the direct release of long-acting oxidants, toxic oxygen compounds, and proteolytic enzymes that may result in damage to Bruch's membrane (Ernst et al. 1987). It is also possible that the inflammatory cells are chemotactically attracted to the site due to injury in Bruch's membrane and do not play a role in the pathogenesis of AMD.

Few epidemiological studies to date have shown a relationship of systemic inflammatory disease and markers of inflammation with AMD (AREDS 2000; Blumenkranz et al. 1986;

EDCCS 1992; Klein et al. 1995b, 1999a, 2003a, 2003c; Smith et al. 1998; Klein 1999; Snow and Seddon 1999; Vingerling et al. 1995a). In Beaver Dam, higher white blood cell count at baseline, independent of smoking, was associated with the 10-year incidence of drusen larger than 125 µm in diameter (risk ratio, RR, per kilogram per microliter, 1.10, 95% confidence interval, CI, 1.03, 1.17), retinal pigment epithelial depigmentation (RR 2.08, 95% CI 1.01, 1.16), and progression of age-related maculopathy (RR 1.09, 95% CI 1.03, 1.15; Klein et al. 2003c). This association has also been found in a small case-control study ($n=14$) but not in other populations studied (Klein et al. 1995b, 1999a, 2003a).

In Beaver Dam, serum albumin, another marker of systemic inflammation (and debilitating conditions), was found to be associated with higher incidence of exudative macular degeneration (RR per milligram per deciliter, 3.22, 95% CI 1.32, 7.69; Klein et al. 2003c). Another marker of systemic inflammation, high plasma fibrinogen levels, is associated with prevalent late but not early AMD in the Blue Mountains Eye Study (Smith et al. 1998) and with exudative macular degeneration in the NHANES III (Klein et al. 1995b). However, plasma fibrinogen levels are not associated with AMD in three other studies (Friedman 2000) and there is no association of C-reactive protein with AMD in one study (Klein et al. 2003a).

Two diseases associated with inflammation, emphysema and gout, have also been found to be associated with late AMD (see above). There is a release of cytokines, such as interleukin-1 and interleukin-8, and an increase in circulating levels of tumor necrosis factor-α (TNF-α) in these conditions, two factors hypothesized to play a role in the pathogenesis of exudative macular degeneration (Barnes 2000; Ghezzi et al. 1991; Hachicha et al. 1995). Anti-inflammatory drug use has not been associated with AMD (Christen et al. 2001; Gilles et al. 2001; Klein et al. 2001c).

## 5.2.4 Lifestyle Behavior

### 5.2.4.1 Smoking

Cigarette smoking has been found to be related to late AMD in some (Paetkau et al. 1978; Hyman et al. 1983; EDCCS 1992; Klein et al. 1993a; Christen et al. 1996; Seddon et al. 1996; Smith et al. 1996, 2001; Vingerling et al. 1996; Delcourt et al. 1998; AREDS 2000), but not all, epidemiological studies (Maltzman et al. 1979; Blumenkranz et al. 1986; Vinding 1989; Klein et al. 1998a, 2002b). Smoking is known to depress serum antioxidant levels and RPE drug-detoxification pathways, alter choroidal blood flow, and reduce macular carotenoid pigments, all pathogenetic mechanisms thought to be associated with AMD (Bettman et al. 1958; Friedman 1970; Pryor et al. 1983; Stryker et al. 1988; Hammond et al. 1996). In Beaver Dam, smoking was associated with the 10-year incidence of large soft drusen and pigmentary abnormalities, signs of early AMD, but not with the incidence of late AMD (Klein et al. 2002b). The reason for the inconsistency between cross-sectional and prospective associations in Beaver Dam is not understood.

### 5.2.4.2 Alcohol

Of the many epidemiological studies that have examined the association of alcohol consumption with AMD (Maltzman et al. 1979; EDCCS 1992; Ritter et al. 1995; Smith and Mitchell 1996; Cruickshanks et al. 1997; Moss et al. 1998; Obisesan et al. 1998; Ajani et al. 1999; Cho et al. 2000; Klein et al. 2002b), few have found a relationship (Smith and Mitchell 1996; Klein et al. 2002b). A cross-sectional association of alcohol consumption with early AMD (RR 1.61, 95% CI 1.07, 2.41) has been reported in the Blue Mountains Eye Study (Smith and Mitchell 1996). In Beaver Dam, people who reported being heavy drinkers at baseline were more likely to develop late AMD (RR 6.94, 95% CI 1.85, 26.1) than people who reported never having been heavy drinkers (Klein et al. 2002b). The wide confidence intervals in this study, resulting from the low frequency of both heavy drinking and the incident late AMD, suggests that this may also be a chance finding.

Heavy alcohol intake increases oxidant stress and affects mechanisms that protect against oxidative damage (Rosenblum et al. 1989; Rikans and Gonzalez 1990). Data from experimental studies in animals suggest that the retina is susceptible to oxidative damage and that this damage may be minimized by the presence of antioxidant nutrients (Katz et al. 1982). Heavy drinkers have also been found to have low serum carotene, vitamin E, and zinc, all factors postulated to protect against the development of AMD (Lieber et al. 1979).

In the National Health and Nutrition Examination Survey I, wine consumption is reported to be protective (RR 0.66, 95% CI 0.55, 0.79) for AMD (Obisesan et al. 1998). The authors of this report speculate that antioxidant phenolic compounds found in high concentrations in red wine may explain their finding. In Beaver Dam, wine consumption was associated with a reduction in the incidence of early AMD; however, the association did not reach statistical significance (Klein et al. 2002b). Neither wine nor beer have been found to be associated with AMD in other studies (Smith and Mitchell 1996). Contrary to baseline findings, consumption of beer is not associated with the incidence of pigmentary abnormalities or exudative macular degeneration in Beaver Dam (Klein et al. 2002b).

### 5.2.4.3
### Caffeine

Caffeine has been hypothesized to be associated with AMD through a possible vasoconstrictive effect on the retinal and choroidal circulation (Lofti and Grunwald 1991; Grunwald et al. 1998). However, data from the Beaver Dam Eye Study showed that coffee or overall caffeine intake was not associated with the 5-year incidence of early AMD or any of its defining lesions (Tomany et al. 2001).

### 5.2.4.4
### Physical Activity

In Beaver Dam, independent of age and other risk factors, physical activity at baseline was associated with a protective effect against the progression of AMD (RR in those who worked up a sweat five times a week compared with those who did not, 0.69; 95% CI 0.47, 1.00; $P=0.05$) and the incidence of neovascular AMD (RR 0.12; 95% CI 0.02, 0.91; $P=0.04$; Klein et al. 2003b). This is consistent with findings from the Eye Disease Case Control Study in which neovascular AMD is associated with less physical activity (EDCCS 1992). This protective relationship of physical activity may be due to obesity (see above), which has been found to be associated with AMD in some studies; however, in Beaver Dam, the association found was independent of body mass index (Klein et al. 2003b).

### 5.2.4.5
### Hormone Replacement Therapy

The loss of the protective effect of estrogens on Bruch's membrane in postmenopausal women has been postulated as a reason for higher incidence of neovascular AMD in older women compared with men (Klein et al. 1997). Female gender in older mice and estrogen deficiency in middle-aged mice is associated with an increase in the severity of sub-RPE deposit formation (Cousins et al. 2003). Cousins et al. have hypothesized that "estrogen deficiency may increase susceptibility to formation of sub-RPE deposits by dysregulating turnover of Bruch's membrane, contributing to collagenous thickening and endothelial changes (Cousins et al. 2003). However, epidemiological data have not shown a consistent protective effect of reproductive history or hormone replacement therapy on AMD (The Eye Disease Case-Control Study Group 1992; Klein BEK et al. 1994; Vingerling et al. 1995c; Smith et al. 1997; van Leeuwen et al. 2004). Data from the Rotterdam Study have shown that women with early menopause after an operation removing one or both ovaries have a significantly increased risk of late AMD (RR 3.8, 95% CI 1.1 to 12.6) compared with women who have their menopause at 45 years of age or older (Vingerling et al. 1995b). Increased number of years of menses is also related to increased risk of early AMD in the Blue Mountains Eye Study (Smith et al. 1997). Data from EDCCS (1992) have shown that the odds of subretinal AMD are reduced by 70% in women who were on hormone replacement therapy (HRT) at the time of the study compared with those who are not. However, HRT is not associated with the 5-year incidence of late AMD in pooled analyses of the Beaver Dam, Blue Mountains, and Rotterdam Eye Study (Klein BEK et al. 1994; van Leeuwen et al. 2004).

The Women's Health Initiative Sight Exam (WHI-SE) was an ancillary study to one arm of the Women's Health Initiative, a randomized trial of hormonal replacement in postmenopausal women aged 50–79 years of age (Rossouw et al. 2002). The purpose of the WHI-SE was to evaluate whether hormone replacement therapy in women would prevent or slow the progression of AMD. The trial was stopped due to the harmful effect of HRT (increased risk of coronary heart disease and stroke mortality). The data from this trial regarding the association between HRT and AMD progression are currently being analyzed.

### 5.2.4.6
**Diet and Vitamin Supplementation**

The reader is referred elsewhere to detailed reviews on the relation of diet to AMD (Christen 1994; Mares-Perlman and Klein 1999). Green leafy vegetables and antioxidants (from diet and multivitamins) have been inconsistently associated with a decreased risk of AMD (EDCCS 1993; Christen 1994; Seddon et al. 1994; Mares-Perlman and Klein 1999). In Beaver Dam, dietary intake of saturated fat was found to be associated with an 80% increased risk of AMD (Mares-Perlman et al. 1995; Delcourt et al. 2001b). Caution must be exercised, because it is difficult to control for other lifestyle choices that might influence or account for these associations.

The AREDS, a large, multicentered randomized controlled clinical trial, has involved 3,640 study participants, 60–80 years of age, with extensive small drusen, intermediate drusen, large drusen, noncentral geographic atrophy, or pigmentary abnormalities in one or both eyes, or advanced AMD or vision loss due to AMD in one eye (AREDS 1999). Participants were randomly assigned to receive daily tablets containing antioxidants (500 mg vitamin C; 400 IU vitamin E; and 15 mg beta-carotene); 80 mg zinc and 2 mg copper; antioxidants plus zinc; or a placebo. Comparisons with the placebo group show a significant reduction of incident late AMD with antioxidants plus zinc (OR 0.72, 95% CI 0.52–0.98; AREDS 2001). The treatment effect was more marked in eyes with larger areas of intermediate-sized drusen, large drusen, or noncentral geographic atrophy. The authors of the report conclude that there is no benefit from the study formulations for persons with no or minimal AMD. Such multivitamin supplementation does not appear to have an effect until more advanced stages of early AMD are apparent.

### 5.2.5
**Environmental Factors**

### 5.2.5.1
**Light**

Data from animal studies and case reports in humans have suggested that exposure to intense bright sunlight or ultraviolet (UV) radiation may cause damage to the RPE similar to that seen in AMD (Borges et al. 1990; Winkler et al. 1999; Roberts 2001; Augustin et al. 2002; Taylor et al. 1992). However, data from epidemiological studies have only weak association of light and no association of UV exposure with AMD (Vinding 1990; Cruickshanks et al. 1993, 2001; Darzins et al. 1997; Mitchell et al. 1998; Delcourt et al. 2001a; Tomany et al. 2004a, 2004b). This may be due to the difficulty in assessing these relations using questionnaires. It is difficult to accurately obtain information about confounding factors such as use of hats and sunglasses, cataract presence and severity, and levels of macular pigment in the eyes that may affect the relation of UV or sunlight exposure at the retina to the incidence of AMD.

In Beaver Dam there were few significant relationships between environmental exposure to light and the 10-year incidence and progression of AMD (Tomany et al. 2004b). Participants exposed to summer sun for more than 5 h a day during their teens, 30s, and at the baseline examination were more likely to develop increased retinal pigment (RR 3.17; 95% CI 1.24,8.11) and early AMD (RR 2.14; 95% CI 0.99,4.61) by 10 years than those exposed less than 2 h per day in the same time periods. In participants reporting the highest summer sun exposure levels in their teens and 30s, use of hats and sunglasses at least half the time in the same time periods was associated with a decreased risk of developing soft, indistinct drusen (RR 0.55; 95% CI 0.33,0.90) and RPE depigmentation (RR 0.51; 95% CI 0.29,0.91). In Beaver Dam, UV-B exposure, winter leisure time spent outdoors, skin sun-sensitivity, or number of bad sunburns experienced at the baseline examination were not associated with AMD.

### 5.2.5.2
### X-rays

There are few data regarding relationships between diagnostic X-rays and AMD. Hyman et al. (1983) have found no relation between occupational exposures to X-rays and AMD. In Beaver Dam a history of computed tomography scans of the head or dental X-rays were not related to the 5-year incidence and progression of AMD (Klein et al. 2000).

### 5.2.6
### Ocular Factors

### 5.2.6.1
### Refractive Error

The relationship of hyperopia to AMD has been inconsistent (Maltzman et al. 1979; Ferris 1983; Hyman et al. 1983; EDCCS 1992; Sandberg et al. 1993; Chaine et al. 1998; AREDS 2000; Wong et al. 2002). In the NHANES, persons who were hyperopic were more likely to have AMD than those who were emmetropic (OR 1.61; 95% CI 1.15–2.25; Goldberg et al. 1988). In EDCCS a refractive error of more than 1+ is associated with neovascular AMD (OR 1.5; 95% CI 0.9 to 2.4, $P=0.05$; The Eye Disease Case-Control Study Group 1992). In Rotterdam, hyperopia of 1 diopter or more is associated with increased risk of AMD compared with myopia but was not associated with neovascular AMD (Klein 1999). In the Blue Mountains Eye Study, early AMD is more likely in eyes with moderate hyperopia (more than 3.00 to 6.00 diopters, OR 1.8, 95% CI 1.1–3.1) and high hyperopia (more than 6.00D, OR 8.6, 95% CI 2.1 to 35.5) compared with eyes that are emmetropic (−1.00 to more than +1.00 diopter; Wang et al. 1998). In Beaver Dam there was no association between hyperopia (+1.00 Diopter or greater spherical equivalent) and the 10-year incidence of early (RR 1.0; 95% CI 0.7–1.1) and late AMD (OR 1.2; 95% CI 0.6–2.3; Wong et al. 2002). Moderate and high hyperopia, as defined in the Blue Mountains, are not associated with incident or progressed AMD (Weiter et al. 1985). There is also no association of myopia with AMD in either study. Boker et al. (1993) have hypothesized that the association between hyperopia and AMD may be related to a reduction in choroidal blood flow in eyes with shorter axial lengths, which may predispose these eyes to development of choroidal neovascularization. Alternatively, hyperopia and AMD may both be related to aging (Pierscionek and Weale 1996).

### 5.2.6.2
### Iris Color

Lighter iris pigmentation has been found to be associated with AMD in some (Ferris 1983; Weiter et al. 1985; Holz et al. 1994a, 1994b; Chaine et al. 1998; Mitchell et al. 1998) but not all studies (Vinding 1990; EDCCS 1992; Klein et al. 1998a, 1998b; Frank et al. 2000). In Beaver Dam, people with brown eyes are more likely to develop soft, indistinct drusen at 10 years than people with blue eyes (Tomany et al. 2003). This may be due to an easier detection of such drusen in a more darkly pigmented fundus in brown- versus blue-eyed persons. In Beaver Dam, people with blue eyes were found to have significantly higher risk of developing increased retinal pigment and RPE depigmentation. An age interaction was found: the 10-year incidence of RPE depigmentation was significantly lower in participants under the age of 65 years with brown rather than blue eyes, but not in participants over 65 years. There was no relationship of iris color with the 10-year-incidence late AMD. In cross-sectional data from the Blue Mountains Eye Study, persons with lightly pigmented eyes are found to have a significantly higher incidence of early AMD (OR 1.45, 95%CI 1.09–1.92; Mitchell et al. 1998) than those with darkly pigmented eyes. Holz et al. (1994b) have shown that the increased risk of AMD is found only in persons who report having light-colored irides in youth (RR 2.4, 95% CI 1.2–4.7) and in those who report having dark-colored irides in youth that changed to a light color by adulthood (RR 9.4, 95% CI 2.9–32.0) compared with persons reporting dark colored irides that did not change. They postulate that in some individuals there may be a decrease in RPE melanin (as indirectly measured by developing lighter-colored irides with age), as well as a loss of the protective effect against factors associated with AMD.

The reasons for the inconsistent relationships of iris color and lesions associated with AMD are not known. It has been hypothesized that melanin, found in the tissues of the eye, may protect the retina from direct sunlight exposure, reducing direct oxidative damage and thereby reducing the risk of AMD (Jampol and Tielsch 1992). A study by Menon et al. (1992), measuring the amount of melanin in brown and blue eyes, has found no statistically significant difference between the type or content of melanin in the irides of brown and blue eyes. However, the study has found the RPE-choroid from brown eyes to have more melanin than corresponding tissues from blue eyes. It is possible that this increased melanin protects against the development of RPE depigmentation but not of soft, indistinct drusen.

### 5.2.6.3
### Cataract and Cataract Surgery

Data from a number of studies have shown cataract and cataract surgery to be related to developing AMD (Sperduto and Seigel 1980; Sperduto et al. 1981; Liu et al. 1989; Vinding 1989; van der Schaft et al. 1992; Klein R et al. 1994a; Wang et al. 1999). In Beaver Dam there was a positive cross-sectional association between nuclear sclerosis or cataract surgery and early AMD (Klein R et al. 1994a). The association with nuclear sclerosis is consistent with findings from the Chesapeake Bay Watermen Study (Bressler et al. 1989), but not with findings from the Framingham Eye Study (Sperduto and Seigel 1980; Sperduto et al. 1981) or the Blue Mountains Eye Study (Wang et al. 1999). In Beaver Dam, while controlling for age, gender, systolic blood pressure, and history of heavy drinking, smoking, and vitamin use, cataract at baseline was associated with the 10-year incidence of early AMD (RR 1.30; 95% CI 1.04–1.63), soft indistinct drusen (RR 1.38; 95% CI 1.08–1.75), increased retinal pigment (RR 1.38; 95% CI 1.07–1.79), and progression of AMD (OR 1.37; 95% CI 1.06–1.77), but not with the incidence of late AMD (Klein et al. 2002c). This has been postulated to be due to environmental factors such as diet (Mares-Perlman and Klein 1999), light exposure (Cruickshanks et al. 1993, 2001; Vingerling et al. 1997), or genetic factors (Heiba et al. 1993, 1994), which may be related to both cataract and AMD.

Most reports regarding the association of cataract surgery and AMD have come from clinical case series, case-control studies, and cross-sectional studies, and the results have been inconsistent (Blair and Ferguson 1979; Sperduto and Seigel 1980; Sperduto et al. 1981; Lui et al. 1989; Vinding 1989; Klein R et al. 1994a; Klein BEK 1994; Pollack et al. 1996, 1998; Vingerling et al. 1997; Chaine et al. 1998; Shuttleworth et al. 1998; Wang et al. 1999; Armbrecht et al. 2000). In the National Health and Nutrition Examination Survey, a significant association has been found between aphakia and AMD (OR 2.00; 95% CI 1.44, 2.78; Klein et al. 1995b). In a clinic-based study, Pollack et al. (1998) have reported a higher risk of progression in eyes with moderate AMD after cataract surgery. In that study, exudative macular degeneration developed during the first postoperative year in 19% of the eyes that had surgery compared with 4% in the fellow eyes in patients older than 65 years of age. These findings are also consistent with a higher frequency of neovascular AMD found by histopathologic study of eyes that had implantation of an intraocular lens compared with phakic eyes (van der Schaft et al. 1994). On the other hand, the Age-Related Eye Disease Study has shown that cataract surgery does not significantly accelerate progression to advanced late AMD at 4 years (RR 1.06; 95% CI 0.85, 1.32; Gensler et al. 2004). In addition, no association has been found between cataract surgery and AMD prevalence in the Rotterdam Study (Vingerling et al. 1997).

In the Beaver Dam Eye Study, cataract surgery before baseline was associated with the 10-year incidence of late AMD (RR 3.81; 95% CI 1.89–7.69), increased retinal pigment (RR 1.89; 95% CI 1.18–3.03), RPE depigmentation (RR 1.95; 95% CI 1.17–3.25), pure geographic atrophy (RR 3.18; 95% CI 1.33–7.60), neovascular AMD (RR 4.31; 95% CI 1.71–10.9), and progression of AMD (RR 1.97; 95% CI 1.29–3.02), but not with the incidence of early AMD (Klein et al. 2002c). Similar findings have been reported in the Blue Mountains Eye Study for the 5-year incidence of late AMD (Wang et al. 2003b). These findings have broad clinical implications if the relationship between cataract surgery and AMD is more than the relationship between nuclear cataract and AMD. It is possible that eyes that had undergone cataract surgery prior to baseline in Beaver Dam and in the Blue Mountains were ones in which early AMD was progressing and causing additional visual symptoms to warrant cataract surgery. The association may also have been a result of easier visualization and detection of AMD lesions after cataract surgery. It is also possible that some of the signs of AMD in eyes that had undergone cataract surgery prior to the baseline examination may be related to photic retinal injuries (increased retinal pigmentation and RPE depigmentation) due to the operating microscope (Jaffe 1982; McDonald and Irvine 1983; Kleinmann et al. 2002). Inflammatory changes, with or without the development of transient cystoid macular edema, that may occur in eyes after cataract surgery may be related to the development of late AMD (van der Schaft et al. 1994).

### 5.2.6.4
### Other

In Beaver Dam, no relationship has been reported between intraocular pressure, perfusion pressure, open angle glaucoma, generalized and focal retinal arteriolar narrowing, retinal arteriovenous nicking or retinopathy and the incidence or progression of AMD (Klein R, unpublished data).

### 5.2.7
### Socioeconomic Factors and Work Exposures

Low income and little education are related to increased morbidity and mortality from a number of diseases (Marmot et al. 1987). These associations have been attributed to under use of healthcare resources, exposure to noxious work or adverse home environments, high-risk behaviors, and poor diet. Most observations regarding socioeconomic factors and AMD have been from case-control and cross-sectional studies (Klein 1999).

In Beaver Dam, education and type of work, independent of vitamin supplementation and smoking, were associated with the incidence of early AMD (Klein R et al. 2001c). Similarly, in EDCCS, persons with AMD were 30% less likely to have completed 12 years of school compared with those without AMD (EDCCS 1992). In the AREDS, while controlling for age, smoking, and hypertension status, persons with higher education were at lower risk of having large drusen (OR 0.72), geographic atrophy (OR 0.44), and neovascular AMD (OR 0.44; AREDS 2000). An inverse relation of education with AMD has also been found in the Health and Nutrition Examination Survey (Klein and Klein 1982). On the other hand, no association of education with AMD has been found cross-sectionally in other studies (Kahn et al. 1977; Klein R et al. 1994b, 1999c). The reasons for the differences among studies are not known. Education may be a marker of unknown lifelong exposures not directly measured in these studies. No relation of income to incident or progressed AMD was found in Beaver Dam.

In Beaver Dam there was a higher incidence of AMD in service workers. The associations remained in waiter/waitresses, cooks, and bartenders after controlling for smoking and drinking history (Klein et al. 2001c). However, passive smoking was not measured and might have accounted for the association found in these subgroups of service workers.

In one case-control study, men with AMD had higher odds of 3.8 (95% CI 1.0–14.5) of responding positively to the question, "Did you ever work around chemicals which caused your eyes to burn, on a regular basis?" (Hyman et al. 1983). This association was not found in women. There was no relationship in this study between AMD and working around extremely bright lights, dim lights, X-rays, radioactive materials, bright flames, radar, or bright sunlight.

## 5.3
## Age-Related Macular Degeneration and Survival

Few data are available regarding whether persons with AMD are at higher risk of morbidity and mortality than persons without AMD. While controlling for other risk factors, AMD has not been found to be related to mortality in population-based studies (Klein et al. 1995a; Wang et al. 2001; Borger et al. 2003).

## 5.4
## Public Health Issues

Review of epidemiological data provided herein suggests inconsistencies in the associations of all intervenable risk factors and AMD. Cigarette smoking, hypertension, and cataract surgery emerge as factors likely to increase the risk of progression to late AMD. Cessation of smoking has been advocated by some to prevent progression to late AMD (Mitchell et al. 1999). Epidemiological data suggest that if there is a significant cataract requiring cataract surgery and moderate-to-severe early AMD is present, the surgeon should discuss the possibility of progression of the AMD following cataract surgery. The cataract surgery is indicated when the cataract has reached a stage leading to visual impairment affecting quality of life. There are no clinical trial data showing that blood pressure reduction will result in less risk of neovascular AMD. Thus, unlike for diabetic retinopathy, there have been no public health approaches developed for early detection and prevention of AMD. Only recommendations for care of individual patients exist.

## 5.5
## Conclusions

Epidemiologic data have quantified the high prevalence and incidence of early AMD and its natural history in the general population. These data have also been helpful in quantifying the strong relationship of AMD with age and a family history of the disease and the relationships of other putative risk factors with AMD (Klein et al. 1999a, 1999b, 1999c; Wang et al. 2001). However, there are a number of limitations that make interpretation of the epidemiological findings difficult. One limitation is that the disease occurs late in life. This may preclude measurement of exposures (e.g., light) that occur earlier in life. In addition, persons at risk of late AMD are often too ill to participate in an epidemiological study or die before they can be studied, limiting the ability to find relationships when they exist. Furthermore, although the lesions characterizing the disease have been described, the specific phenotypic expression of early stages of the disease has yet to be fully explored. It is possible that what is currently grouped together as early or late AMD is an expression of combinations of different genes that respond differently to different environmental exposures such as light and systemic conditions such as hypertension and inflammation. It would be of benefit to characterize persons at risk of the disease prior to the appearance of the first signs of early AMD, and to characterize persons with the same phenotypic expression of the disease. It is hoped that advances in genetic epidemiology and the development of more sensitive, computer-assisted noninvasive measurement approaches of drusen, the RPE, Bruch's membrane, and the choroidal vasculature will help achieve this.

## References

Age-Related Eye Disease Study Research Group (1999) Design implications. AREDS Report No. 1. Control Clin Trials 20:573–600

Age-Related Eye Disease Study Research Group (2000) Risk factors associated with age-related macular degeneration. A case-control study in the Age-Related Eye Disease Study. AREDS Report No 3. Ophthalmology 107:2224–2232

Age-Related Eye Disease Study Research Group (2001) A randomized, placebo-controlled, clinical trial of high-dose supplementation with vitamins C and E, beta carotene, and zinc for age-related macular degeneration and vision loss. AREDS Report No. 8. Arch Ophthalmol 119:1417–1436

Ajani UA, Christen WG, Manson JE, Glynn RJ, Schaumberg D, Buring JE, Hennekens CH (1999) A prospective study of alcohol consumption and the risk of age-related macular degeneration. Ann Epidemiol 9:172–177

Armbrecht AM, Findlay C, Kaushal S, Aspinall P, Hill AR, Dhillon B (2000) Is cataract surgery justified in patients with age related macular degeneration? A visual function and quality of life assessment. Br J Ophthalmol 84:1343–1348

Augustin AJ, Dick HB, Offermann I, Schmidt-Erfurth U (2002) The significance of oxidative mechanisms in diseases of the retina. Klin Monatsbl Augenheilkd 219:631–643

Barnes PJ (2000) Chronic obstructive pulmonary disease. N Engl J Med 343:269–280

Barondes MJ, Pagliarini S, Chisholm IH, Hamilton AM, Bird AC (1992) Controlled trial of laser photocoagulation of pigment epithelial detachments in the elderly: 4-year review. Br J Ophthalmol 76:5–7

Beatty S, Koh H, Phil M, Henson D, Boulton M (2000) The role of oxidative stress in the pathogenesis of age-related macular degeneration. Surv Ophthalmol 45:115–134

Bettman JW, Fellows V, Chao P (1958) The effect of cigarette smoking on the intraocular circulation. Arch Ophthalmol 59:481–488

Bird AC, Bressler NM, Bressler SB, Chisholm IH, Coscas G, Davis MD, Jong PT de, Klaver CC, Klein BE, Klein R et al. (1995) An international classification and grading system for age-related maculopathy and age-related macular degeneration. The International ARM Epidemiological Study Group. Surv Ophthalmol 39:367–374

Bischoff PM, Flower RW (1983) High blood pressure in choroidal arteries as a possible pathogenetic mechanism in senile macular degeneration. Am J Ophthalmol 96:398–399

Blair CJ, Ferguson J Jr (1979) Exacerbation of senile macular degeneration following cataract extraction. Am J Ophthalmol 87:77–83

Blumenkranz MS, Russell SR, Robey MG, Kott-Blumenkranz R, Penneys N (1986) Risk factors in age-related maculopathy complicated by choroidal neovascularization. Ophthalmology 93:552–558

Boker T, Fang T, Steinmetz R (1993) Refractive error and choroidal perfusion characteristics in patients with choroidal neovascularization and age-related macular degeneration. Ger J Ophthalmol 2:10–13

Borger PH, Leeuwen R van, Hulsman CA, Wolfs RC, Kuip DA van der, Hofman A, Jong PT de (2003) Is there a direct association between age-related eye diseases and mortality? The Rotterdam Study. Ophthalmology 110:1292–1296

Borges J, Li ZY, Tso MO (1990) Effects of repeated photic exposures on the monkey macula. Arch Ophthalmol 108:727–733

Bressler NM, Bressler SB, West SK, Fine SL, Taylor HR (1989) The grading and prevalence of macular degeneration in Chesapeake Bay watermen. Arch Ophthalmol 107:847–852

Bressler SB, Maguire MG, Bressler NM, Fine SL (1990) Relationship of drusen and abnormalities of the retinal pigment epithelium to the prognosis of neovascular macular degeneration. The Macular Photocoagulation Study Group. Arch Ophthalmol 108:1442–1447

Bressler NM, Muñoz B, Maguire MG, Vitale SE, Schein OD, Taylor HR, West SK (1995) Five-year incidence and disappearance of drusen and retinal pigment epithelial abnormalities. Waterman study. Arch Ophthalmol 113:301–308

Bressler SB, Muñoz B, Phillips D, West S, the SEE Project Team (1998) Prevalence of age-related macular degeneration in a population-based study: SEE project (Abstract). Presented at the Macula Society Meeting, Boca Raton, FL, p 86

Chaine G, Hullo A, Sahel J, Soubrane G, Espinasse-Berrod MA, Schutz D, Bourguignon C, Harpey C, Brault Y, Coste M, Moccatti D, Bourgeois H (1998) Case-control study of the risk factors for age related macular degeneration. France DMLA Study Group. Br J Ophthalmol 82:996–1002

Chakrabarti S, Prashar S, Sima AAF (1990) Augmented polyol pathway activity and retinal pigment epithelial permeability in the diabetic BB rat. Diabet Res Clin Pract 8:1–11

Cho E, Hankinson SE, Willett WC, Stampfer MJ, Spiegelman D, Speizer FE, Rimm EB, Seddon JM (2000) Prospective study of alcohol consumption and the risk of age-related macular degeneration. Arch Ophthalmol 118:681–688

Christen WG Jr (1994) Antioxidants and eye disease. Am J Med 97:14S–17S; 22S–28S

Christen WG, Glynn RJ, Manson JE, Ajani UA, Buring JE (1996) A prospective study of cigarette smoking and risk of age-related macular degeneration in men. J Am Med Assoc 276:1147–1151

Christen WG, Glynn RJ, Ajani UA, Schaumberg DA, Chew EY, Buring JE, Manson JE, Hennekens CH (2001) Age-related maculopathy in a randomized trial of low-dose aspirin among US physicians. Arch Ophthalmol 119:1143–1149

Cousins SW, Marin-Castano ME, Espinosa-Heidmann DG, Alexandridou A, Striker L, Elliot S (2003) Female gender, estrogen loss, and sub-RPE deposit formation in aged mice. Invest Ophthalmol Vis Sci 44:1221–1229

Cruickshanks KJ, Klein R, Klein BEK (1993) Sunlight and age-related macular degeneration. The Beaver Dam Eye Study. Arch Ophthalmol 111:514–518

Cruickshanks KJ, Hamman RF, Klein R, Nondahl DM, Shetterly SM (1997) The prevalence of age-related maculopathy by geographic region and ethnicity: the Colorado-Wisconsin Study of Age-Related Maculopathy. Arch Ophthalmol 115:242–250

Cruickshanks KJ, Klein R, Klein BEK, Nondahl DM (2001) Sunlight and the 5-year incidence of early age-related maculopathy. The Beaver Dam Eye Study. Arch Ophthalmol 119:246–250

Curcio CA, Millican CL, Bailey T, Kruth HS (2001) Accumulation of cholesterol with age in human Bruch's membrane. Invest Ophthalmol Vis Sci 42:265–274

Darzins P, Mitchell P, Heller RF (1997) Sun exposure and age-related macular degeneration. An Australian case-control Study. Ophthalmology 104:770–776

Delcourt C, Diaz JL, Ponton-Sanchez A, Papoz L (1998) Smoking and age-related macular degeneration. The POLA Study. Pathologies Oculaires Liees a l'Age. Arch Ophthalmol 116:1031–105

Delcourt C, Cristol JP, Tessier F, Leger CL, Descomps B, Papoz L (1999) Age-related macular degeneration and antioxidant status in the POLA study. POLA Study Group. Pathologies Oculaires Liees a l'Age. Arch Ophthalmol 117:1384–1390

Delcourt C, Carriere I, Ponton-Sanchez A, Fourrey S, Lacroux A, Papoz L; POLA Study Group (2001a) Light exposure and the risk of age-related macular degeneration: the Pathologies Oculaires Liees a l'Age (POLA) Study. Arch Ophthalmol 119:1463–1468

Delcourt C, Michel F, Colvez A, Lacroux A, Delage M, Vernet MH; POLA Study Group (2001b) Associations of cardiovascular disease and its risk factors with age-related macular degeneration: the POLA study. Ophthalmic Epidemiol 8:237–249

Dithmar S, Sharara NA, Curcio CA, Le NA, Zhang Y, Brown S, Grossniklaus HE (2001) Murine high-fat diet and laser photochemical model of basal deposits in Bruch membrane. Arch Ophthalmol 119:1643–1649

Dubina TL, Mints AYa, Zhuk EV (1984) Biological age and its estimation. III. Introduction of a correction to the multiple regression model of biologicalal age in cross-sectional and longitudinal studies. Exp Gerontol 19:133–143

Ernst E, Hammerschmidt DE, Bagge U, Matrai A, Dormandy JA (1987) Leukocytes and the risk of ischemic diseases. J Am Med Assoc 257:2318–2324

Evans JR (2001) Risk factors for age-related macular degeneration. Prog Retin Eye Res 20:227–253

The Eye Disease Case-Control Study Group (1992) Risk factors for neovascular age-related macular degeneration. Arch Ophthalmol 110:1701–1708

The Eye Disease Case-Control Study Group (1993) Antioxidant status and neovascular age-related macular degeneration. Arch Ophthalmol 111:104–109

Feeney-Burns L, Ellersieck MR (1985) Age-related changes in the ultrastructure of Bruch's membrane. Am J Ophthalmol 100:686–697

Ferris FL 3rd (1983) Senile macular degeneration: review of epidemiological features. Am J Epidemiol 118:132–151

Frank RN, Puklin JE, Stock C, Canter LA (2000) Race, iris color, and age-related macular degeneration. Trans Am Ophthalmol Soc 98:109–115

Friedman E (1970) Choroidal blood flow. Pressure-flow relationships. Arch Ophthalmol 83:95–99

Friedman E (2000) The role of the atherosclerotic process in the pathogenesis of age-related macular degeneration. Am J Ophthalmol 130:658–663

Fryczkowski AW, Sato SE, Hodes BL (1988) Changes in the diabetic choroidal vasculature: scanning electron microscopy findings. Ann Ophthalmol 20:299–305

Gass JD (1973) Drusen and disciform macular detachment and degeneration. Arch Ophthalmol 90:206–217

Gensler G, Klein BEK, Klein R, Chew EY, the AREDS Research Group (2004) The effect of cataract surgery on progression to advanced AMD. Abstract Invest Ophthalmol Vis Sci (in press)

Ghezzi P, Dinarello CA, Bianchi M, Rosandich ME, Repine JE, White CW (1991) Hypoxia increases production of interleukin-1 and tumor necrosis factor by human mononuclear cells. Cytokine 3:189–194

Gibson JM, Rosenthal AR, Lavery J (1985) A study of the prevalence of eye disease in the elderly in an English community. Trans Ophthalmol Soc UK 104:196–203

Gilles MC, Luo W, Chua W, et al. (2001) The efficacy of a single intravitreal injection of triamcinolone for neovascular age-related macular degeneration. One-year results of a randomized clinical trial: IVTAS (abstract). Invest Ophthalmol Vis 42:S522

Goldberg J, Flowerdew G, Smith E, Brody JA, Tso MO (1988) Factors associated with age-related macular degeneration. An analysis of data from the first National Health and Nutrition Examination Survey. Am J Epidemiol 128:700–710

Goldstein JL, Hobbs HH, Brown MS (2001) Familial hypercholesterolemia. In: Scriver CR, Beadet AL, Valle D, Sly WS (eds) The metabolic and molecular basis of inherited disease, 8th edn, vol 2. McGraw-Hill, New York, pp 2863–2913

Gorin MB, Breitner JC, De Jong PT, Hageman GS, Klaver CC, Kuehn MH, Seddon JM (1999) The genetics of age-related macular degeneration. Mol Vis 5:29

Green WR, Enger C (1993) Age-related macular degeneration histopathologic studies. The 1992 Lorenz E Zimmerman Lecture. Ophthalmology 100:1519–1535

Gregor Z, Bird AC, Chisholm IH (1977) Senile disciform macular degeneration in the second eye. Br J Ophthalmol 61:141–147

Grunwald JE, Hariprasad SM, DuPont J, Maguire MG, Fine SL, Brucker AJ, Maguire AM, Ho AC (1998) Foveolar choroidal blood flow in age-related macular degeneration. Invest Ophthalmol Vis Sci 39:385–390

Hachicha M, Naccache PH, McColl SR (1995) Inflammatory microcrystals differentially regulate the secretion of macrophage inflammatory protein 1 and interleukin 8 by human neutrophils: a possible mechanism of neutrophil recruitment to sites of inflammation in synovitis. J Exp Med 182:2019–2025

Hageman GS, Luthert PJ, Victor Chong NH, Johnson LV, Anderson DH, Mullins RF (2001) An integrated hypothesis that considers drusen as biomarkers of immune-mediated processes at the RPE-Bruch's membrane interface in aging and age-related macular degeneration. Prog Retin Eye Res 20:705–732

Hall NF, Gale CR, Syddall H, Phillips DI, Martyn CN (2001) Risk of macular degeneration in users of statins: cross sectional study. Br Med J 323:375–376

Hammond BR Jr, Wooten BR, Snodderly DM (1996) Cigarette smoking and retinal carotenoids: implications for age-related macular degeneration. Vis Res 18:3003–3009

Hayes KC, Lindsey S, Stephan ZF, Brecker D (1989) Retinal pigment epithelium possesses both LDL and scavenger receptor activity. Invest Ophthalmol Vis Sci 30:225–232

Heiba IM, Elston RC, Klein BEK, Klein R (1993) Genetic etiology of nuclear cataract: Evidence for a major gene. Am J Med Genet 47:1208–1214

Heiba IM, Elston RC, Klein BEK, Klein R (1994) Sibling correlations and segregation analysis of age-related maculopathy: The Beaver Dam Eye Study. Genet Epidemiol 11:51–67

Hidayat AA, Fine BS (1985) Diabetic choroidopathy. Light and electron microscopic observations of seven cases. Ophthalmology 92:512–522

Hirvela H, Luukinen H, Laara E, Sc L, Laatikainen L (1996) Risk factors of age-related maculopathy in a population 70 years of age or older. Ophthalmology 103:871–877

Holz FG, Piguet B, Minassian DC, Bird AC, Weale RA (1994a) Decreasing stromal iris pigmentation as a risk factor for age-related macular degeneration. Am J Ophthalmol 117:19–23

Holz FG, Wolfensberger TJ, Piguet B, Gross-Jendroska M, Wells JA, Minassian DC, Chisholm IH, Bird AC (1994b) Bilateral macular drusen in age-related macular degeneration. Prognosis and risk factors. Ophthalmology 101:1522–1528

Hyman LG, Lilienfeld AM, Ferris FL 3rd, Fine SL (1983) Senile macular degeneration: a case-control study. Am J Epidemiol 118:213–227

Hyman L, Schachat AP, He Q, Leske MC (2000) Hypertension, cardiovascular disease, and age-related macular degeneration. Age-Related Macular Degeneration Risk Factors Study Group. Arch Ophthalmol 118:351–358

Jaffe NS (1982) The intracapsular-extracapsular controversy. Aust J Ophthalmol 10:115–119

Jampol LM, Tielsch J (1992) Race, macular degeneration, and the Macular Photocoagulation Study. Arch Ophthalmol 110:1699–1700

Jonasson F, Thordarson K (1987) Prevalence of ocular disease and blindness in a rural area in the eastern region of Iceland during 1980 through 1984. Acta Ophthalmol (Suppl) 182:40–43

Kahn HA, Leibowitz HM, Ganley JP, Kini MM, Colton T, Nickerson RS, Dawber TR (1977) The Framingham Eye Study II. Association of ophthalmic pathology with single variables previously measured in the Framingham Heart Study. Am J Epidemiol 106:33–41

Katz ML, Parker KR, Handelman GJ, Bramel TL, Dratz EA (1982) Effects of antioxidant nutrient deficiency on the retina and retinal pigment epithelium of albino rats: a light and electron microscopic study. Exp Eye Res 34:339–369

Klaver CC, Kliffen M, Duijn CM van, Hofman A, Cruts M, Grobbee DE, Broeckhoven C van, Jong PT de (1998) Genetic association of apolipoprotein E with age-related macular degeneration. Am J Hum Genet 63:200–206 (erratum in Am J Hum Genet 63:1252)

Klaver CC, Assink JJ, Leeuwen R van, Wolfs RC, Vingerling JR, Stijnen T, Hofman A, Jong PT de (2001) Incidence and progression rates of age-related maculopathy: the Rotterdam Study. Invest Ophthalmol Vis Sci 42:2237–2241

Klein BE, Klein R (1982) Cataracts and macular degeneration in older Americans. Arch Ophthalmol 100:571–573

Klein BEK, Klein R, Jensen SC, Ritter LL (1994) Are sex hormones associated with age-related maculopathy in women? The Beaver Dam Eye Study. Trans Am Ophthalmol Soc 92:289–297

Klein BEK, Klein R, Moss SE (2000) Exposure to diagnostic X-rays and incident age-related eye disease. Ophthalmic Epidemiol 7:61–65

Klein BEK, Klein R, Lee KE, Jensen SC (2001a) Measures of obesity and age-related eye diseases. Ophthalmic Epidemiol 8:251–262

Klein BEK, Klein R, Lee KE, Moore EL, Danforth L (2001b) Risk of incident age-related eye diseases in people with an affected sibling. The Beaver Dam Eye Study. Am J Epidemiol 154:207–211

Klein R (1999) Epidemiology. In: Berger JW, Fine SL, Maguire MG (eds) Age-related macular degeneration. Mosby, St. Louis, pp 31–55

Klein R, Davis MD, Magli YL, Segal P, Klein BE, Hubbard L (1991a) The Wisconsin Age-Related Maculopathy Grading System. Ophthalmology 98:1128–1134

Klein R, Klein BEK, Linton KL, De Mets DL (1991b) The Beaver Dam Eye Study: visual acuity. Ophthalmology 98:1310–1315

Klein R, Klein BEK, Linton KL (1992) Prevalence of age-related maculopathy. The Beaver Dam Eye Study. Ophthalmology 99:933–943

Klein R, Klein BEK, Linton KL, De Mets DL (1993a) The Beaver Dam Eye Study: the relation of age-related maculopathy to smoking. Am J Epidemiol 137: 190–200

Klein R, Klein BEK, Franke T (1993b) The relationship of cardiovascular disease and its risk factors to age-related maculopathy. The Beaver Dam Eye Study. Ophthalmology 100:406–414

Klein R, Klein BEK, Wang Q, Moss SE (1994a) Is age-related maculopathy associated with cataracts? Arch Ophthalmol 112:191–196

Klein R, Klein BEK, Jensen SC, Moss SE, Cruickshanks KJ (1994b) The relation of socioeconomic factors to age-related cataract, maculopathy, and impaired vision. The Beaver Dam Eye Study. Ophthalmology 101:1969–1979

Klein R, Klein BEK, Moss SE (1995a) Age-related eye disease and survival. The Beaver Dam Eye Study. Arch Ophthalmol 113:333–339

Klein R, Rowland ML, Harris MI (1995b) Racial/ethnic differences in age-related maculopathy. Third National Health and Nutrition Examination Survey. Ophthalmology 102:371–381

Klein R, Klein BEK, Lee KE (1996) Changes in visual acuity in a population. The Beaver Dam Eye Study. Ophthalmology 103:1169–1178

Klein R, Klein BEK, Jensen SC, Meuer SM (1997) The five-year incidence and progression of age-related maculopathy: The Beaver Dam Eye Study. Ophthalmology 104:7–21

Klein R, Klein BEK, Moss SE (1998a) Relation of smoking to the incidence of age-related maculopathy. The Beaver Dam Eye Study. Am J Epidemiol 147:103–110

Klein R, Klein BE, Jensen SC, Cruickshanks KJ (1998b) The relationship of ocular factors to the incidence and progression of age-related maculopathy. Arch Ophthalmol 116:506–513

Klein R, Clegg L, Cooper LS, Hubbard LD, Klein BE, King WN, Folsom AR, for the Atherosclerosis Risk in Communities Study Investigators (1999a) Prevalence of age-related maculopathy in the Atherosclerosis Risk in Communities Study. Arch Ophthalmol 117:1203–1210

Klein R, Klein BEK, Cruickshanks KJ (1999b) The prevalence of age-related maculopathy by geographic region and ethnicity. Prog Retin Eye Res 18:371–389

Klein R, Klein BEK, Jensen SC, Mares-Perlman JA, Cruickshanks KJ, Palta M (1999c) Age-related maculopathy in a multiracial United States population. The National Health and Nutrition Examination Survey III. Ophthalmology 106:1056–1065

Klein R, Klein BEK, Lee KE, Cruickshanks KJ, Chappell RJ (2001a) Changes in visual acuity in a population over a 10-year period. The Beaver Dam Eye Study. Ophthalmology 108:1757–1766

Klein R, Klein BEK, Jensen SC, Cruickshanks KJ, Lee KE, Danforth LG, Tomany SC (2001b) Medication use and the 5-year incidence of early age-related maculopathy: The Beaver Dam Eye Study. Arch Ophthalmol 119:1354–1359

Klein R, Klein BEK, Jensen SC, Moss SE (2001c) The relation of socioeconomic factors to the incidence of early age-related maculopathy: The Beaver Dam Eye Study. Am J Ophthalmol 132:128–131

Klein R, Klein BEK, Tomany SC, Meuer SM, Huang GH (2002a) Ten-year incidence and progression of age-related maculopathy. The Beaver Dam Eye Study. Ophthalmology 109:1767–1779

Klein R, Klein BEK, Tomany SC, Moss SE (2002b) The ten-year incidence of age-related maculopathy and smoking and drinking. The Beaver Dam Eye Study. Am J Epidemiol 156:589–598

Klein R, Klein BEK, Wong TY, Tomany SC, Cruickshanks KJ (2002c) The association of cataract and cataract surgery with the long-term incidence of age-related maculopathy. The Beaver Dam Eye Study. Arch Ophthalmol 120:1551–1558

Klein R, Klein BE, Marino EK, Kuller LH, Furberg C, Burke GL, Hubbard LD (2003a) Early age-related maculopathy in the Cardiovascular Health Study. Ophthalmology 110:25–33

Klein R, Klein BEK, Tomany SC, Cruickshanks KJ (2003b) The association of cardiovascular disease with long-term incidence of age-related maculopathy. The Beaver Dam Eye Study. Ophthalmology 110:1273–1280

Klein R, Klein BEK, Tomany SC, Cruickshanks KJ (2003c) Association of emphysema, gout, and inflammatory markers with long-term incidence of age-related maculopathy. Arch Ophthalmol 121: 674–678

Klein R, Klein BEK, Tomany SC, Danforth LG, Cruickshanks KJ (2003d) Relation of statin use to the 5-year incidence and progression of age-related maculopathy. Arch Ophthalmol 121:1151–1155

Kleinmann G, Hoffman P, Schechtman E, Pollack A (2002) Microscope-induced retinal phototoxicity in cataract surgery of short duration. Ophthalmology 109:334–338

Kornzweig AL (1977) Changes in the choriocapillaris associated with senile macular degeneration. Ann Ophthalmol 9:753–756, 759–762

Leibowitz HM, Krueger DE, Maunder LR, Milton RC, Kini MM, Kahn HA, Nickerson RJ, Pool J, Colton TL, Ganley JP, Loewenstein JI, Dawber TR (1980) The Framingham Eye Study monograph: an ophthalmological and epidemiological study of cataract, glaucoma, diabetic retinopathy, macular degeneration, and visual acuity in a general population of 2,631 adults, 1973–1975. Surv Ophthalmol 24:335–610

Leeuwen R van, Klaver CC, Vingerling JR, Hofman A, Jong PT de (2003) The risk and natural course of age-related maculopathy: follow-up at 6.5 years in the Rotterdam Study. Arch Ophthalmol 121:519–526

Leeuwen R van, Tomany SC, Wang JJ, et al. (2004) Is medication use associated with the incidence of early age-related maculopathy? Pooled findings from three continents. Ophthalmology

Lip PL, Blann AD, Hope-Ross M, Gibson JM, Lip GY (2001) Age-related macular degeneration is associated with increased vascular endothelial growth factor, hemorheology and endothelial dysfunction. Ophthalmology 108:705–710

Liu IY, White L, LaCroix AZ (1989) The association of age-related macular degeneration and lens opacities in the aged. Am J Public Health 79:765–769

Lieber CS, Seitz HK, Garro AJ, et al. (1979) Alcohol-related diseases and carcinogenesis. Cancer Res 39:2863–2886

Lotfi K, Grunwald JE (1991) The effect of caffeine on human macular circulation. Invest Ophthalmol Vis Sci 32:3028–3032

Macular Photocoagulation Study Group (1993) Five-year follow-up of fellow eyes of patients with age-related macular degeneration and unilateral extrafoveal choroidal neovascularization. Arch Ophthalmol 111:1189–1199

Macular Photocoagulation Study Group (1997) Risk factors for choroidal neovascularization in the second eye of patients with juxtafoveal or subfoveal choroidal neovascularization secondary to age-related macular degeneration. Arch Ophthalmol 115:741–747

Malek G, Li CM, Guidry C, Medeiros NE, Curcio CA (2003) Apolipoprotein B in cholesterol-containing drusen and basal deposits of human eyes with age-related maculopathy. Am J Pathol 162:413–425

Maltzman BA, Mulvihill MN, Greenbaum A (1979) Senile macular degeneration and risk factors: a case-control study. Ann Ophthalmol 11:1197–1201

Marano CW, Matschinsky FM (1989) Biochemical manifestations of diabetes mellitus in microscopic layers of the cornea and retina. Diabetes Metab Rev 5:1–15

Mares-Perlman JA, Klein R (1999) Diet and age-related macular degeneration. In: Taylor A (ed) Nutritional and environmental influences on the eye. CRC Press, Boca Raton, FL, pp 181–214

Mares-Perlman JA, Brady WE, Klein R, Vanden Langenberg GM, Klein BEK, Palta M (1995) Dietary fat and age-related maculopathy. Arch Ophthalmol 113:743–748

Marmot MG, Kogevinas M, Elston MA (1987) Social/economic status and disease. Annu Rev Public Health 8:111–135

Martinez GS, Campbell AJ, Reinken J, Allan BC (1982) Prevalence of ocular disease in a population study of subjects 65 years old and older. Am J Ophthalmol 94:181–189

McCarty CA, Mukesh BN, Guymer RH, Baird PN, Taylor HR (2001) Cholesterol-lowering medications reduce the risk of age-related maculopathy progression. Med J Aust 175:340

McDonald HR, Irvine AR (1983) Light-induced maculopathy from the operating microscope in extracapsular cataract extraction and intraocular lens implantation. Ophthalmology 90:945–951

Menon IA, Wakeham DC, Persad MA, Avaria M, Trope GE, Basu PK (1992) Quantitative determination of the melanin contents in ocular tissues from human blue and brown eyes. J Ocul Pharmacol 8:35–42

Miceli MV, Newsome DA (1991) Cultured retinal pigment epithelium cells from donors with type I diabetes show an altered insulin response. Invest Ophthalmol Vis Sci 32:2847–2853

Mitchell P, Smith W, Attebo K, Wang JJ (1995) Prevalence of age-related maculopathy in Australia. The Blue Mountains Eye Study. Ophthalmology 102:1450–1460

Mitchell P, Smith W, Wang JJ (1998) Iris color, skin sun sensitivity, and age-related maculopathy. The Blue Mountain Eye Study. Arch Ophthalmol 105:1359–1363

Mitchell P, Chapman S, Smith W (1999) Smoking is a major cause of blindness. Med J Aust 171:173–174

Mitchell P, Wang JJ, Foran S, Smith W (2002) Five-year incidence of age-related maculopathy lesions: the Blue Mountains Eye Study. Ophthalmology 109:1092–1097

Moss SE, Klein R, Klein BE, Jensen JC, Meuer SM (1998) Alcohol consumption and the 5-year incidence of age-related maculopathy: the Beaver Dam Eye Study. Ophthalmology 105:789–794

Muñoz BE, West SK, Klein R (2003) Prevalence and risk factors for age related macular degeneration in a population based sample of US Hispanics: Proyecto VER (abstract). Presented at the May 2003 ARVO meeting in Ft. Lauderdale, FL

Newsome DA, Huh W, Green WR (1987) Bruch's membrane age-related changes vary by region. Curr Eye Res 6:1211–1221

Obisesan TO, Hirsch R, Kosoko O, Carlson L, Parrott M (1998) Moderate wine consumption is associated with decreased odds of developing age-related macular degeneration in NHANES-1. J Am Geriatr Soc 46:1–7

O'Rourke MF (1982) Arterial Function in Health and Disease. Churchill Livingstone, Edinburgh, p 211

Paetkau ME, Boyd TA, Grace M, Bach-Mills J, Winship B (1978) Senile disciform macular degeneration and smoking. Can J Ophthalmol 13:67–71

Pang CP, Baum L, Chan WM, Lau TC, Poon PM, Lam DS (2000) The apolipoprotein E epsilon4 allele is unlikely to be a major risk factor of age-related macular degeneration in Chinese. Ophthalmologica 214:289–291

Pauleikhoff D, Chen JC, Chisholm IH, Bird AC (1990) Choroidal perfusion abnormality with age-related Bruch's membrane change. Am J Ophthalmol 109:211–217

Pauleikhoff D, Wormald RP, Wright L, Wessing A, Bird AC (1992) Macular disease in an elderly population. Ger J Ophthalmol 1:12–15

Penfold PL, Killingsworth MC, Sarks SH (1986) Senile macular degeneration. The involvement of giant cells in atrophy of the retinal pigment epithelium. Invest Ophthalmol Vis Sci 27:364–371

Penfold PL, Madigan MC, Gillies MC, Provis JM (2001) Immunological and aetiological aspects of macular degeneration. Prog Retin Eye Res 20:385–414

Pierscionek BK, Weale RA (1996) Risk factors and ocular senescence. Gerontology 42:257–269

Poirier J, Minnich A, Davignon J (1995) Apolipoprotein E, synaptic plasticity and Alzheimer's disease. Ann Med 27:663–670

Pollack A, Marcovich A, Bukelman A, Oliver M (1996) Age-related macular degeneration after extracapsular cataract extraction with intraocular lens implantation. Ophthalmology 103:1546–1554

Pollack A, Bukelman A, Zalish M, Leiba H, Oliver M (1998) The course of age-related macular degeneration following bilateral cataract surgery. Ophthalmic Surg Lasers 29:286–294

Pryor WA, Hales BJ, Premovic PI, Church DF (1983) The radicals in cigarette tar: their nature and suggested physiological implications. Science 220:425–427

Rikans LE, Gonzalez LP (1990) Antioxidant protection systems of rat lung after chronic ethanol inhalation. Alcohol Clin Exp Res 14:872–877

Ritter LL, Klein R, Klein BEK, Mares-Perlman JA, Jensen SC (1995) Alcohol use and age-related maculopathy in the Beaver Dam Eye Study. Am J Ophthalmol 120:190–196

Roberts JE (2001) Ocular phototoxicity. J Photochem Photobiol B 64:136–143

Rohan RM, Fernandez A, Udagawa T, et al. (2000) Genetic heterogeneity of angiogenesis in mice. FASEB J 14:871–876

Rosenblum ER, Gavaler JS, Van Thiel DH (1989) Lipid peroxidation: a mechanism for alcohol-induced testicular injury. Free Radic Biol Med 7:569–577

Rossouw JE, Anderson GL, Prentice RL, LaCroix AZ, Kooperberg C, Stefanick ML, Jackson RD, Beresford SA, Howard BV, Johnson KC, Kotchen JM, Ockene J, Writing group for the Women's Health Initiative Investigators (2002) Risks and benefits of estrogen plus progestin in healthy postmenopausal women: principal results from the Women's Health Initiative randomized controlled trial. J Am Med Assoc 288:321–333

Roy M, Kaiser-Kupfer M (1990) Second eye involvement in age-related macular degeneration: a four-year prospective study. Eye 4:813–818

Sandberg MA, Tolentino MJ, Miller S, Berson EL, Gaudio AR (1993) Hyperopia and neovascularization in age-related macular degeneration. Ophthalmology 100:1009–1013

Sarks JP, Sarks SH, Killingsworth MC (1988) Evolution of geographic atrophy of the retinal pigment epithelium. Eye 2:552–577

Sarks SH (1976) Ageing and degeneration in the macular region: a clinico-pathological study. Br J Ophthalmol 60:324–341

Sarks SH, Van Driel D, Maxwell L, Killingsworth M (1980) Softening of drusen and subretinal neovascularization. Trans Ophthalmol Soc UK 100:414–422

Sarks SH, Penfold PL, Killingsworth MC, Driel D van (1985) Patterns in macular degeneration. In: Ryan SJ, Dawson AK, Little HL (eds) Retinal diseases. Grune & Stratton, New York, pp 87–93

Schachat AP, Hyman L, Leske MC, Connell AM, Wu SY (1995) Features of age-related macular degeneration in a Black population. The Barbados Eye Study Group. Arch Ophthalmol 113:728–735

Schaft TL van der, Mooy CM, Bruijn WC de, Oron FG, Mulder PG, Jong PT de (1992) Histologic features of the early stages of age-related macular degeneration. Ophthalmology 99:278–286

Schaft TL van der, Mooy CM, Bruijn WC de, Mulder PG, Pameyer JH, Jong PT de (1994) Increased prevalence of disciform macular degeneration after cataract extraction with implantation of an intraocular lens. Br J Ophthalmol 78:441–445

Schaumberg DA, Christen WG, Hankinson SE, Glynn RJ (2001) Body mass index and the incidence of visually significant age-related maculopathy in men. Arch Ophthalmol 119:1259–1265

Schmidt S, Saunders AM, De La Paz MA, Postel EA, Heinis RM, Agarwal A, Scott WK, Gilbert JR, McDowell JG, Bazyk A, Gass JD, Haines JL, Pericak-Vance MA (2000) Association of the apolipoprotein E gene with age-related macular degeneration: possible effect modification by family history, age, and gender. Mol Vis 6:287–293

Schultz DW, Klein ML, Humpert A, Majewski J, Schain M, Weleber RG, Ott J, Acott TS (2003) Lack of an association of apolipoprotein E gene polymorphisms with familial age-related macular degeneration. Arch Ophthalmol 121:679–683

Seddon JM, Ajani UA, Sperduto RD, Hiller R, Blair N, Burton TC, Farber MD, Gragoudas ES, Haller J, Miller DT, Yannuzzi LA, Willett WC (1994) Dietary carotenoids, vitamins A, C, and E, and advanced age-related macular degeneration. Eye Disease Case-Control Study Group. J Am Med Assoc 272:1413–1420

Seddon JM, Willett WC, Speizer FE, Hankinson SE (1996) A prospective study of cigarette smoking and age-related macular degeneration in women. J Am Med Assoc 276:1147–1151

Seddon JM, Cote J, Davis N, Rosner B (2003) Progression of age-related macular degeneration: association with body mass index, waist circumference, and waist-hip ratio. Arch Ophthalmol 121:785–792

Shuttleworth GN, Luhishi EA, Harrad RA (1998) Do patients with age-related maculopathy and cataract benefit from cataract surgery? Br J Ophthalmol 82:611–616

Smiddy WE, Fine SL (1984) Prognosis of patients with bilateral macular drusen. Ophthalmology 91:271–277

Smith W, Mitchell P (1996) Alcohol intake and age-related maculopathy. Am J Ophthalmol 122:743–745

Smith W, Mitchell P, Leeder SR (1996) Smoking and age-related maculopathy. The Blue Mountains Eye Study. Arch Ophthalmol 114:1518–1523

Smith W, Mitchell P, Wang JJ (1997) Gender, oestrogen, hormone replacement and age-related macular degeneration: results from the Blue Mountains Eye Study. (Suppl 1) 25:S13–S15

Smith W, Mitchell P, Leeder SR, Wang JJ (1998) Plasma fibrinogen levels, other cardiovascular risk factors, and age-related maculopathy: the Blue Mountains Eye Study. Arch Ophthalmol 116:583–587

Smith W, Assink J, Klein R, Mitchell P, Klaver CC, Klein BE, Hofman A, Jensen S, Wang JJ, Jong PT de (2001) Risk factors for age-related macular degeneration: Pooled findings from three continents. Ophthalmology 108:697–704

Snow KK, Seddon JM (1999) Do age-related macular degeneration and cardiovascular disease share common antecedents? Ophthalmic Epidemiol 6:125–143

Souied EH, Benlian P, Amouyel P, Feingold J, Lagarde JP, Munnich A, Kaplan J, Coscas G, Soubrane G (1998) The epsilon4 allele of the apolipoprotein E gene as a potential protective factor for exudative age-related macular degeneration. Am J Ophthalmol 125:353–359

Sparrow JM, Dickinson AJ, Duke AM, Thompson JR, Gibson JM, Rosenthal AR (1997) Seven year follow-up of age-related maculopathy in an elderly British population. Eye 11:314–324

Sperduto RD, Seigel D (1980) Senile lens and senile macular changes in a population-based sample. Am J Ophthalmol 90:86–91

Sperduto RD, Hiller R, Seigel D (1981) Lens opacities and senile maculopathy. Arch Ophthalmol 99:1004–1008

Strahlman ER, Fine SL, Hillis A (1983) The second eye of patients with senile macular degeneration. Arch Ophthalmol 101:1191–1193

Stryker WS, Kaplan LA, Stein EA, Stampfer MJ, Sober A, Willett WC (1988) The relation of diet, cigarette smoking, and alcohol consumption to plasma beta-carotene and alpha-tocopherol levels. Am J Epidemiol 127:283–296

Taylor HR, West S, Muñoz B, Rosenthal FS, Bressler SB, Bressler NM (1992) The long-term effect of visible light on the eye. Arch Ophthalmol 110:99–104

Tielsch JM (1995) Vision problems in the US: a report on blindness and vision impairment in adults aged 40 and older. Prevent Blindness America. Schaumburg, IL, pp 1–20

Tomany SC, Cruickshanks KJ, Klein R, Klein BEK, Knudtson MD (2003) Sunlight and the 10-year incidence of age-related maculopathy. The Beaver Dam Eye Study. Arch Ophthalmol

Tomany SC, Klein R, Klein BEK (2001) The relation of coffee and caffeine to the 5-year incidence of early age-related maculopathy: The Beaver Dam Eye Study. Am J Ophthalmol 132:271–273

Tomany SC, Wang JJ, Klein R, Mitchell P, Jong PT de, Vingerling JR, Klein BEK, Smith W (2004a) Risk factors for incident age-related macular degeneration: pooled findings from three continents. Ophthalmology (in press)

Tomany SC, Klein R, Klein BEK (2004b) The relationship between iris color, hair color, and skin sun sensitivity to the 10-year incidence of age-related maculopathy. The Beaver Dam Eye Study. Ophthalmology (in press)

Tong L, Lai M, Klein R, Varma R, LALES Group (2003) Prevalence of age-related Maculopathy (ARM) in a population-based cohort of adult Latinos. The Los Angeles Latino Eye Study (LALES) (Abstract). Presented at the 2003 ARVO Meeting in Ft. Lauderdale, FL

VanNewKirk MR, Nanjan MB, Wang JJ, Mitchell P, Taylor HR, McCarty CA (2000) The prevalence of age-related maculopathy: The Visual Impairment Project. Ophthalmology 107:1593–1600

Vinding T (1989) Age-related macular degeneration. Macular changes, prevalence and sex ratio. An epidemiological study of 1000 aged individuals. Acta Ophthalmol (Copenh) 67:609–616

Vinding T (1990) Pigmentation of the eye and hair in relation to age-related macular degeneration. An epidemiological study of 1000 aged individuals. Acta Ophthalmol (Copenh) 68:53–58

Vinding T, Appleyard M, Nyboe J, Jensen G (1992) Risk factor analysis for atrophic and exudative age-related macular degeneration. An epidemiological study of 1000 aged individuals. Acta Ophthalmol (Copenh) 70:66–72

Vingerling JR, Dielemans I, Bots ML, Hofman A, Grobbee DE, de Jong PT (1995a) Age-related macular degeneration is associated with atherosclerosis. The Rotterdam Study. Am J Epidemiol 142:404–409

Vingerling JR, Dielemans I, Hofman A, Grobbee DE, Hijmering M, Kramer CF, Jong PT de (1995b) The prevalence of age-related maculopathy in the Rotterdam Study. Ophthalmology 102:205–210

Vingerling JR, Dielemans I, Witteman JC, Hofman A, Grobbee DE, Jong PT de (1995c) Macular degeneration and early menopause: a case control study. Br Med J 310:1570–1571

Vingerling JR, Hofman A, Grobbee DE, Jong PT de (1996) Age-related macular degeneration and smoking. The Rotterdam Study. Arch Ophthalmol 114:1193–1196

Vingerling JR, Klaver CC, Hofman A, Jong PT de (1997) Cataract extraction and age-related macular degeneration: The Rotterdam Study. Abstract presented at ARVO S472, Investigative Ophthalmology

Vinores SA, Campochiaro PA, May EE, Blaydes SH (1988) Progressive ultrastructural damage and thickening of the basement membrane of the retinal pigment epithelium in spontaneously diabetic BB rats. Exp Eye Res 46:545–558

Wang JJ, Mitchell P, Smith W (1998) Refractive error and age-related maculopathy: the Blue Mountains Eye Study. Invest Ophthalmol Vis Sci 39:2167–2171

Wang JJ, Mitchell PG, Cumming RG, Lim R (1999) Cataract and age-related maculopathy: the Blue Mountains Eye Study. Ophthalmic Epidemiol 6:317–326

Wang JJ, Mitchell P, Simpson JM, Cumming RG, Smith W (2001) Visual impairment, age-related cataract, and mortality. Arch Ophthalmol 119:1186–1190

Wang JJ, Foran S, Smith W, Mitchell P (2003a) Risk of age-related macular degeneration in eyes with macular drusen or hyperpigmentation: the Blue Mountains Eye Study cohort. Arch Ophthalmol 121:658–663

Wang JJ, Klein R, Smith W, Klein BEK, Tomany S, Mitchell P (2003b) Cataract surgery and development of age-related macular degeneration in 5 years: pooled findings from the Beaver Dam and Blue Mountains Eye studies. Ophthalmology (in press)

Weiter JJ, Delori FC, Wing GL, Fitch KA (1985) Relationship of senile macular degeneration to ocular pigmentation. Am J Ophthalmol 99:185–187

Winkler BS, Boulton ME, Gottsch JD, Sternberg P (1999) Oxidative damage and age-related macular degeneration. Mol Vis 5:32

Wong TY, Klein R, Klein BEK, Tomany SC (2002) Refractive errors and 10-year incidence of age-related maculopathy. Invest Ophthalmol Vis Sci. 43:2869–2873

# Prevalence and Risk Factors for Age-Related Macular Degeneration in China

Zheng Qin Yin, Meidong Zhu, Wen Shan Jiang

## Contents

6.1 Introduction  103
6.2 Regional Prevalence and Relationship of Morbidity to Age and Gender  104
6.2.1 Population-Based Studies  105
6.2.2 Clinical Review Studies  106
6.2.3 The Relationship of Morbidity to Age and Gender  107
6.3 Prevalence of 'Wet' and 'Dry' AMD  107
6.4 Visual Acuity in AMD  107
6.5 Risk Factors of AMD in China  107
6.5.1 Race/Ethnicity  109
6.5.2 Light Exposures/Occupation  109
6.5.3 Smoking  109
6.5.4 Drinking  110
6.5.5 Systemic Diseases  111
6.6 Conclusions  111

References  111

## 6.1 Introduction

Racial/ethnic differences in the incidence and prevalence of age-related macular degeneration (AMD) in China compared with Western countries have been noted in previous studies (Gregor and Joffe 1978; Wu 1987; Klein et al. 1999; see also Chap. 5). Despite this, in China AMD is still considered one of the most important causes of blindness in the population older than 50 years of age (Wu 1987; National Bureau of Statistics of China 2000; Wang and Hu 2001).

With the improvement of economic conditions in China, the most common causes of blindness such as cataract, corneal diseases, trachoma and glaucoma have been largely controlled. Today, AMD has an increased prevalence, now being fourth on the list of causes of blindness in the age group of 60 years and older, and draws more and more attention from ophthalmologists, medical researchers, and society (Sun 1999). Epidemiological study of AMD in China has shown that 6.98% of the people 65 years and over have AMD (National Bureau of Statistics of China 2000). The rate increases to 17.6% in persons 70 years or older (Wang and Hu 2001). Considering the large population of China, it has been estimated that AMD currently affects at least 20 million individuals.

According to WHO statistics published in October 2003, China has a total population of 1.292 billion, of whom 130 million people are over 60 years of age (WHO 2000). The life expectancy of Chinese people in 2003 was 69.8 years for men and 72.7 years for women (WHO 2003). Studies also show that the population is aging, and three stages of the aging of China's population have been identified. The first stage, from 1990 to 2003, is considered as one of "preaging of the population"; during this period, the number of persons aged over 60 years in China increased from 97,190,000 (8.6% of the population) to 136,000,000 (10.2%). The second stage is the "developing stage" (from 2003 to 2020). During this period, the proportion of the population over 60 years will increase from 10.2% to 15.6%. In the third, "peak stage" (from 2020 to 2050), the proportion of people in China over 60 years will increase to 27.4% and reach 240 million people (Sun 1999; WHO 2003). At that stage, AMD is expected to be a much more severe problem in China than at present.

There have been few published studies describing the incidence and prevalence of AMD from large population-based cohorts of older people in China. The purpose of this chapter is to review and examine the prevalence of AMD in China for the past 20 years based on population-based studies. The investigation covers eight regions, representing the eastern, western, northern, southern, and central areas of China (Fig. 6.1). More than 250 million Oriental/Asian persons, including Tibetans, Moslems, and Hans (Chinese) have been involved in those studies. The risk factors for AMD and their impacts in China have also been investigated and analyzed.

## 6.2 Regional Prevalence and Relationship of Morbidity to Age and Gender

The diagnoses of AMD reported below have been made according to criteria established by the National Academic Group of Fundus Disease of China (Table 6.1; National Academic Group of Fundus Disease of China 1987) combined with the visual criteria of the Framingham Eye Study (Leibowitz et al. 1980).

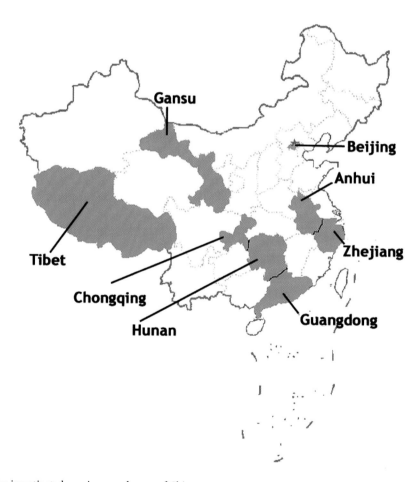

Fig. 6.1. The investigated provinces and areas of China

Table 6.1. The clinical diagnostic criteria of AMD

|  | Nonexudative (dry) type | Exudative (wet) type |
| --- | --- | --- |
| Age | Over age 45 years | Over age 45 years |
| Eye | Two eyes affected | Two eyes affected, one after another |
| Visual acuity | Decreases slowly | Decreases fast |
| Fundus findings | Earlier period: pigment dispersed and proliferation throughout the macular; light reflex of the fovea is unclear or disappeared; discrete drusen are the most typical features | Earlier period: pigment dispersed and proliferation throughout the macular; light reflex of the fovea is unclear or disappeared; drusen are often fused |
|  | Late stage: gold foil appearance in macula, geographic pigment epithelial atrophy, macular cystoid degeneration or lamellar hole are present | Metaphase: serous and/or hemorrhagic detachment of macula, subretinal hemorrhage, intraretinal hemorrhage and vitreous hemorrhage can been seen in serious patients |
|  |  | Late stage: scar formation |
| FFA | Transmitted fluorescence or low fluorescence in macular area, no fluorescence leakage | Subretinal neovascularization in macular area, fluorescence leakage, hemorrhage obstructs fluorescence |

## 6.2.1
### Population-Based Studies

Population-based studies have been carried out in six provinces/areas of China, including Guangdong, Hunan, Zhejiang and Gansu provinces, Tibet, and Shunyi – a county of Beijing. In all cases the examinations were almost identical. After randomized clustering sampling, all selected individuals were enumerated according to work places or the village residence register. Personal and family histories including nationality, occupation, smoking and drinking habits, and ocular and systemic diseases were collected. Visual acuity was measured, and a comprehensive ophthalmic examination of eyelid, cornea, lens, and fundus with dilated pupils were also carried out.

## 6.2.1.1
### Guangdong Province

Guangdong province is located in the southeastern coastal area of China, with a population of about 70.5 million. The investigation was carried out in two different areas, including Guangzhou (a modern city) and Doumen County (a rural area). A sample of 6,745 people, aged from 40 to 93 years old, were investigated (Wu et al. 1987; Guan et al. 1989; He et al. 1998).

In rural Doumen County, 5,342 people (2,421 men and 2,921 women), 50 years or older were examined in 1997 (He et al. 1998). The overall prevalence of AMD in the group is 8.4%, with prevalence increasing with age; 2.9%, 7.8%, and 12.9% in the age groups 50–59, 60–69, and 70 years or older (respectively). Prevalence is similar in men and women (8.5% and 8.4%, respectively). Of the eyes diagnosed with AMD, 5% are legally blind (visual acuity equal or less than 20/400) and 49% had low vision.

In Guangzhou city, two studies were conducted in 1987 (Guan et al. 1989; Wu et al. 1987). In the first, 1,091 people, including 666 men and 425 women, were investigated. Participants were divided into four age groups: 40–49, 50–59, 60–69, and 70–93 years of age. Overall prevalence in Guangzhou is 4.95% and there is a steady increase in prevalence with age (0.87%, 5.05%, 7.77%, and 15.33% in the four groups, respectively). Again, the prevalence of AMD in men and women is similar (4.95% and 4.94%, respectively; Guan et al. 1987).

In the second study, 312 Guangzhou citizens, including 224 men and 88 women, were examined (Wu et al. 1987). Prevalence in this study is

6.41%, with a slightly higher proportion of women affected (6.82% women compared with 6.25% men). Prevalence in the age groups 50–59, 60–69, and over 70 years were 5.79%, 5.28%, and 15.00%, respectively.

### 6.2.1.2
### Hunan Province

Hunan province is located in the middle south area of China with a population of approximately 63.1 million. The investigation, carried out in the province from December 1986 to March 1987, covered five basic sampling units, including city, lake, hilly land, low mountain, and mountain areas (Huang et al. 1992). After randomized clustering 11,182 citizens were identified; of these 9,902 were examined, of whom 1,589 subjects were 50 years or older. In the sample group, the overall prevalence of AMD is 1.46%. However, in the age groups 50–59, 60–69, and over 70 years, prevalence of AMD is 4.30%, 11.20%, and 17.63%, respectively.

### 6.2.1.3
### Zhejiang Province

This province is located in the eastern coastal region of China and has a population of approximately 42.7 million. The investigation was carried out in three areas in 1989, and 730 people, whose ages ranged from 50 to 92 years, were examined (Xuan et al. 1994). The sample includes 342 men and 388 women. The overall prevalence of AMD is 7.4%, with 8.18% men and 6.70% women affected. Prevalence in the age groups 50–59, 60–69, 70–79, and 80 years and over is 3.39%, 6.25%, 13.51%, and 13.64%, respectively.

### 6.2.1.4
### Gansu Province

Gansu Province is located in the northwest of China with a population of 23.8 million. A total of 7,563 residents, including 598 Tibetan, 886 Moslem, and 6,079 Han, were investigated in 2001 (Tian et al. 1980). Of these 4,128 are men and 3,435 women, and ages range between 45 and 89 years. The overall prevalence of AMD is 7.66%; prevalence in men is 8.09% and in women, 7.13%. Prevalence in the age groups 45–55, 56–65, 66–75, and over 75 is 5.48%, 6.09%, 10.85%, and 18.89%, respectively.

### 6.2.1.5
### Tibet

The capital city of Tibet, Lhasa, is located more than 3,700 m above sea level. A sample population of 186 Tibetans from Lhasa, aged between 50 and 90 years old, were investigated in 1987 (Wu 1987; Wu et al. 1987). Overall prevalence of AMD in Lhasa Tibetans is 15.59% and increases from 9.88% in the sixth decade to 13.70%, and 34.38% in the seventh and eighth decades, respectively.

### 6.2.1.6
### Beijing

Shunyi County is located in north area of China (30 km north of Beijing) with a population of about 560,000. After randomized clustering sampling of the whole population, 10,854 citizens were selected and 10,414 people were examined (Hu et al. 1988). Only 40 subjects, comprising 20 men and 20 women, have been diagnosed with AMD, an overall prevalence in the population of 0.38%.

### 6.2.2
### Clinical Review Studies

Clinical review studies have been conducted in Chongqing city and Anhui province.

### 6.2.2.1
### Chongqing

Chongqing city is located in the southwest of China and has a population about 30 million people. The review study was carried out on patients who were examined using fluorescence angiography (FFA) in Southwest Eye Hospital between January 1998 and the end of May 2003 (Meng et al. 2003). During this period 183,274 patients were registered as outpatients of Southwest Eye Hospital of whom 5,410 had FFA examination. Of those, 240 AMD cases have been diagnosed in 3,361 patients who had FFA and are 40 years or older. The detected rate of AMD among the outpatient is 0.13%. The prevalence of AMD among FFA-examined patients is 4.4% and is 7.14% in the patients of 40 years or more. In the age groups 40–49, 50–59, 60–69, and 70 or more, the prevalence of AMD is 1.82%, 5.61%, 10.22%, and 18.65%, respectively.

### 6.2.2.2
### Anhui Province

Anhui province is located in the east of China with a population of about 59.86 million. The clinical review study was carried out in the Department of Ophthalmology, Fengtai Hospital (Fang et al. 2000). Between January 1997 and December 1998, 2,798 patients were registered in the outpatient clinic. Of these, 158 have been diagnosed with AMD, according to the diagnosis criteria referred to (National Academic Group of Fundus Disease of China 1987). The prevalence of AMD in the registered patients is 1.13% and in the age groups 40–49, 50–59, 60–69, and 70 or more is 0.90%, 3.66%, 6.04%, and 16.23%, respectively.

The prevalence of AMD in the different age groups in the regions of China discussed is listed in Table 6.2. The data indicate that AMD morbidity is the highest (15.59%) in the capital city of Tibet, Lhasa. The average prevalence over the eight regions is 7.60%.

### 6.2.3
### The Relationship of Morbidity to Age and Gender

The data in Table 6.2 also indicate that the prevalence of AMD in China is strongly associated with age. The prevalence reaches from 13.09% (Guangdong province) to 34.38% (Lhasa city) in the 70 years or older group. There is no significantly statistical difference of prevalence between men and women. Overall, the findings are similar to those of the US Beaver Dam Eye Study and the Australian Blue Mountains Eye Study (reviewed in Chap. 5). The risk factors in relation to higher prevalence in different regions of China are discussed below.

## 6.3
## Prevalence of 'Wet' and 'Dry' AMD

Epidemiological findings of the relative incidence of the 'wet' and 'dry' forms of AMD are summarized in Table 6.3 (Wu 1987; Wu et al. 1987; Zhao et al. 1987; Hu et al. 1988; Guan et al. 1989; Huang et al. 1992; Wu et al. 1992; Xuan et al. 1994; Fang et al. 2000; Tian et al. 2002; Meng et al. 2003). The data indicate that a higher proportion of patients suffer from atrophic (dry) AMD than those who suffer from the exudative (wet) AMD in China. The overall ratio of the two types in China is 5.33:1. In Gansu province the ratio of dry to wet is the highest (114.8:1), and it is lowest in Chongqing city (1.38:1). It is likely that the low ratio of dry to wet AMD found in clinical review data-groups (including Chongqing) is due to patients being selected for FFA examination, since patients with wet AMD are more likely to require FFA.

## 6.4
## Visual Acuity in AMD

The AMD visual acuity data are summarized in Table 6.4. (Wu 1987; Wu et al. 1987; Zhao et al. 1987; Hu et al. 1988; Guan et al. 1989; Wu et al. 1992; He et al. 1998; Fang et al. 2000; Meng et al. 2003). The table shows that 60.76% of AMD patients have decreased visual acuity (between 20/60 and 20/30). 32.96% of patients are in the low-vision range and 6.28% patients, legally blind. It appears that, because of the rather high prevalence of atrophic AMD in China, the proportion of blindness associated with AMD is lower in China than in Australia (Blue Mountains Eye Study) and the USA (Beaver Dam Study).

## 6.5
## Risk Factors of AMD in China

There are a variety of risk factors associated with AMD. The only factors consistently found to be associated with AMD are age and a family history (see Chap. 5). Epidemiological data of

Table 6.2. Prevalence of AMD in China

| Region | Age | No. of subjects | No. of cases | Detected rate |
|---|---|---|---|---|
| Guangdong province | 40–49 years | 462 | 4 | 0.87% |
| | 50–59 years | 2,017 | 72 | 3.57% |
| | 60–69 years | 2,035 | 157 | 7.71% |
| | >70 years | 2,231 | 292 | 13.09% |
| | Total | 6,745 | 525 | 7.78% |
| Hunan province | 50–59 years | 768 | 33 | 4.30% |
| | 60–69 years | 509 | 57 | 11.20% |
| | >70 years | 312 | 55 | 17.63% |
| | Total | 1,589 | 145 | 9.13% |
| Zhejiang province | 50–59 years | 177 | 6 | 3.39% |
| | 60–69 years | 368 | 23 | 6.25% |
| | >70 years | 185 | 25 | 13.51% |
| | Total | 730 | 54 | 7.40% |
| Gansu province | 45–55 years | 3,614 | 198 | 5.48% |
| | 56–65 years | 2,151 | 131 | 6.09% |
| | 66–75 years | 1,115 | 121 | 10.85% |
| | >75 years | 683 | 129 | 18.89% |
| | Total | 7,563 | 579 | 7.66% |
| Capital city of Tibet (Lhasa) | 50–59 years | 81 | 8 | 9.88% |
| | 60–69 years | 73 | 10 | 13.70% |
| | >70 years | 32 | 11 | 34.38% |
| | Total | 186 | 29 | 15.59% |
| Shunyi county of Beijing | Natural population | 10,414 | 40 | 0.38% |
| Chongqing city[a] | 40–49 years | 1,042 | 19 | 1.82% |
| | 50–59 years | 1,052 | 59 | 5.61% |
| | 60–69 years | 881 | 90 | 10.22% |
| | >70 years | 386 | 72 | 18.65% |
| | Total | 3,361 | 240 | 7.14% |
| Anhui province[a] | 45–49 years | 666 | 6 | 0.90% |
| | 50–59 years | 819 | 30 | 3.66% |
| | 60–69 years | 894 | 54 | 6.04% |
| | >70 years | 419 | 68 | 16.23% |
| | Total | 2,798 | 158 | 5.65% |

[a] Clinical review studies

Table 6.3. The incidence/prevalence of two types of AMD in China

| Region | Dry AMD | Wet AMD | Total | Rate of dry/wet |
|---|---|---|---|---|
| Guangdong province | 72 | 2 | 74 | 36:1 |
| Capital city of Tibet | 28 | 1 | 29 | 28:1 |
| Hunan province | 399 | 54 | 453 | 7.39:1 |
| Gansu province | 574 | 5 | 579 | 114.8:1 |
| Chongqing city[a] | 207 | 150 | 357 | 1.38:1 |
| Anhui province[a] | 143 | 55 | 198 | 2.6:1 |
| Total | 1,423 | 267 | 1,690 | 5.33:1 |

[a] Clinical review studies

Table 6.4. The visual acuity of AMD patients in China

| Visual acuity | <20/400 | 20/400–20/60 | >20/60–20/60 | Total |
|---|---|---|---|---|
| Eyes | 156 | 819 | 1,510 | 2,485 |
| Rate | 6.28% | 32.96% | 60.76% | 100 |

family history of AMD in China were not available in the cases reviewed here. Other risk factors which may impact the prevalence of AMD in China, including race/ethnicity, occupation, light exposure, smoking, alcohol intake, and systemic diseases are reviewed below.

## 6.5.1
### Race/Ethnicity

China is a multiracial/ethnic country. Little published population-based data are available regarding the prevalence of AMD in different racial and ethnic groups. Table 6.5 summarizes data showing the prevalence of AMD in three racial/ethnic groups (Wu et al. 1987; Tian et al. 2002). The data show a prevalence of 13.39%, 8.01% and 7.07% in Tibetan, Moslem, and Han ethnic groups, respectively. The prevalence of AMD in Tibetans is significantly higher than in Moslem and Han ethnic groups. Although all three groups are Oriental/Asian, their environments, dietary intake, and living habits are each quite different. Tibet has been considered to be a hypoxic environment, with high levels of ultraviolet radiation that may contribute the high prevalence of AMD amongst Tibetans (Wu 1987; Wu et al. 1987).

Table 6.5. The prevalence of AMD in different racial/ethnic groups in China

| Racial/Ethnic Group | Subjects (n) | AMD cases (n) | Detected rate |
|---|---|---|---|
| Tibetans | 784 | 105 | 13.39% |
| Moslems | 886 | 71 | 8.01% |
| Hans (Chinese) | 6,391 | 452 | 7.07% |

## 6.5.2
### Light Exposures/Occupation

Guan et al. (1989) and Xuan et al. (1994) have divided occupations of their AMD patients into two 'light exposure' groups: *direct* and *nondirect* light exposure. Direct light exposure occupations are defined to include agricultural workers, fishermen, welders, and other groups who work outdoors. *Nondirect* light exposure occupations include clerks, persons with household duties, and others working indoors. Table 6.6 lists the prevalence of AMD in two different sample populations with *direct* and *nondirect* occupations in China. In both sample populations, the incidence of AMD is higher in *direct* light exposure occupations than in *nondirect* light exposure occupations. The data imply that the chronic light damage is one of the important risk factors of AMD.

Wu et al. (1992) have investigated the prevalence of AMD in different occupations, including miners, agricultural workers, and clerks in Chenzhou District of Hunan province. The results (Table 6.7) show light intensities that workers in different occupations are exposed to along with the prevalence of AMD. Except in the cases of miners, the prevalence of AMD correlates positively with intensity of light exposure (regression analysis, $P<0.05$). The high prevalence of AMD in miners may be related to other risk factors which need further investigation.

Other reports, however, have revealed different results. Table 6.8 shows light impact on prevalence of AMD in Hunan and Anhui provinces (Huang et al. 1992; Fang et al. 2000). The data show the highest prevalence of AMD associated with household duties, while agricultural populations have the lowest prevalence (significant, $P<0.05$). The specific reasons for the results are unknown.

## 6.5.3
### Smoking

It has been suggested that, in addition to age, cigarette smoking is a risk factor associated with AMD and is most pronounced for the dry

Table 6.6. The impact of light exposure to AMD

|  | Guangzhou | | | | Zhejiang | | | |
|---|---|---|---|---|---|---|---|---|
|  | AMD | Non-AMD | Total | Detected rate | AMD | Non-AMD | Total | Detected rate |
| Direct light exposure occupations | 17 | 163 | 180 | 9.44% | 22 | 177 | 199 | 11.06% |
| Nondirect light exposure occupations | 37 | 874 | 911 | 4.06% | 32 | 499 | 531 | 6.03% |
| $P$ | $P<0.005$ | | | | $P<0.05$ | | | |

Table 6.7. The prevalence of AMD with intensity of light (Wu et al. 1992)

| Group | Intensity of light | Incidence of AMD |
|---|---|---|
| Miners | 0–520 lx | 7.89% |
| Agricultural workers | 100,000 lx | 7.33% |
| Workers | 80–2,600 lx | 4.94% |
| Clerks | 80–1,800 lx | 2.78% |

Table 6.8. Prevalence of AMD with light exposure in Hunan and Anhui provinces

| Occupation | Subjects (n) | Cases (n) | Detected rate |
|---|---|---|---|
| Workers | 2,525 | 62 | 2.46% |
| Agricultural workers | 10,714 | 105 | 0.98% |
| Clerks | 2,721 | 80 | 2.94% |
| Household-duties workers | 997 | 53 | 5.32% |
| Others | 3,070 | 8 | 0.26% |

form of AMD (Guan et al. 1989; Klein et al. 1993; Smith et al. 1996; Mitchell et al. 1999; see also Chap. 5). Smoking is known to depress serum antioxidant levels and to reduce choroidal blood flow. Studies indicate that around 70% of men and 20–30% of women in China smoke, and a recent investigation has reported that there is an increase in female smokers, especially in the younger age groups. The relationship between smoking and AMD in China have been analyzed by Wu (1987) and Guan et al. (1989) and the results are shown in Table 6.9. In Guan's study, the prevalence of AMD in smokers is higher than in nonsmokers (Guan et al. 1989). However, the results from Wu et al. (1987) do not show a difference.

### 6.5.4
### Drinking

Some studies have examined the association of alcohol consumption with the prevalence of AMD. It has been reported that heavy drinkers

Table 6.9. The relationship between smoking and AMD

|  |  | Guan et al. 1989 | | Wu et al. 1987 | |
|---|---|---|---|---|---|
|  |  | AMD | Non-AMD | AMD | Non-AMD |
| Smoking | Yes | 28 | 362 | 8 | 44 |
|  | No | 26 | 657 | 20 | 104 |
| $P$ |  | $P<0.05$ | | $P>0.05$ | |

Table 6.10. The relationship between drinking and AMD

|  |  | Guan et al. 1989 | | Wu et al. 1987 | |
|---|---|---|---|---|---|
|  |  | AMD | Non-AMD | AMD | Non-AMD |
| Drinking | Yes | 9 | 113 | 15 | 84 |
|  | No | 45 | 924 | 12 | 62 |
| $P$ |  | $P>0.05$ | | $P>0.05$ | |

Table 6.11. The relationship between systematic diseases and AMD

|  |  | Guan et al. 1989 | | Xuan et al. 1994 | |
| --- | --- | --- | --- | --- | --- |
|  |  | AMD | Non-AMD | AMD | Non-AMD |
| Hypertension | Yes | 10 | 53 | 18 | 135 |
|  | No | 44 | 984 | 36 | 541 |
|  | P | $P<0.01$ |  | $P<0.05$ |  |
| Heart diseases | Yes | 4 | 23 |  |  |
|  | No | 50 | 1014 |  |  |
|  | P | $P>0.05$ |  |  |  |
| Respiratory systematic diseases | Yes | 2 | 11 |  |  |
|  | No | 52 | 1026 |  |  |
|  | P | $P>0.05$ |  |  |  |
| Chronic bronchitis | Yes |  |  | 14 | 102 |
|  | No |  |  | 40 | 574 |
|  | P |  |  | $P<0.05$ |  |

at baseline are more likely to develop late AMD than those who have never been heavy drinkers (Klein et al. 2002; see also Chap. 5), and there may be an association of alcohol consumption with early AMD (Smith and Mitchell 1996). It has also been suggested (Obisesan et al. 1998) that moderate wine consumption may be protective for AMD due to the high concentration of antioxidant phenolic compounds found in red wine. However, data from population-based studies in China, (Table 6.10) suggest that drinking is not a risk factor for AMD in that population (Wu et al. 1987; Guan et al. 1989).

## 6.5.5
### Systemic Diseases

Guan et al. (1989) have analyzed the relationship between hypertension, heart disease, respiratory systemic disease, and AMD in China, while Xuan et al. (1994) has reported the relationship between hypertension, chronic bronchitis, and AMD. The data indicate (Table 6.11) that hypertension is significantly correlated with AMD and that chronic bronchitis might also be a risk factor for AMD (Guan et al. 1989; Xuan et al. 1994). These results are similar to the results observed in other studies (Hyman et al. 2000; Klein et al. 2003; see also Chap. 5).

## 6.6
### Conclusions

Epidemiological studies carried out over the past 20 years in China show that the prevalence of AMD is similar in most areas of China, except in Lhasa, where prevalence is much higher. As expected, the prevalence of AMD is found to increase with age; there are no significant differences with gender. Also in China the prevalence of 'dry' AMD is higher than 'wet' AMD. Chronic light damage, smoking, hypertension, and chronic bronchitis might also be risk factors for AMD in China.

### References

Fang Y, Zhao CL, Hu DR (2000) Epidemiologic analysis of 158 cases of age-related macular degeneration. Chin J Dis Control Prevent 4:127–128

Gregor Z, Joffe L (1978) Senile macular changes in the black African. Br J Ophthalmol 62:547–550

Guan GH, Zhan YJ, Huang ZW, Yu YG, Li JH (1989) Investigation of epidemiology of age-related macular degeneration. Chin J Ocul Fund Dis 5:208–111

He MG, Xu JJ, Wu KL, Li SZ (1998) The prevalence of age-related macular degeneration in Doumen county, Guanding Province. Chin J Ocul Fund Dis 14:122–124

Hu Z, Zhao JL, Dong FT, Liu XL, Peng YE, Li P, Wang SF, Li YJ, Chen XH (1988) Investigation of epidemiology of ocular fundus diseases in Shunyi county of Beijing. Chin J Ocul Fund Dis 4:193–195

Huang P, He RX, He GZ, Huang LH, Xie JY (1992) Investigation of epidemiology of age-related macular degeneration in Hunan Province. Chin Ophthalmic Res 10:60–61

Hyman L, Schachat AP, He Q, Leske MC (2000) Hypertension, cardiovascular disease, and age-related macular degeneration. Age-Related Macular Degeneration Risk Factors Study Group. Arch Ophthalmol 118:351–358

Klein R, Klein BE, Linton KL, DeMets DL (1993) The Beaver Dam Eye Study: the relation of age-related maculopathy to smoking. Am J Epidemiol 137:190–200

Klein R, Klein BEK, Jensen SC, Mares-Perlman JA, Cruickshanks KJ, Palta M (1999) Age-related maculopathy in a multiracial United States population. The National Health and Nutrition Examination Survey III. Ophthalmology 106:1056–1065

Klein R, Klein BEK, Tomany SC, Moss SE (2002) Ten-year incidence of age-related maculopathy and smoking and drinking. The Beaver Dam Eye Study. Am J Epidemiol 156:589–598

Klein R, Klein BEK, Tomany SC, Cruickshanks KJ (2003) The association of cardiovascular disease with the long-term incidence of age-related maculopathy. The Beaver Dam Eye Study. Ophthalmology 110:1273–1280

Leibowitz HM, Krueger DE, Maunder LR, Milton RC, Kini MM, Kahn HA, Nickerson RJ, Pool J, Colton TL, Ganley JP, Loewenstein JI, Dawber TR (1980) The Framingham eye study monograph: an ophthalmological and epidemiological study of cataract, glaucoma, diabetic retinopathy, macular degeneration, and visual acuity in a general population of 2631 adults, 1973–1975. Surv Ophthalmol 24:335–610

Meng XH, Yin ZQ, Jiang WS (2003) Clinical analysis of age-related macular degeneration. (Unpublished data)

Mitchell P, Chapman S, Smith W (1999) Smoking is a major cause of blindness. Med J Aust 171:173–174

National Academic Group of Fundus Disease of China (1987) The clinical diagnostic criteria of age-related macular degeneration. Chin J Ocul Fund Dis 3: Cover 3

National Bureau of Statistics of China (2000) The bulletin of the 5th National population mass survey

Obisesan TO, Hirsch R, Kosoko O, Carlson L, Parrott M (1998) Moderate wine consumption is associated with decreased odds of developing age-related macular degeneration in NHANES-1. J Am Geriatr Soc 46:1–7

Smith W, Mitchell P (1996) Alcohol intake and age-related maculopathy. The Blue Mountains Eye Study. Am J Ophthalmol 122:743–745

Smith W, Mitchell P, Leeder SR (1996) Smoking and age-related maculopathy. The Blue Mountains Eye Study. Arch Ophthalmol 114:1518–1523

Sun BC (1999) Clinical low vision, 2nd edition. Hua Xia, Beijing, pp 177–180

Tian MN, Li L, Zhang CA, Mu YZ, Wie ZC, Bai CX, Wang HY (2002) Investigation of epidemiology of age-related macular degeneration. The 8th Ophthalmology Meeting Paper Assembler, pp 128–129

Wang X, Hu S (2001) A review of research actuality on age-related macular degeneration. Yan Ke Xue Bao (Eye Science) 17(4):245–251

WHO (2000) Life Tables for 191 countries. World mortality in 2000. WHO Geneva, Switzerland

WHO (2003) World Health Organization statistics. WHO, Geneva, Switzerland

Wu L (1987) Study of aging macular degeneration in China. Jpn J Ophthalmol 31:349–367

Wu LZ, Chen YZ, Cao XY, Huang ZS, Xu XJ, Xin DY, Tian N, Wu FL, Liu AQ, Tang SP, Zhu SP, Bao GQ, Sheng XF (1987) Aging macular degeneration in Tibetan and Han. Eye Sci 3:179–181

Wu ZQ, Cao JE, Yao XH, Long XG, Zheng QL (1992) Investigation of epidemiology of age-related macular degeneration. Chin J Ophthalmol 28:246–247

Xuan MZ, Wan J, Zhao J, Jiang JK, Wu LL (1994) Investigation of epidemiology of age-related macular degeneration. J Zhejiang Med Univ 23:78–79; 82

Zhao GY, Shu HL, Xia YC, Zhan YS, Xia HX (1987) Investigation of epidemiology of age-related macular degeneration. J Chin Pract Ophthalmol 5:633–635

## Experimental Models of Macular Degeneration

Ray F. Gariano

### Contents

7.1 Introduction   113
7.2 Aging   113
7.3 Laser-Induced CNV   114
7.4 Growth Factor-Induced CNV   114
7.5 Genetically Defined Animal Models   116
7.6 Conclusions   118
References   118

## 7.1
## Introduction

The ideal experimental model of age-related macular degeneration (AMD) would exhibit in its early phases features of atrophic degeneration such as drusen and pigment epithelial changes, in the central retina. Over time, a minority of animals would develop choroidal neovascularization (CNV) with exudation and decreasing vision, culminating in subretinal fibrosis. The animal of choice might be the mouse, because of the ease of genetic manipulation, and the disease would respond to therapies that show partial effectiveness in humans.

Numerous difficulties hinder development of such an experimental model. First, as outlined in Chap. 1, the macula is unique to the primate retina (see also Hendrickson and Yuodelis 1984). Thus the possible contributions of cellular and functional specializations of the macula to the pathogenesis of AMD cannot be assessed in less expensive and more easily manipulated and genomically well-defined non-primates.

Second, AMD by definition develops in older persons. Certain pathologic alterations of AMD are associated with senescence independent of life-span– and therefore occur in laboratory animals at a corresponding stage of life– while other alterations may require cumulative pathogenic activity over a number of years too large to be practical for most species and investigations.

Creation of in vivo models of AMD is further complicated by the protean presentations of the disease. Animal models of AMD typically exhibit one, or at most a few, of the many signs of human disease. Furthermore, it is uncertain which findings of AMD are desirable to incorporate into an experimental model, because it is unclear which are essential to progression of the disease and to vision loss and which are harmless, secondary, or epiphenomenal (Csaky 2003). Finally, detection of vision loss in animals is challenging.

Despite these difficulties, experimental systems that capture selected features of AMD have contributed greatly to our understanding of the disease pathogenesis and to assessment of potential therapeutic interventions. Conversely, insights into the etiology of AMD have directed the search for experimental models in promising directions.

## 7.2
## Aging

Experimental animals exhibit several aging changes reminiscent of those found in AMD eyes. In particular, older rhesus monkeys develop RPE alterations such as pigment mottling and hypopigmentation, and increasing numbers of drusen and drusen-like deposits; these are especially prominent in females (Ulshafer et al. 1987; Engel et al. 1988; Monaco and Wormington 1990). However, sub-RPE deposits in senescent monkeys are biochemically distinct from those in humans (Hirata and Feeney-Burns 1992), and CNV does not occur.

Certain strains of mice with accelerated senescence have been examined for evidence of retinal degeneration. In SAM P8 mice, for example, progressive age-related pathology includes loss of RPE cells, depositions in the sub-RPE space that resemble basal laminar deposits, mild choriocapillaris atrophy, and marked thickening of Bruch's membrane (Majji et al. 2000). Interestingly, intra-Bruch's membrane vascular invasion was detected in SAM P8 mice at age 11 months, though this finding could not be confirmed by inspection of choroidal vascular casts. It will be of interest to determine if these mutant strains are particularly susceptible to experimental manipulations that stimulate CNV in wild type mice, and whether their ocular findings can be therapeutically delayed.

## 7.3
### Laser-Induced CNV

Most efforts to replicate features of AMD in animals have focused on subretinal angiogenesis, since the exudative form of the disease is responsible for most severe vision loss in patients. Early investigators described a variety of methods to induce CNV in experimental animals, including penetrating trauma (Hsu et al. 1989), enzymatic degradation of Bruch's membrane (Ryan et al. 1980), systemic naphthalene administration, (Orzalesi et al. 1994), *Histoplasma capsulatum* inoculation (Jester and Smith 1985), subretinal vitreous injection (Zhu et al. 1989), and induction of local inflammation (Sakamoto et al. 1994). Although such techniques could be effective, they induced CNV at relatively low rates and often with associated ocular pathology unrelated to AMD.

The first reliable animal model of CNV was based on the clinical observation that subretinal neovessels may rarely occur in patients receiving retinal laser therapy for various indications (Lewen 1988). Ryan and colleagues described CNV in monkey eyes at the site of previously placed laser burns; the occurrence of CNV was more likely if the laser application was in the macular region, and if Bruch's membrane was focally disrupted (Ryan 1982). Similar to CNV in humans, laser-induced CNV showed leakage on fluorescein angiography, and evolved to a fibrotic stage. A role for leukocytes in CNV was suggested by their accumulation adjacent to neovascular growths. Also, participation of the RPE in permeability was shown by the reduction in angiographic leakage as mature lesions became enveloped by RPE cells.

The application of laser-induced CNV to rabbits (elDirini et al. 1991), rats (Frank et al. 1989) and mice (Tobe et al. 1998) greatly increased its utility and practicality, such that today it is the most common technique employed to assay therapeutic approaches to CNV (Fig. 7.1). Advantages of the laser-CNV technique include relative ease to create the laser burns via slit-lamp delivery, and to detect CNV with fluorescence angiography, and partial quantifiability in terms of the fraction of burns that give rise to CNV, the size of the neovascular lesion, and the degree of angiographic leakage. A limitation of this model is that it appears to have a more prominent inflammatory component than is seen in AMD; indeed laser-induced experimental CNV is eliminated with steroidal or non-steroidal anti-inflammatory medication, (Ishibashi et al. 1985; Sakamoto et al. 1995), while these agents induce a much more modest effect on CNV in AMD (Danis et al. 2000; Ciulla et al. 2003; Gillies et al. 2003).

## 7.4
### Growth Factor-Induced CNV

The central macula, and the outer retina in general, are normally avascular. Even during in utero development, when the foveal and perifoveal depression are yet to form by translocation of the inner retinal layers, the presumptive central macula is devoid of retinal blood vessels (En-

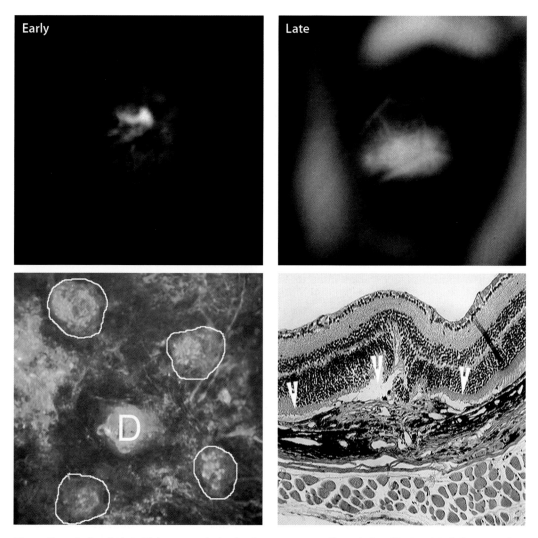

Fig. 7.1. Laser-induced choroidal neovascularization in the mouse eye. Early (*top left*) and late (*top right*) phase fluorescein angiograms taken 4 weeks after diode laser photocoagulation; the progressive increase in fluorescence between frames indicates leakage. FITC-dextran perfusion (*lower left*) visualizes the intravascular compartment of laser-induced lesions (*circled*) surrounding the optic nerve head (*D*). Histopathology section of a choroidal neovascular lesion (*lower right*) obtained 4 weeks after laser shows a fibrovascular infiltration into the subretinal space (*arrowheads*). (Courtesy of Dr. Diego Espinosa-Heidmann and Dr. Scott Cousins)

german 1976; Gariano et al. 1994). As developing retinal vessels extend from the optic nerve towards the periphery, they abruptly stop as they approach the central macula, and course around this "no-go" zone (Provis et al. 2000). Similarly, the outer retina remains avascular both in macular and extramacular regions. The growth of subretinal vessels in AMD may thus be viewed as a breach of the avascular privilege of the central macula and outer retina.

This perspective focuses future research on factors that normally preclude vascular invasion in retinal compartments. Such factors may include higher relative tissue oxygen levels in the outer retina (Braun et al. 1995), absence of astrocytes from central and outer retina (Gariano et al. 1996; Provis et al. 2000), and differential distribution of pro- and anti-angiogenic growth factors and extracellular matrix components between vascular and avascular retinal compartments.

Pigment derived epithelial factor (PEDF) is a candidate angiogenesis inhibitor to prevent vascularization of the central macula and outer retina (Bouck 2002). PEDF is synthesized by retinal ganglion cells and by RPE cells (Karakousis et al. 2001; Ogata et al. 2002), is present in the interphotoreceptor and subretinal spaces, and binds receptors in the outer retina and the ganglion cell layer (Aymerich et al. 2001). PEDF levels are lower in patients with exudative AMD than in age-matched controls, raising the possibility that loss of PEDF is a permissive factor in the development of CNV (Holekamp et al. 2002).

PEDF immunolocalization correlates inversely with CNV formation in a rat laser-induced CNV model (Renno et al. 2002), and adeno-associated viral vector-induced subretinal PEDF expression inhibits laser-induced CNV in mice (Mori et al. 2002, 2003). Subretinal transplantation of PEDF-expressing iris pigment epithelium also inhibits laser-induced CNV lesions (Semkova et al. 2002). Mice born without detectable PEDF are viable and appear healthy (Bouck 2002), but it is unknown whether they provide a model of spontaneous CNV, or more readily develop CNV in response to experimental manipulations such as laser or growth factors.

Several pro-angiogenic growth factors are present within CNV in AMD. While a causative role is not established for any such molecules, two of them, vascular endothelial growth factor (VEGF; Kvanta et al. 1996) and fibroblast growth factors (FGF; Frank et al. 1996), have been employed to create reliable experimental models of subretinal neovascularization.

Pharmacologic administration of basic FGF or VEGF to the subretinal space within pellets or microspheres causes growth of new vessels into the subretinal space (Kimura et al. 1995; Cui et al. 2000). The new vessels arise most often from the deep inner retinal vessels, extend into the photoreceptor layer and subretinal space, and regress soon after the inciting growth factor is depleted. This latter feature limits the usefulness of these models to evaluate potential long-term therapies.

More persistent delivery of specific growth factors has been attained in photoreceptors and RPE cells by targeted gene expression. When VEGF was expressed in photoreceptor cells, subretinal neovascularization arose from deep inner retinal vessels (Okamoto et al. 1997). Transgenic expression of VEGF in the RPE resulted in intrachoroidal neovascularization that leaked, but did not penetrate, into the subretinal space (Schwesinger et al. 2001). Subretinal injection of an adenoviral-associated vector encoding VEGF in rats and mice appears to provide a superior model because in this model the resulting CNV penetrated Bruch's membrane to proliferate within the subretinal space, and persisted for over 20 months (Baffi et al. 2000; Wang et al. 2003). As noted above, transgenic methods used in combination with the laser-induced CNV model may facilitate screening for genes that inhibit, stimulate, or stabilize CNV. In this way, PEDF has been shown to suppress experimental CNV. Similarly, laser-induced CNV is inhibited by angiostatin (Lai et al. 2001) and endostatin (Mori et al. 2001), and occurs less frequently and aggressively in knockout mice deficient in matrix metalloproteinases 2 and 9 (Lambert et al. 2003).

Growth factor-stimulated CNV may also be elicited indirectly by immune mechanisms. Lipid hydroperoxides, for example, accumulate in Bruch's membrane with age (Spaide et al. 1999). When injected into the subretinal space of rabbits, linoleic hydroperoxide stimulates CNV, possibly through an immune response that recruits or liberates angiogenic factors (Tamai et al. 2002).

## 7.5
## Genetically Defined Animal Models

Fundus ophthalmoscopy, fluorescein angiography, and electroretinography in mice (Hawes et al. 1999) allow rapid screening of normal murine strains and those with spontaneous or induced genetic mutations. Ocular findings in the

Bst (belly spot tail) murine strain were discovered in this way. Bst mice exhibit spontaneous thickening and discontinuities in Bruch's membrane, RPE alterations, and invasion of choriocapillaris vessels into the subretinal space (Smith et al. 2000). The utility of this strain is limited, however, by the relatively low penetrance of the CNV phenotype and by associated ocular defects, including coloboma and retinal detachment.

Mice lacking a functional very low density lipoprotein receptor (vldlr) exhibit striking retinal vasculature abnormalities that mimic a subset of patients with exudative ARMD (Heckenlively et al. 2003). In the first 2 weeks after birth in vldlr$^{-/-}$ mice, vessels develop normally to form three vascular laminae within the inner retina (Connolly et al. 1988; Dorrell et al. 2003). Soon thereafter, however, neovessels arising from inner retinal vessels extend into the normally avascular outer retina, and then into the subretinal space. Unlike normal intraretinal vessels, this outer retinal neovascularization is hyper-permeable.

These features resemble the abnormal intra- and sub-retinal neovascularization in patients with Retinal angiomatous Proliferation (RAP). RAP, which responds particularly poorly to both conventional thermal laser ablation and photodynamic therapy, may be present in up to 20% of patients with exudative ARMD (Yannuzzi et al. 2001).

How the defect in vldlr leads to retinal vascular abnormalities is under investigation. Vldlr contributes guidance cues to developing neurons as they migrate to their appropriate cortical laminae (Trommsdorff et al. 1999); possibly, similar guidance cues may pertain to retinal vascular patterning. Vldlr also regulates activity of the urokinase system, whose extracellular matrix degrading capacity is critical to angiogenesis. Regardless of the mechanism of angiogenesis in vldlr mutant mice, this mutation may help to understand the bases of RAP, and to test anti-angiogenic and anti-permeability agents in vivo.

Cellular and humoral immune and inflammatory mechanisms have been implicated in the development and progression of ARMD since early descriptions of macrophages in CNV membranes (Lopez et al. 1996). Anti-inflammatory therapy shows promise for patients with ARMD (Danis et al. 2000). However, it has been unclear whether immune processes actually cause aspects of ARMD or are secondary responses to degeneration.

Recent studies with animal models of ARMD suggest that the immune system is a critical participant in development of CNV. First, anti-inflammatory therapies markedly reduce experimental CNV (Ciulla et al. 2001; Bora et al. 2003). Second, chemical depletion of circulating macrophages inhibits formation of laser-induced CNV in mice (Espinosa-Heidemann et al. 2003; Sakurai et al. 2003a). Interference with leukocyte function by targeted disruption of genes encoding leukocyte adhesion molecules CD18 or ICAM-1 also inhibits growth of experimental CNV (Sakurai et al. 2003b).

Ambati and colleagues presented immune-defective murine strains that exhibit several features of atrophic macular degeneration as well as CNV (Ambati et al. 2003). Mice lacking monocyte chemoattractant protein (MCP-1) or its receptor CCR2 exhibit impaired macrophage function, and develop subretinal lipofuscin and drusenoid deposits, focal and diffuse RPE atrophy, and choroidal neovascularization. Depositions contain complement component C5 and advanced glycation end products, both of which induce RPE production of VEGF. The authors propose that MCP-1 and CCR2 deficiency prevents the normal clearance of deposited proteins, whose accumulation may stimulate VEGF-dependent CNV.

Gene mutations have been identified in several inherited macular degenerations. For example, the photoreceptor disc rim protein ABCA4 is defective in Stargardt's disease, bestrophin in Best's disease, peripherin/RDS in adult vitelliform dystrophy, and tissue inhibitor of metalloproteinase-3 (TIMP3) in Sorsby's fundus dystrophy (see Michaelides et al. 2003). With the possible exception of ABCA4, mutations from monogenic macular dystrophies do not appear involved in ARMD.

In Sorsby's fundus dystrophy (SFD), a rare autosomal dominantly inherited disease that phenotypically resembles both atrophic and exudative ARMD, mutations in TIMP3 (Weber et al. 1994; Jacobson et al. 2002) may cause aberrant extracellular matrix remodeling and lead to thickening of Bruch's membrane. Also, TIMP3 inhibits angiogenesis (Qi et al. 2003), and its absence may enable growth of CNV in patients with SFD.

Defects in TIMP3 have not been detected in patients with ARMD (De La Paz et al. 1997). However, insights gained from a mouse knock-in model of SFD allow some role for TIMP3 in ARMD. In mice with a mutation in the orthologous murine TIMP3 gene similar to mutations reported in humans, the mutant TIMP3 protein appears to maintain protease inhibiting activities (Weber et al. 2002). Site-specific excess– rather than deficiency– of functional TIMP3 in Bruch's membrane and the outer retina may result. TIMP3 levels are elevated in retinitis pigmentosa without evident coding mutations. Thus, TIMP3 may yet participate in ARMD even in the absence of mutations in the TIMP3 gene (see Tymms 1999).

Putative genetic susceptibility loci for ARMD have been localized to several chromosomes, but specific gene products that confer risk are yet to be characterized (Michaelides et al. 2003). Identification of these genes is likely to spur generation of animal models with homologous mutations.

## 7.6
## Conclusions

Several experimental systems replicate pathological aspects of AMD, including (1) natural or accelerated senescence; (2) structural injury (e.g., laser-induced CNV); (3) pharmacologic administration of angiogenic molecules; (4) tissue and cell-specific altered expression of growth factors; (5) outer retinal invasion by misguided inner retinal vessels (e.g., vldlr mutation); and (5) mutant strains with defined molecular defects (e.g., in immune function). Not discussed in this chapter are several in vitro systems – such as RPE cell cultures – that allow investigators to focus on and control limited aspects of AMD, for example, RPE-retinal interactions, and oxidative stress.

These diverse experimental approaches parallel the clinical impression that (1) atrophic findings of AMD occur in several settings, including normal aging, inherited retinal dystrophies, drug toxicities and inflammatory disorders; and (2) neovascular findings similar to those of exudative AMD occur in association with diverse pathological insults, such as structural damage to Bruch's membrane (e.g., laser burns), inherited dystrophies, inflammatory diseases (e.g., ocular histoplasmosis), tumors, and trauma. Given the polygenic and multifactorial nature of AMD pathogenesis, it seems likely that insights gained from several animal models will be required to understand and treat the spectrum of findings in patients with AMD.

## References

Ambati J, Anand A, Sakurai E, Fernandez S, Lynn, BC, Kuziel WA, Rollins BJ, Ambati BK. (2003) An animal model of age-related macular degeneration in senescent Ccl-2 or Ccr-2 deficient mice. Nat Med 9: 1390–1397

Aymerich MS, Alberdi EM, Martinez A, Becerra SP (2001)Evidence for pigment epithelium-derived factor receptors in the neural retina. Invest Ophthalmol Vis Sci 42:3287–3293

Baffi J, Byrnes G, Chan CC, Csaky KG (2000) Choroidal neovascularization in the rat induced by adenovirus mediated expression of vascular endothelial growth factor. Invest Ophthalmol Vis Sci 41:3582–3589

Bora PS, Hu Z, Tezel TH, Sohn JH, Kang SG, Cruz JM, Bora NS, Garen A, Kaplan HJ (2003) Immunotherapy for choroidal neovascularization in a laser-induced mouse model simulating exudative (wet) macular degeneration. Pro Natl Acad Sci USA 100:2679–2684

Bouck N (2002) PEDF: anti-angiogenic guardian of ocular function. Trends Mol Med 8:330–334

Braun RD, Linsenmeier RA, Goldstick TK (1995) Oxygen consumption in the inner and outer retina of the cat. Invest Ophthalmol Vis Sci 36:542–554

Ciulla TA, Criswell MH, Danis RP, Hill TE (2001) Intravitreal triamcinolone acetonide inhibits choroidal neovascularization in a laser-treated rat model. Arch Ophthalmol 119:399–404

Ciulla TA, Criswell MH, Danis RP, Fronheiser M, Yuan P, Cox TA, Csaky KG, Robinson MR (2003) Choroidal neovascular membrane inhibition in a laser treated rat model with intraocular sustained release triamcinolone acetonide microimplants. Br J Ophthalmol 87:1032–1037

Connolly SE, Hores TA, Smith LE, D'Amore PA (1988) Characterization of vascular development in the mouse retina. Microvasc Res 36:275–290

Csaky K (2003) Anti-vascular endothelial growth factor therapy for neovascular age-related macular degeneration: promises and pitfalls. Ophthalmology 110: 879–881

Cui JZ, Kimura H, Spee C, Thumann G, Hinton DR, Ryan SJ (2000) Natural history of choroidal neovascularization induced by vascular endothelial growth factor in the primate. Graefes Arch Clin Exp Ophthalmol 238:326–333

Danis RP, Ciulla TA, Pratt LM, Anliker W (2000) Intravitreal triamcinolone acetonide in exudative age-related macular degeneration. Retina 20:244–250

De La Paz MA, Pericak-Vance MA, Lennon F, Haines JL, Seddon JM (1997) Exclusion of TIMP3 as a candidate locus in age-related macular degeneration. Invest Ophthalmol Vis Sci 38:1060–1065

Dorrell MI, Aguilar E, Friedlander M (2002) Retinal vascular development is mediated by endothelial filopodia, a preexisting astrocytic template and specific R-cadherin adhesion. Invest Ophthalmol Vis Sci 43: 3500–3510

elDirini AA, Ogden TE, Ryan SJ (1991) Subretinal endophotocoagulation. A new model of subretinal neovascularization in the rabbit. Retina 11:244–249

Engel HM, Dawson WW, Ulshafer RJ, Hines MW, Kessler MJ (1988) Degenerative changes in maculas of rhesus monkeys. Ophthalmologica 196:143–150

Engerman RL (1976) Development of the macular circulation. Invest Ophthalmol 15:835–840

Espinosa-Heidmann DG, Suner IJ, Hernandez EP, Monroy D, Csaky KG, Cousins SW (2003) Macrophage depletion diminishes lesion size and severity in experimental choroidal neovascularization. Invest Ophthalmol Vis Sci 44:3586–3592

Frank RN, Das A, Weber ML (1989) A model of subretinal neovascularization in the pigmented rat. Curr Eye Res 8:239–247

Frank RN, Amin RH, Eliott D, Puklin JE, Abrams GW (1996) Basic fibroblast growth factor and vascular endothelial growth factor are present in epiretinal and choroidal neovascular membranes. Am J Ophthalmol 122:393–403

Gariano RF, Iruela-Arispe ML, Hendrickson AE (1994) Vascular development in primate retina: comparison of laminar plexus formation in monkey and human. Invest Ophthalmol Vis Sci 35:3442–3455

Gariano RF, Sage EH, Kaplan HJ, Hendrickson AE (1996) Development of astrocytes and their relation to blood vessels in fetal monkey retina. Invest Ophthalmol Vis Sci 37:2367–2375

Gillies MC, Simpson JM, Luo W, Penfold P, Hunyor AB, Chua W, Mitchell P, Billson F (2003) A randomized clinical trial of a single dose of intravitreal triamcinolone acetonide for neovascular age-related macular degeneration: one-year results. Arch Ophthalmol 121:667–673

Hawes NL, Smith RS, Chang B, Davisson M, Heckenlively JR, John SW (1999) Mouse fundus photography and angiography: a catalogue of normal and mutant phenotypes. Mol Vis 5:22

Heckenlively JR, Hawes NL, Friedlander M, Nusinowitz S, Hurd R, Davisson M, Chang B (2003) Mouse model of subretinal neovascularization with choroidal anastomosis. Retina 23:518–522

Hendrickson AE, Yuodelis C (1984) The morphological development of the human fovea. Ophthalmology 91:603–612

Hirata A, Feeney-Burns L (1992) Autoradiographic studies of aged primate macular retinal pigment epithelium. Invest Ophthalmol Vis Sci 33:2079–2090

Holekamp NM, Bouck N, Volpert O (2002) Pigment epithelium-derived factor is deficient in the vitreous of patients with choroidal neovascularization due to age-related macular degeneration. Am J Ophthalmol 134:220–227

Hsu HT, Goodnight R, Ryan SJ (1989) Subretinal choroidal neovascularization as a response to penetrating retinal injury in the pigmented rabbit. Jpn J Ophthalmol 33:358–366

Ishibashi T, Miki K, Sorgente N, Patterson R, Ryan SJ (1985) Effects of intravitreal administration of steroids on experimental subretinal neovascularization in the subhuman primate. Arch Ophthalmol 103: 708–711

Jacobson SG, Cideciyan AV, Bennett J, Kingsley RM, Sheffield VC, Stone EM (2002) Novel mutation in the TIMP3 gene causes Sorsby fundus dystrophy. Arch Ophthalmol 120:376–379

Jester JV, Smith RE (1985) Subretinal neovascularization after experimental ocular histoplasmosis in a subhuman primate. Am J Ophthalmol 100:252–258

Karakousis PC, John SK, Behling KC, Surace EM, Smith JE, Hendrickson A, Tang WX, Bennett J, Milam AH (2001) Localization of pigment epithelium derived factor (PEDF) in developing and adult human ocular tissues. Mol Vis 7:154–163

Kimura H, Sakamoto T, Hinton DR, Spee C, Ogura Y, Tabata Y, Ikada Y, Ryan SJ (1995) A new model of subretinal neovascularization in the rabbit. Invest Ophthalmol Vis Sci 36:2110–2119

Kvanta A, Algvere PV, Berglin L, Seregard S (1996) Subfoveal fibrovascular membranes in age-related macular degeneration express vascular endothelial growth factor. Invest Ophthalmol Vis Sci 37: 1929–1934

Lai CC, Wu WC, Chen SL, Xiao X, Tsai TC, Huan SJ, Chen TL, Tsai RJ, Tsao YP (2001) Suppression of choroidal neovascularization by adeno-associated virus vector expressing angiostatin. Invest Ophthalmol Vis Sci 42:2401–2407

Lambert V, Wielockx B, Munaut C, Galopin C, Jost M, Itoh T, Werb Z, Baker A, Libert C, Krell HW, Foidart JM, Noel A, Rakic JM (2003) MMP-2 and MMP-9 synergize in promoting choroidal neovascularization. FASEB J DOI 10.1096/fj.03-0113fje

Lewen RM (1988) Subretinal neovascularization complicating laser photocoagulation of diabetic maculopathy. Ophthalmic Surg 19:734–737

Lopez PF, Sippy BD, Lambert HM, Thach AB, Hinton DR (1996) Transdifferentiated retinal pigment epithelial cells are immunoreactive for vascular endothelial growth factor in surgically excised age-related macular degeneration-related choroidal neovascular membranes. Invest Ophthalmol Vis Sci 37:855–868

Majji AB, Cao J, Chang KY, Hayashi A, Aggarwal S, Grebe RR, De Juan E Jr.(2000) Age-related retinal pigment epithelium and Bruch's membrane degeneration in senescence-accelerated mouse. Invest Ophthalmol Vis Sci 41:3936–3942

Michaelides M, Hunt DM, Moore AT (2003) The genetics of inherited macular dystrophies. J Med Genet 40:641–650

Monaco WA, Wormington CM (1990) The rhesus monkey as an animal model for age-related maculopathy. Optom Vis Sci 67:532–537

Mori K, Ando A, Gehlbach P, Nesbitt D, Takahashi K, Goldsteen D, Penn M, Chen CT, Mori K, Melia M, Phipps S, Moffat D, Brazzell K, Liau G, Dixon KH, Campochiaro PA (2001) Inhibition of choroidal neovascularization by intravenous injection of adenoviral vectors expressing secretable endostatin. Am J Pathol 59:313–320

Mori K, Gehlbach P, Ando A, McVey D, Wei L, Campochiaro PA (2002) Regression of ocular neovascularization in response to increased expression of pigment epithelium-derived factor. Invest Ophthalmol Vis Sci 43:2428–2434

Mori K, Gehlbach P, Yamamoto S, Duh E, Zack DJ, Li Q, Berns KI, Raisler BJ, Hauswirth WW, Campochiaro PA (2003) AAV-mediated gene transfer of pigment epithelium-derived factor inhibits choroidal neovascularization. Invest Ophthalmol Vis Sci 43:1994–2000

Ogata N, Wada M, Otsuji T, Jo N, Tombran-Tink J, Matsumura M (2002) Expression of pigment epithelium-derived factor in normal adult rat eye and experimental choroidal neovascularization. Invest Ophthalmol Vis Sci 43:1168–1175

Okamoto N, Tobe T, Hackett SF, Ozaki H, Vinores MA, LaRochelle W, Zack DJ, Campochiaro PA (1997) Transgenic mice with increased expression of vascular endothelial growth factor in the retina: a new model of intraretinal and subretinal neovascularization. Am J Pathol 151:281–291

Orzalesi N, Migliavacca L, Miglior S (1994) Subretinal neovascularization after naphthalene damage to the rabbit retina. Invest Ophthalmol Vis Sci 35:696–705

Provis JM, Sandercoe T, Hendrickson AE (2000) Astrocytes and blood vessels define the foveal rim during primate retinal development. Invest Ophthalmol Vis Sci 41:2827–2836

Qi JH, Ebrahem Q, Moore N, Murphy G, Claesson-Welsh L, Bond M, Baker A, Anand-Apte B (2003) A novel function for tissue inhibitor of metalloproteinases-3 (TIMP3): inhibition of angiogenesis by blockage of VEGF binding to VEGF receptor-2. Nat Med 9:407–415

Renno RZ, Youssri AI, Michaud N, Gragoudas ES, Miller JW (2002) Expression of pigment epithelium-derived factor in experimental choroidal neovascularization. Invest Ophthalmol Vis Sci 3:1574–1580

Ryan SJ (1982) Subretinal neovascularization. Natural history of an experimental model. Arch Ophthalmol 100:1804–1809

Ryan SJ, Mittl RN, Maumenee AE (1980) Enzymatic and mechanically induced subretinal neovascularization. Graefes Arch Clin Exp Ophthalmol 215:21–27

Sakamoto T, Sanui H, Ishibashi T, Kohno T, Takahira K, Inomata H, Kohen L, Ryan SJ (1994) Subretinal neovascularization in the rat induced by IRBP synthetic peptides. Exp Eye Res 58:155–160

Sakamoto T, Soriano D, Nassaralla J, Murphy TL, Oganesian A, Spee C, Hinton DR, Ryan SJ (1995) Effect of intravitreal administration of indomethacin on experimental subretinal neovascularization in the subhuman primate. Arch Ophthalmol 113:222–226

Sakurai E, Anand A, Ambati BK, Rooijen N van, Ambati J (2003a) Macrophage depletion inhibits experimental choroidal neovascularization. Invest Ophthalmol Vis Sci 44:3578–3585

Sakurai E, Taguchi H, Anand A, Ambati BK, Gragoudas ES, Miller JW, Adamis AP, Ambati J (2003b) Targeted disruption of the CD18 or ICAM-1 gene inhibits choroidal neovascularization. Invest Ophthalmol Vis Sci 44:2743–2749

Schweisinger C, Yee C, Rohan RM, Joussen AM, Fernandez A, Meyer TN, Poulaki V, Ma JJ, Redmond TM, Liu S, Adamis AP, D'Amato RJ (2001) Intrachoroidal neovascularization in transgenic mice overexpressing vascular endothelial growth factor in the retinal pigment epithelium. Am J Pathol 158:1161–1172

Semkova I, Kreppel F, Welsandt G, Luther T, Kozlowski J, Janicki H, Kochanek S, Schraermeyer U (2002) Autologous transplantation of genetically modified iris pigment epithelial cells: a promising concept for the treatment of age-related macular degeneration and other disorders of the eye. Proc Natl Acad Sci USA 99:13090–13095

Smith RS, John SW, Zabaleta A, Davisson MT, Hawes NL, Chang B (2000) The bst locus on mouse chromosome 16 is associated with age-related subretinal neovascularization. Proc Natl Acad Sci USA 97:2191–2195

Spaide RF, Ho-Spaide WC, Browne RW, Armstrong D (1999) Characterization of peroxidized lipids in Bruch's membrane. Retina 19:141–147

Tamai K, Spaide RF, Ellis EA, Iwabuchi S, Ogura Y, Armstrong D (2002) Lipid hydroperoxide stimulates subretinal choroidal neovascularization in the rabbit. Exp Eye Res 74:301–308

Tobe T, Ortega S, Luna JD, Ozaki H, Okamoto N, Derevjanik NL, Vinores SA, Basilico C, Campochiaro PA (1998) Targeted disruption of the FGF2 gene does not prevent choroidal neovascularization in a murine model. Am J Pathol 153:1641–1646

Trommsdorff M, Gotthardt M, Hiesberger T, Shelton J, Stockinger W, Nimpf J, Hammer RE, Richardson JA, Herz J (1999) Reeler/Disabled-like disruption of neuronal migration in knockout mice lacking the VLDL receptor and ApoE receptor 2. Cell 97:689–701

Tymms MJ (1999) Sorsby's fundus dystrophy: what does TIMP3 tell us about general mechanisms underlying macular degeneration? Clin Exp Optometry 82:124–129

Ulshafer RJ, Engel HM, Dawson WW, Allen CB, Kessler MJ (1987) Macular degeneration in a community of rhesus monkeys. Ultrastructural observations. Retina 7:198–203

Wang F, Rendahl KG, Manning WC, Quiroz D, Coyne M, Miller SS (2003) AAV-mediated expression of vascular endothelial growth factor induces choroidal neovascularization in rat. Invest Ophthalmol Vis Sci 44:781–790

Weber BH, Vogt G, Pruett RC, Stohr H, Felbor U (1994) Mutations in the tissue inhibitor of metalloproteinases-3 (TIMP3) in patients with Sorsby's fundus dystrophy. Nat Genet 8:352–356

Weber BH, Lin B, White K, Kohler K, Soboleva G, Herterich S, Seeliger MW, Jaissle GB, Grimm C, Reme C, Wenzel A, Asan E, Schrewe H (2002) A mouse model for Sorsby fundus dystrophy. Invest Ophthalmol Vis Sci 43:2732–2740

Yannuzzi LA, Negrao S, Iida T, Carvalho C, Rodriguez-Coleman H, Slakter J, Freund KB, Sorenson J, Orlock D, Borodoker N (2001) Retinal angiomatous proliferation in age-related macular degeneration. Retina 21:416–434

Zhu ZR, Goodnight R, Sorgente N, Ogden TE, Ryan SJ (1989) Experimental subretinal neovascularization in the rabbit. Graefes Arch Clin Exp Ophthalmol 227:257–262

# Chapter 8

## Transporters and Oxidative Stress in AMD

David V. Pow, Robert K.P. Sullivan, Susan M. Williams, Elizabeth WoldeMussie

## Contents

8.1 Redox Reactions, Health, Oxidative Damage and Disease  123
8.2 Does Epidemiology Support a Role for Oxidation in AMD?  124
8.3 Exogenous Factors Influencing AMD Incidence  125
8.4 Hallmark Features of Oxidative Damage and Free Radical Damage  125
8.4.1 Oxidized Proteins  125
8.4.2 Oxidized Lipids  125
8.4.3 DNA Damage  125
8.5 Oxidative Challenges in the Eye  125
8.5.1 Endogenous Antioxidants, Enzymes and Related Molecules in the Retina  126
8.5.2 Classes of Antioxidants  128
8.5.3 The Glutathione System  128
8.5.4 GSH in the Retina: Intrinsic Synthesis and Possible Interplay with the RPE  128
8.5.5 Synthesis of GSH  128
8.5.6 The GSH-GSSG Cycle  129
8.5.7 Elimination of Superoxides by Superoxide Dismutases  131
8.5.8 Ascorbic Acid  132
8.5.9 Vitamin E  132
8.5.10 Taurine  133
8.5.11 Metallothionein  136
8.5.12 Zinc, Copper, Selenium and Manganese  136
8.5.13 Selenium Compounds  137
8.5.14 Manganese  137
8.5.15 Macular Pigments  138
8.5.16 Ascorbate and Recycling of Carotenoids  138
8.5.17 A Summary of Mechanisms for Antioxidant Protection  138
8.6 Cellular Interactions and Photoreceptor Death  138
8.6.1 Light-Mediated Damage: Experimental Lesions and AMD  139
8.6.2 Mechanisms of Cell Death  140
8.6.3 Cell Death in Response to Oxidative Damage  140
8.7 A Role for Glial Cells in AMD?  141
8.7.1 Metabolic Functions of Müller Cells in the AMD Retina  141
8.7.2 Cell Death and Glutamate Toxicity  144
8.8 Conclusions  144
References  145

## 8.1 Redox Reactions, Health, Oxidative Damage and Disease

Redox reactions that involve the linked oxidation of one molecule and the concomitant reduction of another are normal and vital elements of eukaryotic biology that drive both the transfer of energy and biosynthetic activity. The normal function of tissues such as the eye is dependent upon the balance between the generation of free radicals and oxidative species, and the availability of antioxidants and free radical scavengers. Even apparently damaging species such as hydrogen peroxide are, at appropriate concentrations, biologically important molecules. Thus, hydrogen peroxide may directly or indirectly regulate the activity of enzymes as diverse as tyrosine kinases (Tokano et al. 2002) and guanylate cyclases (White et al. 1976). However, dysregulation of the production and removal of oxidative species and free radicals may be detrimental to the body. In the past 15 years, there has been a massive promotion of the notion that free radicals and oxidant species are potential causal agents in disease states as divergent as Parkinson's disease (Simpkins and Jankovic 2003), motor neuron disease (Rothstein et al. 1994), Alzheimer's disease (Aliev et al. 2003), cardiovascular disease (Burke and Fitzgerald 2003), cataracts (Bush and Goldstein 2001), and cancer (Kelly 1998). Studies on the eyes of cynomolgus monkeys, which exhibit an age-related macular degeneration (AMD)-like condition, including the formation of drusen, reveal the presence of greatly reduced levels of the antioxidant enzymes catalase and glutathione peroxidase, and greatly reduced levels of retinal (but not plasma) zinc (Nicolas et al. 1996). The reduced zinc levels in retina might be interpreted as indicating that low zinc causes AMD. However, the most parsimonious interpretation is that if zinc levels are not depressed in plasma then depressed retinal levels are due to a reduction in the level of zinc-binding proteins (or more likely GSH), such that this lack in turn causes a lowering of retinal zinc. Thus one observation can lead to two diametrically opposed interpretations if all the data are not taken in context completely. Despite the potential for misinterpretation or multiple interpretations, the data of Nicholas et al. (1996) have been widely assumed to demonstrate the presence of oxidative stress in these affected retinas. Obviously these types of biochemical measurements are difficult to perform on human retinas, due to inevitable postmortem delays in obtaining tissues; thus most human studies have relied on epidemiological evaluation of measures such as plasma antioxidants (Cohen et al. 1994; Samiec et al. 1998) and the clinical consequences of attempts to manipulate these parameters.

## 8.2
## Does Epidemiology Support a Role for Oxidation in AMD?

As the epidemiology of AMD is reviewed elsewhere in this book, only a brief examination is undertaken here, to clarify the recent findings that have analysed the relationship between antioxidant availability and AMD. Two distinct themes emerge: the potential for deficits in levels of antioxidants, especially those derived from the diet, and the oxidising potential of extrinsic factors such as light and smoking.

Many small antioxidant molecules are derived from the diet; accordingly one premise might be that deficiencies are present in modern diets. The recommendation of a healthy diet to prevent or cure visual dysfunction may seem like a modern concept. However, the Ebers papyrus, written around 1550 BC (Ebers and Stern 1875) recommended the consumption of ox liver (a rich source of antioxidants such as taurine and glutathione) as a cure for night blindness. Conversely, the notion that Western diets may be deficient in critical antioxidants, and thus contribute to diseases such as AMD, has become a mainstream mantra. A series of dietary supplementation trials have examined molecules such as zinc (Newsome et al. 1988; Cho et al. 2001) and vitamin E (Delcourt et al. 1999a, 1999b). The results of many of these trials have been somewhat ambiguous. Evans (2002) has suggested that most of the studies to date have not indicated a significant protective effect of antioxidants upon the incidence of new cases of AMD, but there may well be retardative effects on the further progression of established cases of AMD. The most parsimonious interpretation of these findings is that, for the majority of the Western populations that have been evaluated, dietary substrates are not normally limiting factors in the initiation of AMD, but that progression of the disease is associated with the production of higher levels of oxidative species, which can then exacerbate the disease state. Accordingly this would point not to a global deficiency in antioxidant availability as a causal factor of AMD, but rather to deficits in the way that the antioxidant network is functioning, possibly because of abnormalities in nondietary-derived antioxidants (e.g. the glial cell-derived antioxidant glutathione). This may account for some of the paradoxical findings such as the POLA. study (see Delcourt et al. 1999a), which shows that higher levels of plasma glutathione peroxidase are evident in late-stage (but not early) AMD. Similarly Nowak et al. (2003) have noted that increased lipid peroxidation in the blood of AMD patients is also associated with an increase in antioxidants such as GSH. We suggest that differences between early and late stages of AMD are indicative of a dynamic adaptive response in the production of antioxidants in response to an oxidative insult.

## 8.3 Exogenous Factors Influencing AMD Incidence

The strongest risk factor identified to date is smoking, which is linked to the development of both cataracts and AMD (Hyman and Neborsky 2002). Tobacco smoke contains several thousand potential oxidants, or molecules that can be bioactivated to form oxidizing species. Tobacco-related chemicals may arrive at the retina via the vasculature or, more likely, may enter via the anterior chamber of the eye, due to the permeable nature of the cornea. Accordingly, behaviours such as smoking may cause oxidative stress. Whether this oxidative stress is then causal of AMD or is merely a coincident bystander in biochemical terms, remains to be proven. Thankfully, the drinking of red wine seems to be a marginally protective vice (Bastianetto 2002).

## 8.4 Hallmark Features of Oxidative Damage and Free Radical Damage

Free radicals and oxidative species can damage tissues by a variety of means, including reacting with nucleotides in DNA, polyunsaturated fatty acids in cellular membranes, and sulphhydryl residues in proteins.

### 8.4.1 Oxidized Proteins

Accumulation and aggregation of abnormal proteins is a common feature of the major neurodegenerative diseases. Sometimes the proteins may have an abnormal sequence, as in the case of the SOD1 (superoxide dismutase 1) mutants found in some familial cases of motor neuron disease, but usually the proteins are normal in terms of primary sequence but have become modified by processes such as oxidation or nitration. Drusen are large proteinaceous bodies present both in normal aged eyes and in eyes afflicted with AMD. Drusen are variable in composition but typically are characterized by the presence of more than a hundred proteins within the aggregates. There is increasing evidence for the presence of oxidative modifications to many of these proteins, suggesting that drusen deposition may be a consequence of oxidative damage (Crabb et al. 2002), which causes malfolding of the proteins, rendering them insoluble.

### 8.4.2 Oxidized Lipids

Peroxidation of lipids, in particular polyunsaturated lipids (such as are abundant in photoreceptors), is a frequent occurrence in stressed tissues, being particularly common in response to challenges such as reperfusion of the brain after a period of hypoxia (Fig. 8.1).

### 8.4.3 DNA Damage

DNA exists in two cellular locations in cells, in the nucleus and in mitochondria. Both pools of DNA may be damaged, especially in response to intense light. Damage appears to be mediated by activated oxygen species that induce the formation of random breaks in DNA (Specht et al. 1999).

## 8.5 Oxidative Challenges in the Eye

Three distinct types of pro-oxidant or free radical-generating challenge confront the normal functioning of the eye. The eye may be subject to chemical factors that are intrinsic in origin, extrinsic in origin, or subject to physical factors such as bombardment by high-energy and pos-

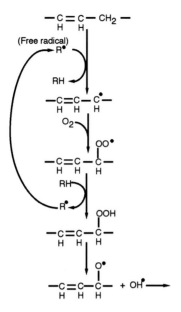

Fig. 8.1. Peroxidation of unsaturated lipids normally involves the initial attack by a free radical species (R), such as might be generated photochemically in a photoreceptor. In the presence of oxygen, this leads to a series of reactions that result in the peroxidation of the lipid and the concomitant regeneration of a free radical that can re-attack the lipid if it is polyunsaturated. Thus, a small number of free radicals can, in the presence of oxygen, cause a disproportionate amount of damage to polyunsaturated lipids unless another molecule (such as vitamin E) acts as a free radical acceptor, to terminate the chain reaction

sibly high-intensity light. In this latter respect, the eye is similar to the human skin and some comparison is appropriate. The skin has evolved defences, including the presence of layers of dead, keratinised cells that can absorb part of the UV spectrum, and variable deposits of melanin. Thus, in the skin, only a small proportion of light penetrates to the live cells in the deeper, well-oxygenated layers. In these deeper layers, the combination of high-energy photons and abundant oxygen presents a dangerous combination that permits the formation of free radicals and a variety of oxidizing species, which leads to damage of cellular constituents, including DNA, leading in turn to apoptosis (Kulms and Schwarz 2002). The retina does not have a thick defensive barrier of light-absorb-

ing dead cells, thus light is free to interact with any light-absorbing molecules in the live cells of the retina and RPE in a well-oxygenated environment. The review by Winkler et al. (1999) provides a seminal overview of the area that has served as a template for most subsequent studies. The basic tenets of the oxidative hypothesis are that photoreceptors and the RPE exist in a highly oxidizing environment due to an unusual combination of factors. The antioxidant defence system of the photoreceptors and RPE needs to quench reactive oxidative species such as singlet oxygen, generated when photons enter the eye and interact with a variety of molecules.

The partial pressure of oxygen around the outer segments of the photoreceptors and RPE is extremely high, based upon measurements such as choroidal blood and saturation of blood in retinal vessels outside the macula (Schweitzer et al. 2001), a saturation that is required because of the high-energy demands of the photoreceptors (Stone et al. 1999). The photoreceptor outer segments contain a high proportion of polyunsaturated fatty acids that are readily oxidized. Simultaneously, the inner segments of the photoreceptors contain a considerable number of mitochondria, to generate ATP. The step-wise reduction of oxygen in mitochondria by electron transfer inevitably leads to production of superoxide; this superoxide is normally inactivated by the mitochondrial form of glutathione peroxidase and catalase, but some leakage will inevitably occur. The conjoint function of these enzyme systems and an abundance of small molecular reductants such as ascorbic acid, vitamin E, lutein, GSH, hypotaurine and urea normally form an effective antioxidant defence. However, if this co-ordinated antioxidant system is overwhelmed, lipid peroxidation within the outer segments may ensue.

### 8.5.1
### Endogenous Antioxidants, Enzymes and Related Molecules in the Retina

The broad tendency in the literature is to discuss individual antioxidants in isolation. However, antioxidant systems participate in a homeostatic web of interlinked pathways. Studies regarding the role of antioxidants in AMD have typically focused on molecules that are normally present at high levels in the macula. The premise is that they allow the normal function of the retina to proceed and that any reduction in levels of these molecules will then predispose the retina to disease. From a philosophical perspective, this may be an incorrect premise, and an alternate view might be that the very susceptibility of the macula to retinal disease is dependent on the intrinsic insufficiency of an antioxidant system in the macular region. Indeed many molecules may have not only beneficial effects but also detrimental effects, as typified in the case of zinc, due to its involvement in a vast assay of cellular processes (over 2,000 proteins use zinc in some way, binding to structural motifs such as "zinc fingers" – relatively small protein motifs that fold around one or more zinc ions – or simply the presence of free sulphydryl groups).

Thus the presence of an antioxidant or related molecule at high levels in the macula is not necessarily evidence that it is normally beneficial. Accordingly any discourse on the antioxidant status of the eye needs to consider the functional properties of these antioxidants, and the relationships of these molecules to other members of the antioxidant web.

A number of terms are used to describe interconnected, but sometimes distinct agents that may cause cellular damage, or protect against such. Some are listed below:

- Free radicals: Unstable molecules that contain one or more unpaired electrons in their outer orbits. Free radicals may or may not have a negative charge associated with them; e.g. OH· (hydroxyl radical) and $O_2^-$ (superoxide anion). Free radicals can originate from endogenous metabolic events. Alternatively, free radicals and oxidative species may be derived exogenously as a result of exposure to chemicals such as tobacco smoke and indirectly through the metabolism (mainly via liver enzymes such as P450 systems) of solvents such as trichloromethane, drugs and pesticides. Radiation from many parts of the electromagnetic spectrum, including blue and ultraviolet light may also cause formation of free radicals.
- Free-radical scavengers: Molecules such as ascorbate which accept the free-radical moiety from a donor molecule such as the α-tocopherol radical, thereby converting, in this case the α-tocopherol radical back to α-tocopherol (vitamin E).
- Molecular oxygen: Oxygen in the form $O_2$ (the normal molecular form of oxygen)
- Ozone ($O_3$): Formed by the reaction of molecular oxygen with atomic oxygen. This is commonly formed when light causes the breakdown of species such as $NO_2$, yielding NO and O. This atomic oxygen reacts rapidly with molecular oxygen to form $O_3$. While little if any ozone is likely to be formed in the eye, air pollution can create a biologically significant loading which may impact on ocular function.
- Reactive oxygen species or reactive oxygen intermediates (ROI/ROS): The group of free radicals and nonradicals that constitute all oxidation and excitation states of $O_2$, from superoxide ($O_2^-$) up to but excluding water, that may be generated or present in physiological environments, including singlet oxygen ($^1O_2$), ozone ($O_3$) hydrogen peroxide ($H_2O_2$), hypohalites, and hydroxyl radical (OH).
- Peroxides: Agents which either directly add oxygen to a molecule or result in loss of a hydrogen; thus hydrogen peroxide ($H_2O_2$) loses an oxygen and is converted to water, ($H_2O$) while donating oxygen to an acceptor molecule.
- Singlet oxygen ($^1O_2$): molecular oxygen ($O_2$) with anti-parallel spin, which can damage molecules as it reverts back to normal molecular oxygen.
- Superoxide: Any compound containing the highly reactive superoxide radical, $O_2^-$, which is produced by reduction of molecular oxygen in many biological oxidations; this highly toxic free radical is continuously removed by the enzyme superoxide dismutase.
- Antioxidants: Molecules that reduce the oxidized target molecule and are oxidized in the process; e.g. reduced glutathione, lutein, ascorbate, vitamin E and urates.

- Metal co-factors: Many antioxidant enzymes are metalloenzymes; they require metallic cofactors either for structural stability or in order for the enzymes to be fully functional, in particular selenium (glutathione peroxidases), zinc, manganese or copper (superoxide dismutases).

## 8.5.2
### Classes of Antioxidants

To counter the effects of oxidation, the body accumulates or produces a variety of antioxidants. Antioxidants in the retina fall into three main categories: (1) small molecules synthesised by the body, such as reduced glutathione (GSH) and hypotaurine; (2) small molecules derived from the diet, such as vitamin A, ascorbic acid (vitamin C), α-tocopherol (vitamin E), zinc and selenium, and (3) enzymes which catalyse interconversions such as superoxide dismutase (SOD), glutathione peroxidase (GPX) and catalase (a peroxidase).

## 8.5.3
### The Glutathione System

Glutathione is the keystone of the mammalian antioxidant defence system and may well be subject to abnormal homeostasis in the AMD retina (Cai et al. 2000). Glutathione is a tri-peptide, containing the amino acids glutamate, cysteine and glycine (Meister 1983). It exists as two forms, reduced glutathione, also known as GSH, where "SH" indicates the presence of a free sulphydryl group, and GSSG, which is a dimer formed by two GSH molecules with concomitant loss of two hydrogens (i.e. it is in the oxidized form). In the central nervous system, GSH and the synthetic machinery for forming such is abundant in glial cells; neurons contain much lower concentrations of GSH (Makar et al. 1994). In the retina, a similar distribution has been observed, GSH being abundant in the Müller cells (Pow and Crook 1995; Huster et al. 2000). Similarly GSH is abundant in the RPE.

## 8.5.4
### GSH in the Retina: Intrinsic Synthesis and Possible Interplay with the RPE

GSH is an interesting molecule, because the levels of such in the retina are extremely dependent on the retina being closely associated with the RPE. Detachment of the retina leads to a rapid loss of GSH immunoreactivity in the retina. Two very distinct possibilities arise from this observation: either much of the retinal GSH is derived from the RPE or alternately the general energetic and metabolic perturbations that ensue after the retina is detached from the RPE prevent appropriate synthesis of GSH by the retinal Müller cells (Fig. 8.2).

## 8.5.5
### Synthesis of GSH

Synthesis of GSH is a multistage process, which starts with the acquisition of the substrates for synthesis by the Müller glial cells. Two key

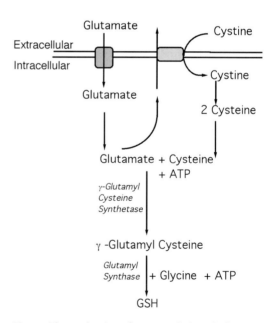

Fig. 8.2. The mechanisms for accumulation of substrates and the synthesis of GSH. Synthesis is an energy-dependent process, energy being needed to drive substrate transport, while synthesis is dependent upon availability of ATP

molecules are glutamate and cysteine. In the retina the Müller cells are the site of de novo glutamate synthesis (Pow and Robinson 1994); thus glutamate supply is not normally a problem if there is sufficient energy to support its synthesis (it is derived from α-ketoglutarate, a product of the Krebs cycle). A problem arises with the supply of cysteine. Cysteine is an essential amino acid; it cannot be synthesized by the Müller cells; accordingly it must be obtained from the extracellular environment. A Müller cell normally accumulates this amino acid not as cysteine, but as the dimeric form, cystine. This accumulation is mediated via a specific transporter, the cystine-glutamate antiporter (CGAP; Sato et al. 1999). The CGAP is a sodium-independent transport system which stoichiometrically couples the influx of cystine to an efflux of glutamate down its concentration gradient. The CGAP is abundantly expressed by the Müller cells, as evinced by the accumulation of α-amino adipic acid (Pow 2001), which is a selective substrate for the CGAP (Bannai 1986; Kato et al. 1993; Tsai et al. 1996). Antibodies against the anti-porter demonstrate that it is localized to those parts of the Müller cell in close apposition to the vasculature (which is the source of the cystine; Fig. 8.3).

One problem that arises in the use of the CGAP is the concomitant export of glutamate from the Müller cell. Glutamate is needed within the cell to make GSH, but when present in the extracellular environment, glutamate can act as an excitotoxin. Glutamate is normally removed from the retinal extracellular space using the Müller cell glutamate transporter GLAST (also called EAAT1; Derouiche and Rauen 1995).

The reaction of cysteine and glutamate to form γ-glutamyl cysteine is catalysed by a zinc-dependent enzyme, γ-glutamylcysteine synthetase (γ-GCS), and driven by the hydrolysis of ATP. γ-GCS is key regulatory enzyme in the synthesis of glutathione. It is a heterodimeric zinc metalloprotein that gains activity owing to formation of a reversible disulphide bond (Soltaninassab et al. 2000). Accordingly the overall antioxidant status of the cell will have a critical regulatory influence on the capacity of the cell to synthesise GSH. Preliminary evidence, in rats at least, suggests that the activity of γ-GCS is down-regulated in aging brains (Liu 2002), concomitant with reduced levels of GSH. It is impossible to verify at this stage if this might be linked to AMD, but it remains a tantalizing possibility. γ-Glutamyl cysteine is then converted to GSH using glutathione synthetase (which is also an oxidation-sensitive enzyme), the reaction again being driven by the hydrolysis of ATP.

## 8.5.6
### The GSH-GSSG Cycle

The GSH-GSSG cycle is a coupled redox cycle, the oxidation of GSH being coupled to the reduction of target molecules. GSSG formed in this cycle is then recycled back to GSH. A measure of the capacity of a molecule to reduce or oxidize another molecule is the standard electrode potential or standard reduction potential, $E°$. The redox potential of the GSH/GSSH redox couple is about −330 mV (measured normally at pH 7, and 25°C), making GSH a potent reductant and thus a good antioxidant.

The key enzymes in the GSH-GSSG cycle are the glutathione peroxidases and glutathione reductase (often called GSSG reductase or, less accurately, GSH reductase). Glutathione peroxidases have been grouped into four main families, GPX1–GPX4, many of which are selenium-dependent metalloenzymes (they require selenium in trace amounts as a co-factor). GPX 1 is located on chromosomes 3, 21 and X, as a number of minor allelic variants which may confer slightly different properties on the enzymes. This form is responsible for the majority of cytoplasmic activity in tissues such as the retina. GPX 2 is specific to the gastrointestinal epithelium, while GPX3 is a secreted plasma glutathione peroxidase. GPX 4 is encoded on chromosome 19. It is unusual, in that it is the only major antioxidant enzyme known to directly reduce phospholipid hydroperoxides within membranes and lipoproteins, acting in conjunction with α-tocopherol (vitamin E) to inhibit lipid peroxidation (Yant et al. 2003). A key finding has been that GPX expression in the retina is enhanced after exposure to bright light

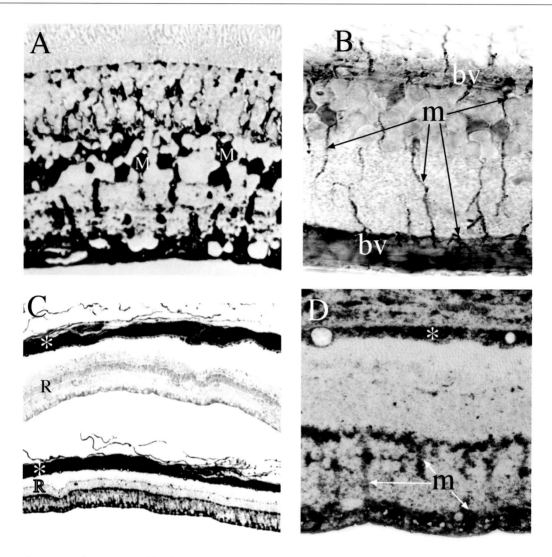

Fig. 8.3A–D. The activity and localisation of the cystine-glutamate anti-porter in the rat retina is revealed using antibodies to a substrate of the transporter, aminoadipic acid (A), or antibodies to the anti-porter itself (B). The anti-porter is localised to Müller cells (*m*), especially those parts close to blood vessels (*bv*). Similarly GSH is localised to Müller cells in the rabbit retina (*R*) and the overlying RPE and choroid (*asterisk, bottom* of C), but is greatly reduced in an experimentally-detached retina (*top* of C). D High-power view of the localisation of GSH in Müller cells (*m*) and the overlying RPE (*asterisk*)

(Ohira et al. 2003), suggesting that the throughput of the GSH-GSSG cycle might be enhanced in response to light-induced oxidative stress. The recycling of GSSG back to GSH is mediated via glutathione reductase. This part of the cycle is driven using the reductive power of NADPH, which is converted to NADP in the process (the actual electron transfer is actually mediated via a further intermediary, FAD, but the fine details of this reaction are outside the scope of this chapter). In turn the production of NADPH is fueled predominantly via the oxidation of glucose-6 phosphate to ribose-6 phosphate. In turn glucose-6 phosphate arises from the gluconeogenic pathway, glycolysis or glycogenolysis. Interestingly, in the frog retina at least, the gluconeogenic pathway is preferentially associated with Müller cells (Goldmann 1990) and indeed the terminal enzyme in that pathway, which would convert glucose-6 phosphate into glucose, is a very good histochemical marker of Müller cells (Hirata et al. 1991). Similarly glycogen is mainly localized to Müller cells rather than neurons (Reichenbach et al. 1993), as is glycogen phosphorylase, the first enzyme involved in breaking down glycogen (Pfeiffer et al. 1994). This glial localization of pathways involved in producing glucose-6 phosphate and thus NADPH reconciles well with the idea that the recycling of GSSG back to GSH is likely to occur in the Müller cells. Whether GSH or GSSG remains in the Müller cells is open to debate. There is significant evidence for the transport of GSH into and possibly out of Müller cells (Kannan et al. 1999), raising the possibility that GSH may be released from Müller cells and thus traffic between neuronal and glial compartments (Schutte and Werner 1998), possibly returning to the Müller cells as GSSG, for recycling back to GSH (Bringmann and Reichenbach 2001).

The GSSG cycle is coupled to a number of other key redox reactions, including superoxide detoxification and recycling of oxidized (dehydro)ascorbate to ascorbate, thereby coupling in turn to the recycling of vitamin E and carotenoids (macular pigments).

## 8.5.7
### Elimination of Superoxides by Superoxide Dismutases

Superoxide dismutases are a family of metalloenzymes which may require copper zinc or manganese to stabilize them. They catalyze the dismutation of superoxide anions, to form hydrogen peroxide. Three mammalian superoxide dismutases have been identified (Zelko et al. 2002): SOD-1, or CuZn-SOD, is a copper- and zinc-containing homodimer that is localised to the cytoplasm. SOD-2, or Mn-SOD, is present in mitochondria, while SOD-3 is a copper- and zinc-containing tetramer that is abundant in plasma. Each of the SODs appears to have distinct physiological roles. Mutations in SOD-1 have been linked to a wide variety of neurological disease states as divergent as motor neuron disease (Rothstein et al. 1994) and sudden infant death syndrome (SIDS; Reid and Tervit 1999). SOD-1 is abundant in photoreceptors (Ogawa et al. 2001); mutations of SOD-1 in the mouse cause a selective degeneration of photoreceptors in response to exposure to bright light (Mittag et al. 1999). Interestingly, SOD-1 is one of the genes encoded by chromosome 21. In Down's syndrome, complete or partial trisomy results in an increase in SOD-1 expression. It has been hypothesized that this may lead to neuronal dysfunction or death in this condition, but curiously there is no particular association between this condition and AMD, perhaps suggesting that, in AMD, if there is any change in SOD it is more likely to be down-regulated than up-regulated.

Abnormalities in SOD-2 may lead to mitochondrial dysfunction, which in the retina are manifest mostly as anatomical abnormalities in the inner layers (Sandbach et al. 2001). Intriguingly, Wu et al. (1994) have noted that low levels of zinc, SOD (presumably SOD-3) and catalase activity are evident in AMD patient sera when compared with controls, suggesting that plasma SOD-3 activity might influence retinal oxidative state and thus function (Fig. 8.4).

The dismutation of superoxide yields hydrogen peroxide. Two key enzymes are thought to be involved in the breakdown of hydrogen peroxide, these are catalase and the glutathione peroxidases. Catalase is a mammalian peroxidase enzyme, which uses iron as a cofactor. It is abundant in endothelial and epithelial cells in the eye (Atalla et al. 1987). Whilst some workers have argued that catalase is present in rod outer segments (Ohia et al. 1994), others have adopted a contrary position (Armstrong et al. 1981). It is probable that catalase is present but not particularly abundant in photoreceptors. As SOD-1 is present in photoreceptors to drive the supply of hydrogen peroxide, the conversion of hydrogen peroxide to water in photoreceptors may be driven by glutathione peroxi-

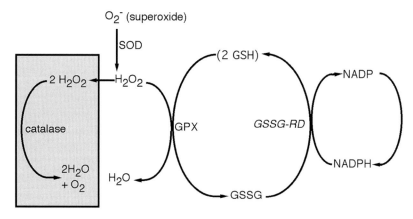

Fig. 8.4. Hydrogen peroxide, produced from superoxide by superoxide dismutase (SOD), is inactivated either by catalase or by glutathione peroxidase (GPX), with the concomitant oxidation of GSH to GSSG. GSSG is converted back to GSH using GSSG reductase (GSSG-RD). The reducing power for this reaction is provided by NADPH

dase. Intriguingly Ohira et al. (2003) have suggested that there is only a low level of expression of glutathione peroxidase in rod photoreceptors and RPE under normal light conditions, as evaluated by immunocytochemistry, but the levels rise in response to a light insult, in RPE and photoreceptors, and prominently in Müller cells. The suggestion that only low levels of GPX are present in the normal retina should, however, be treated with caution, as glutathione peroxidase exists as a number of forms as discussed previously; it is possible that antibody techniques do not identify all the known glutathione peroxidases in the retina.

### 8.5.8
### Ascorbic Acid

Vitamin C or ascorbic acid/ascorbate is a small, extremely effective, water-soluble molecule that is obtained from the diet. Vitamin C is in a unique position to "scavenge" aqueous peroxyl radicals before these destructive substances have a chance to damage the lipids. Ascorbate radicals are relatively harmless as they are neither strong oxidants nor reductants and react poorly with oxygen; thus they produce few superoxide or peroxyl radicals (Buettner and Jurkiewicz 1996). Perhaps the most important feature of ascorbate is its capacity to repair and thus recycle the lipid-soluble tocopheroxyl (chromanoxyl) radical of vitamin E, by virtue of redox reactions at the lipid-aqueous interface (Fig. 8.5).

Oxidized ascorbate (dehydroascorbate) is efficiently converted back to ascorbate using glutathione peroxidase. Because ascorbate can interact both with oxidized vitamin E at the interface of the lipid phase and GSH in the aqueous phase (see below), it is an extremely important antioxidant.

### 8.5.9
### Vitamin E

Vitamin E ($\alpha$-tocopherol) is the primary lipid-soluble, small antioxidant molecule in the plasma membranes of most cells. Its role is to limit the peroxidation of polyunsaturated fatty acids. Vitamin E is derived solely from the diet so dietary supply may be a limiting factor in some instances.

Peroxidation of lipids is a significant problem in the retina because free radicals, which may be generated in a variety of reactions, including photochemical reactions, aggressively attack polyunsaturated fatty acids. A key feature of this process is the regeneration of free radicals and formation of organic peroxides, which can then repeat the damage process. Vi-

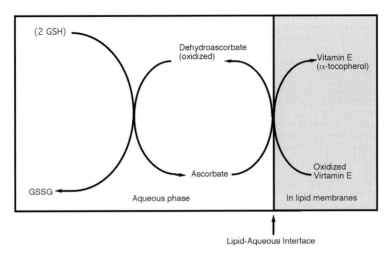

**Fig. 8.5.** The reduction of oxidized α-tocopherol (vitamin E) is mediated by ascorbate, which in turn is oxidized to dehydroascorbate. In turn, dehydroascorbate is recycled back to ascorbate with the concomitant oxidation of GSH to GSSG

tamin E is thought to reduce this damage by acting as an efficient free radical acceptor, forming in the process an α-tocopherol radical (Buettner 1993). The tocopheroxyl radical is formed at the aqueous interface, thus permitting its transfer to ascorbate, with concomitant regeneration of vitamin E and conversion of ascorbate to yield an ascorbate radical. This ascorbate radical in turn feeds back into the glutathione-detoxifying system.

Because of its hydrophobic character, vitamin E requires a transport protein in order to be transported in the blood. Mice that are deficient in α-tocopherol transfer protein are unable to transport this vitamin and thus their tissues are deficient of vitamin E. Mice with this mutation exhibit an age-dependent retinal degeneration (Yokota et al. 2000, 2001) as well as neural degeneration throughout the brain. Transport of vitamin E across the blood-brain barrier or blood-retina barrier may be a limiting factor in its supply to nervous tissues. It is generally thought that transfer across the blood-brain barrier at least may be mediated by its binding to high-density lipoproteins (HDL; Goti et al. 2000), but data are lacking as to the transport mechanism or capacity with respect to the blood-retina barrier.

### 8.5.10
### Taurine

Taurine and hypotaurine are enigmatic amino acids, which are clearly critical for normal visual function, but little is known about how they achieve this protective role. Both are sulphur-containing amino acids; they are abundant in the retina, taurine being present at 10–25 mM concentrations in the rod and cone photoreceptors, and hypotaurine is probably present at millimolar levels. Both molecules have been considered to be potential antioxidants in the retina. Thus, Pasantes-Morales and Cruz (1985) have shown that taurine and hypotaurine inhibit light-induced lipid peroxidation and thus protect rod outer segment structure in the frog retina. Of the two compounds, hypotaurine is the more potent antioxidant; hypotaurine is readily oxidized to form taurine, concomitantly reducing target compounds that are being protected (Fig. 8.6).

Hypotaurine is able to prevent the inactivation of SOD by hydrogen peroxide in a concentration-dependent manner. Hypotaurine probably exerts this effect by reacting with hydroxyl radicals, generated by the enzyme, preventing them from reacting with the active site of the

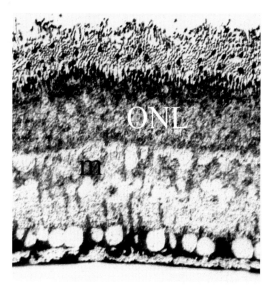

Fig. 8.6. Immunolocalisation of hypotaurine in the rabbit retina. Hypotaurine is localized to the Müller cells (*m*) and is also abundant in the photoreceptors in the outer nuclear layer (*ONL*)

enzyme. (Pecci et al. 2000). Hypotaurine is able to form complexes with zinc (van Gelder 1983). This may account for the interaction that Gottschall-Pass et al. (1997) have described between zinc and taurine, based upon an analysis of oscillatory potentials in the developing rat retina. Taurine deficiency in the cat results in a stereotypic degeneration of both rod and cone photoreceptors (Hayes et al. 1975; Schmidt et al. 1976). Pasantes-Morales et al. (1986) and Leon et al. (1995) have demonstrated that the degenerative lesion starts in the area centralis; intriguingly, the development of the lesion is not dependent on light exposure, as assessed by uniocular occlusion studies. Degeneration is associated initially with the loss of the b-wave, and later, the a-wave of the electroretinogram (ERG), suggesting that signaling changes in the retina precede photoreceptor loss.

Despite the obvious similarity between the gross spatial pattern of photoreceptor loss in taurine depletion and AMD, the fine details of cell loss in taurine-depleted cats have not been fully investigated. In routine histological sections, it is clear that there is a graded loss of photoreceptors, the rod photoreceptors being lost before the cone photoreceptors. A critical feature of this type of degenerative lesion is the lack of concomitant disturbance in the organization of the remaining retina and lack of loss or perturbation of the RPE (Fig. 8.7).

Neurochemically there appears to be little dysfunction, with normal expression of neurotransmitters in bipolar cells and retinal ganglion cells, and glycine and GABA in amacrine cells. This lack of change in the inner retina renders this type of lesion dissimilar to most AMD lesions, where there is growing evidence for changes such as loss of ganglion cells and major changes in the architecture of the glial cells (Wu et al. 2003). Whether this difference is due to the lack of other secondary changes such as the breakdown of the RPE and thus the blood-retina barrier is unclear. Intriguingly, in a rat light-degeneration model of AMD (Sullivan et al. 2003), the expression of at least one of the known taurine transporters (Taut-2) is highly up-regulated at sites proximal to the anatomically defined lesion, possibly in a compensatory attempt to enhance taurine supply in the areas that are about to degenerate (Fig. 8.8).

The mechanism by which taurine depletion causes photoreceptor degeneration is unknown. However, Sturman et al. (1981) have shown that reductions in taurine levels in the cat retina are associated with a concomitant reduction in zinc levels. The change in the ERG in response to taurine deficiency differs from that in zinc deficiency (Jacobson et al. 1986), so it is unclear whether this association between zinc and taurine is instrumental in causing the lesion, but if so it might be expected to cause changes in the inner retina as well, since these cells will also use zinc-dependent enzymes. A simpler explanation is that taurine and hypotaurine are antioxidants, and that lack of antioxidants leads to increased oxidative damage of photoreceptors and thus their death.

A second tangible link between taurine deficiency and oxidative stress is the observation that taurine shares a common metabolic origin with GSH, both being derived from cysteine (Fig. 8.9). While much retinal taurine is probably derived from the diet (especially in carnivorous species such as cats) or synthesized by

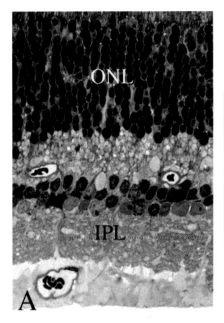
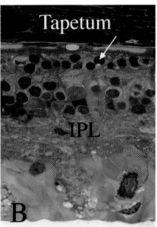

Fig. 8.7A, B. Semithin sections through the retinas of a normal cat (A) and the area centralis of a taurine-depleted cat (B). In the taurine-depleted cat the retina is thin, the overlying tapetum is intact and a few residual photoreceptors (probably cones, based on their morphology) are still present in this example. The retina appears to retain considerable organizational integrity in the inner plexiform layer (IPL) and inner nuclear layer; ONL, outer nuclear layer.

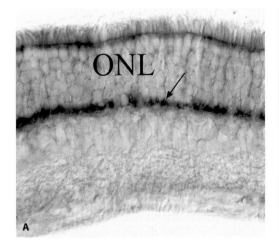
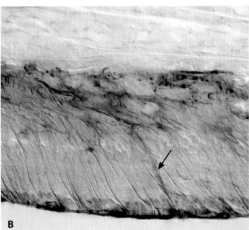

Fig. 8.8A, B. Taurine transporter Taut-1 is localized to photoreceptors, especially their terminals (*arrow*) in the human retina (A). In light-damaged rats (B), Taut-2, which is normally only expressed by the astrocytes (Pow et al. 2002), is strongly expressed by the Müller cells in lesioned areas and areas close to the lesion (*arrow*)

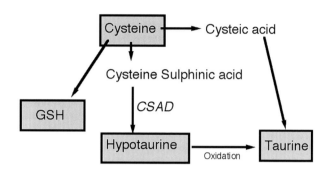

Fig. 8.9.
Taurine, hypotaurine and GSH share a common metabolic origin, in the form of cysteine. Thus there may be a metabolic interdependency, which regulates the relative levels of these molecules

the liver using the enzyme cysteine sulphinic acid decarbocylase (CSAD), some may be synthesized in the retina, at least in species such as rodents and humans. The localization of CSAD in the retina is unclear, but preliminary studies from our laboratory indicate its presence in Müller cells. Given the potential localization of both CSAD and the glutathione-synthesising enzymes in Müller cells, it is plausible that competition for available cysteine may create an interrelationship between the relative amounts of GSH and taurine that can be synthesised in the retina. However, no attempts have been made to date to correlate the relative levels of GSH and taurine under differing physiological conditions.

## 8.5.11
### Metallothionein

Metallothioneins are low molecular weight, zinc-binding proteins, with no disulphide bonds, that exist as four isoforms. Metallothioneins can bind zinc ions to 7 of their 20 cysteines, forming zinc-thiolate "clusters" (Maret 2003). Areas of the brain and retina that contain high levels of zinc typically express high levels of metallothioneins. The metallothioneins are thought to buffer or sequester zinc at presynaptic terminals; or buffer the excess zinc at synaptic junctions. The binding capacity of the metallothoneins is sensitive to the redox state of the environment; thus oxidants cause the release of zinc, whereas reductants enhance the capacity of the cysteine residues to bind zinc. Accordingly metallothioneins are capable of linking the redox state of the cell with zinc availability.

## 8.5.12
### Zinc, Copper, Selenium and Manganese

In discussing these metals, it is impossible to characterise one as being more important than another, because their functions are intimately interconnected. The use of copper, manganese and selenium as cofactors by the superoxide dismutases and glutathione peroxidase, respectively, automatically makes them essential metals for maintenance of normal oxidative function in the retina. Zinc has many divergent roles in the body due to its binding to literally thousand of proteins. Because of this critical position, the homeostasis of zinc is very tightly regulated and attempts to increase levels usually have limited effects on tissue levels in the absence of dysfunction of these homeostatic mechanisms. Conversely experimental dietary depletion of zinc is possible and has been shown to depress the b-wave of the ERG (Jacobson et al. 1986). In the context of AMD, it has been shown that retinas from cynamolgus monkeys with early-onset macular degeneration apparently contain four-fold less zinc than controls (Nicolas et al. 1996).

Maret (2003) has noted that eukaryotes tightly compartment their zinc pools, such that only nanomolar levels are biologically available, most zinc being sequestered by binding to the sulphhydryl groups in proteins, especially metallothioneins, and to the sulphydryl groups of other molecules such as GSH, a reaction which may have functional consequences for these molecules if they are unable to be oxidized to form disulphide bonds. While many of these interactions may be relatively benign, some interactions are likely to have significant biological consequences.

The distribution of zinc within the eye is thought to be somewhat heterogeneous; it is abundant in the inner segments of photoreceptors and the RPE, while little is present in ganglion cells (Akagi et al. 2001). It is important to note that these kinds of distribution maps are based on detection using methods such as Timm's method or zinc chelators such as the fluorimetric Zn probe FluoZin-3 (Kay 2003), which may only detect unbound zinc. Accordingly it is possible that this is not the full extent of zinc distribution on the retina.

High levels of zinc may be stored in synaptic vesicles (Haug 1967). It is often assumed that this zinc is associated with glutamatergic neurons (a nexus made by the assumption that all metallothionein-containing neurons are glutamatergic). However, recent studies have shown that high levels of free zinc may also be present in GABA- and glycine-containing neurons (Birinyi et al. 2001). Zinc can be released from zinc-containing neurons upon stimulation; thus extracellular zinc levels may be determined by retinal activity.

Zinc may have significant modulatory effects both intracellularly and extracellularly. In the extracellular environment, high levels of zinc may inhibit glutamate transport by Müller cells and photoreceptors (Spiridon et al. 1998). Lynch et al. (1998) have demonstrated that zinc modulates binding of glycine to the glycine receptor, while Han and Yang (1999) have suggested that zinc may modulate retinal GABA C receptors. In intracellular locations zinc may modulate the activity of a myriad or proteins. As an example, Permyakov et al. (2003) have shown that recoverin, a calcium-binding protein from retina, which modulates the $Ca^{2+}$-sensitive deactivation of rhodopsin, is a zinc-binding protein, and that zinc modulates its activity. Thus in this instance, zinc may directly influence phototransduction. Zinc deficiencies may directly influence retinal function possibly by reducing levels of retinol-binding protein in plasma and liver (which may limit the translocation of vitamin A from liver stores to the retina) and may also influence the stability of rhodopsin, though the mechanism for this is contentious (reviewed by Grahn et al. 2001).

While a deficiency of zinc may be detrimental, an excess is not always beneficial. Chen and Liao (2003) have demonstrated that, in tissue culture at least, exogenous zinc can be extremely neurotoxic, possibly as a result of chelating GSH. This toxic effect can be counteracted, in astrocyte cultures at least, by addition of GSH (Kim et al. 2003). Accordingly, while it is appropriate to attempt to maintain plasma zinc concentration within the normal physiological range, attempts to enhance or reduce this are likely to have negative biological consequences.

## 8.5.13
## Selenium Compounds

While selenium is critical for the function of glutathione peroxidases, Chen and Maret (2001) have demonstrated that selenium is also capable of reacting with the zinc-thiolate complexes in proteins such as metallothioneins, displacing the bound zinc. Thus selenium may directly influence both the zinc-binding capacity of these molecules and their biological functions.

## 8.5.14
## Manganese

This metal has not been extensively discussed in the literature, but its association with SOD-2 in mitochondria makes it an important metal in any analysis of oxidative function. While it is almost certainly vital for it to be present at low levels, it can in turn be very toxic when present at elevated levels. The recent observation that elevated levels of manganese have a profound inhibitory effect of glutamate transporter (GLAST) expression suggests that glutamate toxicity may represent the primary pathogenic mechanism in response to excessive manganese (Erikson and Aschner 2002).

## 8.5.15
### Macular Pigments

The yellow colouration of the macula derives from the presence of high levels of the carotenoids lutein and zeaxanthin. These lipid-soluble molecules are derived solely from the diet and are abundant in the plasma membranes of the photoreceptors. The presence of these pigments in the macula gives rise to several distinct interpretations. The simplest interpretation is that these molecules act as an optical filter, in much the same way that carotenoid-containing oil droplets filter specific parts of the light spectrum in the photoreceptors of species such as birds and some fishes. While birds may use these physical filters to facilitate colour vision, it is unlikely that this filtration of short wavelengths in the macula is for this purpose, due to the presence of short-, medium- and long-wavelength cones in the normal macula. A more likely possibility is that the macular pigment acts as a biological "sunscreen", absorbing the most energetic, shorter-wavelength photons, thus preventing them from interacting with other molecules to generate free radicals or other oxidizing species. Carotenoids are well suited to this role as most of the light energy that they absorb is radiated away as heat, with only limited formation of free radicals. Lutein and zeaxanthin can also function as lipid-phase antioxidants, quenching free radicals and singlet oxygen, thereby limiting free-radical propagation in lipid membranes (Schalch 1992).

## 8.5.16
### Ascorbate and Recycling of Carotenoids

It is well known that plants use ascorbate to reduce carotenoids that have been oxidized by intense light conditions. (Muller-Moule et al. 2002). This relationship has not been studied extensively in mammals, but it is clear from studies of liposomes containing carotenoids (in the lipid phase) and ascorbate and GSH in the aqueous phase that the presence of ascorbate enhances the antioxidant actions of the carotenoids. These data suggest that ascorbate functions as an acceptor molecule, thereby regenerating the carotenoids, possibly by a mechanism similar to that used to regenerate vitamin E.

## 8.5.17
### A Summary of Mechanisms for Antioxidant Protection

It is clear that any discussion of antioxidants has to accommodate the multiple interactions that different antioxidants have in each cellular compartment. Perhaps the most significant feature of these interactions is the partitioning of many antioxidant pathways in glial cells, rather than those cells which are likely to be the source of most oxidant species, namely the neurons (Fig. 8.10).

## 8.6
### Cellular Interactions and Photoreceptor Death

In AMD, it has been shown that initial photoreceptor losses occur in the perifoveal rods prior to cone degeneration (Curcio et al. 1993). Similar results have been obtained with a variety of different mouse and rat models where abnormal rod genes such as rhodopsin cause not only the death of rods, but also the death of cones (John et al. 2000). Experiments with chimeric mice (Huang et al. 1993; Kedzierski et al. 1998) have demonstrated that cell-to-cell interactions play a critical role in retinal degenerations and indicate that the functional integrity of a cell such as a rod photoreceptor will directly influence the survival of neighbouring cells such as cone photoreceptors, as well as other retinal cells (Wong 1990).

The notion that rod photoreceptors are necessary for the survival of cones (and perhaps other retinal cells) is important, since there is a significant loss of rod photoreceptor cells, without concomitant cone-cell loss, during "normal aging". It is plausible that there is a threshold number of rod photoreceptors needed to sup-

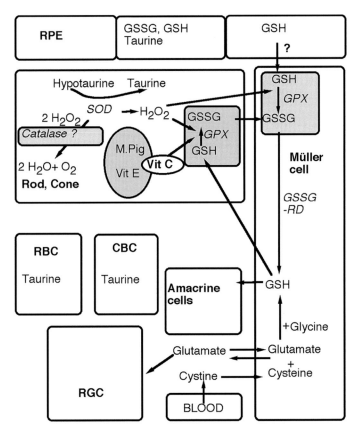

Fig. 8.10. The antioxidant network in the retina is present in multiple cell types. The glutathione system is centred on the Müller cells, while enzymes such as SOD are in neurons. Taurine is present in both neurons and glial cells. GSH released from glial cells can act as a reductant in neurons. Alternatively ROS generated by neurons may be accumulated by the Müller glial cells and detoxified via the GSH system. Ascorbate forms an intermediate system, which links the reduction of lipid phase molecules such as vitamin E to the GSH-based aqueous antioxidant system. Some GSH may be derived from the RPE. Hypotaurine serves as an intrinsic antioxidant, which may protect SOD, and other proteins, especially in photoreceptors. Oxidation of hypotaurine yields taurine, which may also serve as an antioxidant. Any dysfunction in the transport system for accumulating glutamate may lead to elevated levels of extracellular glutamate and thus excitotoxicity, especially in cells with high levels of expression of NMDA receptors such as the retinal ganglion cells (*RGC*). (*CBC* cone bipolar cell, *RBC* rod bipolar cells)

port cones, and that in AMD the loss of rods exceeds a critical threshold in the macula (Curcio et al. 1993, 1996).

## 8.6.1
### Light-Mediated Damage: Experimental Lesions and AMD

Young (1988) originally proposed that light, especially in the blue-to-ultraviolet part of the spectrum is a significant mediator of damage in AMD. Clearly light, especially short-wavelength, high-energy light, has the capacity to cause damage to the retina, via oxidative- and free radical-mediated mechanisms.

Numerous experiments have attempted to recreate the damage seen in AMD, either in short-term experiments that employ short exposures of bright light or constant light exposure or conversely in longer-term experiments, to use animals such as albino rats, which, by virtue of a lack of pigmentation, are very sensitive to normal ambient light levels, but where the retinas degenerate slowly (Sullivan et al. 2003).

A key feature of both slow and fast degenerative paradigms is that rod photoreceptors are lost first (as is the case in AMD), with delayed loss of cone photoreceptors (La Vail 1976). We suggest this initial loss of rod photoreceptors may in turn lead to early loss of other downstream cellular elements in rod pathways, including rod bipolar cells; whereas cells in the cone pathways may be lost later (Fig. 8.11).

## 8.6.2
## Mechanisms of Cell Death

In a variety of model systems of photoreceptor degeneration, such as the RCS rat, photoreceptor cell death is predominantly mediated via the apoptotic pathway (Tso et al. 1994). Similarly, Cai et al. (2000) have demonstrated that RPE cells, when exposed to oxidizing species, die by an apoptotic mechanism, as do Muller glia (Giardino et al. 1998). Photoreceptor death in response to light-induced damage is similarly thought to be mediated via apoptosis. While studies of AMD retinae have been limited, preliminary evidence based on tunnel staining indicates that in AMD, RPE, photoreceptors and inner nuclear layer cells die by apoptosis (Dunaief et al. 2002).

## 8.6.3
## Cell Death in Response to Oxidative Damage

In response to oxidizing damage, the retina normally responds by switching on the expression of DNA repair systems, especially in photoreceptors (Gordon et al. 2002). While these repair mechanisms suffice in most normal situations, they are frequently overwhelmed in experimental light-damage models, leading to photoreceptor death. Mitochondrial DNA is thought to be relatively resistant to free radical-

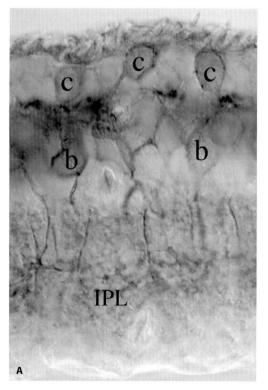

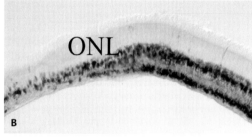

Fig. 8.11A, B. Albino rat retinas exposed to 2 h of bright white light (5,000 lx) and then left in normal ambient cyclical light conditions for 2 weeks. Degeneration is most evident in the central retina, and less conspicuously in the peripheral retina. (A) The antibody marker used in this study (against a glutamate transporter, GLT-1B), demonstrates the retention of cone photoreceptors (*c*) despite the loss of most rod photoreceptors in the central retina. The cone bipolar cells (*b*), which are also labeled, appear to maintain their normal organization in the inner plexiform layer (*IPL*) at this point in time. (B) Immunolabelling for PKC, a rod bipolar cell marker. Reorganization of the rod bipolar cells occurs at an early stage in response to the degeneration of the outer nuclear layer (*ONL*), possibly indicating the early loss of rod photoreceptor inputs to these cells

induced damage (Soong et al. 1996), possibly because of effective free radical defences within the mitochondria. However, as the highest concentrations of mitochondria are in photoreceptor inner segments, it has been proposed that mitochondrial loss or dysfunction may influence their capacity to generate energy and thus lead to impaired photoreceptor function (Liang and Godley 2003). The sequential nature of rod and then cone death may be linked not only to issues of trophic support but may also include a component reflecting simple anatomical features such as the more restricted size of the cone photoreceptor outer segments relative to those of rod photoreceptors, and thus the relative amount of damage due to light-induced formation of free radicals and oxidative species may be less in cones, compared with rods, in response to exposure to a similar amount of light.

by Müller cells in response to insults such as ischaemia or physical trauma (Barnett and Osborne 1995; Kim et al. 1998; Wu et al. 2003). Intriguingly, in aging there is often a modest upregulation of GFAP in Müller cells, perhaps suggestive of low-level physiological stress (Wu et al. 2003). While most observers have restricted their analysis of the expression GFAP in AMD to the central parts of the retina, a more extensive analysis reveals that GFAP may be upregulated over the entire retina. This suggests that the stressor acting on the macula may actually be acting over the entire retina. Intriguingly, Ramirez et al. (2001) have argued on the basis of GFAP labeling and ultrastrucutral studies that in AMD retinae there may be vascular deficits, especially in the inner retina, which may impact on ganglion cell viability (Fig. 8.12).

## 8.7
## A Role for Glial cells in AMD?

While the interactions of rod and cone photoreceptors that govern their survival are generally assumed to be direct, it must be acknowledged that intervening between most rod and cone photoreceptors is a lining of Müller glial cells. Accordingly it is plausible that rod-related trophic effects may in fact be mediated indirectly via the Müller cells

If this is correct, then any anomalies in the Müller cell population could lead to changes in cone photoreceptor survival. Nishikawa and Tamai (2001) have shown that within the fovea the ratio of Müller cells to neurons is lower than elsewhere in the retina and is associated with lower levels of GLAST and glutamine synthetase, both proteins being critical for normal glutamate homeostasis (Pow and Robinson 1994; Pow and Barnett 1999). As the glial cells provide a raft of metabolic support functions, it is plausible that in the macula the limited glial support may be deficient at points of high metabolic stress, including oxidative stress. It is well known that the Müller cells in the macular region appear to be stressed in AMD, as illustrated by the expression of the cytoskeletal marker GFAP, which is normally only expressed

### 8.7.1
### Metabolic Functions of Müller Cells in the AMD Retina

It is difficult to measure the metabolic activities of enzymes in postmortem retinas, but some markers such as neurotransmitter transporters are extremely stable. Recent studies in our laboratories indicate the presence of several distinct but interrelated perturbations in glutamate transporter expression. The expression of GLAST, the dominant Müller cell glutamate transporter, appears to be significantly down-regulated across the entire retina in AMD, an observation which accords with the suggestion that these cells are stressed (as evidenced by GFAP expression; Fig. 8.13).

In concert with the changes observed in retinal Müller cells, examination of control and AMD retinae reveal that the glutamate transporter GLT-1, which is normally expressed only by photoreceptors, is concomitantly down-regulated in the photoreceptors immediately lateral to the lesion site. Curiously, this same splice-variant (probably GLT-1C), which, in the normal human retina is only present in photoreceptors, is also expressed in the retinal ganglion cells across the entirety of the retina in all AMD retinas that we have studied to date (Fig. 8.14).

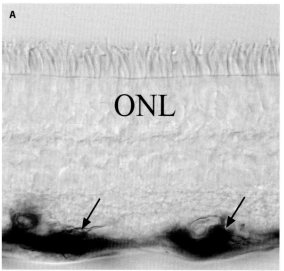
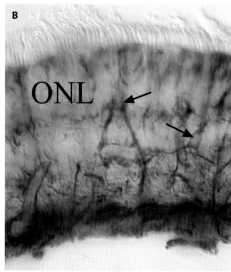

Fig. 8.12A, B. GFAP in mid-peripheral retina of a normal aged human (A) and a retina with AMD. In normal retinae limited GFAP labeling is restricted to astrocytes around blood vessels in the nerve fibre layer (*arrows*), whereas in AMD (B), labeling is frequently also associated with the Müller cells (*arrows*). The outer nuclear layer (*ONL*) is indicated for orientation purposes

Fig. 8.13A, B. Expression of the Müller cell glutamate transporter in mid-peripheral retina from a control eye (A) and an eye with AMD (B). Immunolabelling appears to be consistently weaker in eyes exhibiting symptoms of AMD

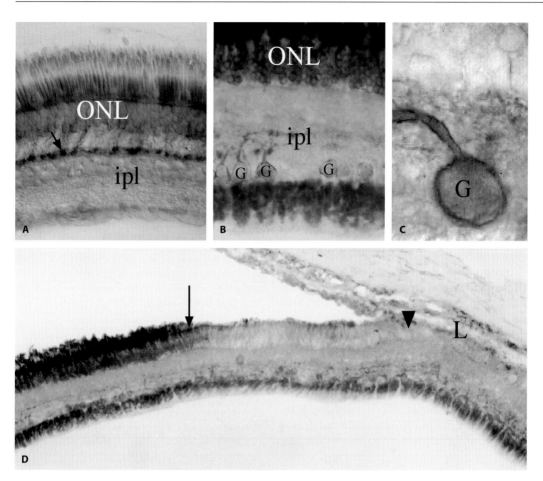

Fig. 8.14A–D. Human retinas labeled with an antibody against an epitope of the glutamate transporter GLT-1 (probably detecting GLT-1C), which labels only the photoreceptors (*arrows*) in control eyes (A), but additionally labels ganglion cells (*G*) in eyes with AMD (B). High-magnification views demonstrate this transporter is localized to the plasma membranes of the ganglion cells and is thus likely to be able to function as a plasmalemmal transporter (C). (D) At the edge of the lesion (*L*) in AMD retinas, there is an area where photoreceptors are present (bounded by an *arrow* and an *arrowhead*), where the photoreceptors switch off their expression of this glutamate transporter, suggesting that these perilesion photoreceptors may not be able to recover glutamate for metabolic or neurotransmitter purposes. Failure to gain energy from accumulated glutamate may result in energetic starvation and subsequent death of these photoreceptors, allowing the lesion to expand

The collective changes that we describe in the glutamate transport system suggest that there are significant imbalances in the homeostasis of glutamate in the AMD retina. We suggest that the novel expression of GLT-1 in the retinal ganglion cells (which has never before been described) may be induced in response to elevated extracellular glutamate levels concomitant with the down-regulation of GLAST. This perturbation of glutamate homeostasis may also explain why retinal ganglion cells have been observed to die in AMD.

The down-regulation of GLAST in the Müller cells has several immediate implications. Reduced GLAST expression will elevate extra-cellular levels of glutamate. Elevated extra-cellular glutamate will plausibly drive the expression of glutamate transporters in the retinal ganglion cells, as a protective mechanism for these cells. Deficiencies in GLAST are also likely to have an effect on GSH production, since Müller cell GSH levels are dependent on maintenance of glutamate transport both out of the cell via the CGAP and then back in via GLAST. Reduced GLAST expression is likely to lead to a reduction in GSH production, especially in areas such as the macula, which have a low glia-to-neuron ratio.

The significance of the down-regulation of the glutamate transporters in the photoreceptors immediately lateral to the lesion sites is less clear. We note that GLT-1 is very prone to oxidative damage and, in our studies on oxidative damage in the brain, we note the rapid internalization and degradation of GLT-1 that has been oxidatively damaged. We suggest that oxidative damage in the perimacular region damages GLT-1, causing its targeting to a degradative pathway. As glutamate accumulated by the photoreceptors using GLT-1 is likely to be fed into the Kreb's cycle to produce energy, any loss of expression would result in an energetic deficit for these photoreceptors, leading to their death. This may be compounded by the well-characterized changes that are evident if the choriocapillaris (Kornzweig et al. 1977), which may reduce blood flow and thus oxygen and energy supply to the photoreceptors. This would provide a mechanism for the expansion of the lesion.

## 8.7.2
### Cell Death and Glutamate Toxicity

We suggest that a final pathway in cell death in the inner retina in AMD may be mediated via glutamate toxicity. Cell death in response to over-activation of glutamate receptors such as the NMDA receptor is usually via an apoptotic mechanism. We suggest that the death of inner retinal neurons in AMD via apoptotic mechanisms (Dunaief et al. 2002) is compatible with the notion that glutamate may cause this death.

## 8.8
### Conclusions

Clearly much is still opaque in the area of AMD research, and the antioxidant hypothesis remains a hypothesis rather than a proven construct. However, many facets of the hypothesis are compatible with experimental and empirical observations. Experimental photoreceptor-degeneration paradigms, including mutations or manipulations such as light exposure or taurine deprivation, cause the sequential degeneration of rod and then cone photoreceptors as is typically observed in AMD. In some cases they may exhibit a centre-periphery gradient, especially in animal models such as the cat, where a large photoreceptor density gradient exists. We suggest that many of these paradigms are characterized by the potential for oxidative stress at some point in the damage process, and that death usually ensues via apoptosis. Also, antioxidant status of the retina may be critical in AMD and it is probable that the web of interactions of antioxidants and oxidant mechanisms makes it essential to study antioxidant systems in a holistic manner. We propose that Müller glial cells and GSH/GSSG status are the pivotal elements in determining overall antioxidant function in the retina. It is also probable that dysfunction of Müller glial cell processes such as the transport of glutamate (which may initially be damaged by oxidizing mechanisms) will have deleterious effects on antioxidant production and may also lead to cell death either via glutamate excitotoxicity, in the case of ganglion cells, or by energy privation, in the case of photoreceptors. Clearly the functions and dysfunctions of glial cells in the AMD retina will provide a rich vein of research in the next few years.

# References

Akagi T, Kaneda M, Ishii K, Hashikawa T (2001) Differential subcellular localization of zinc in the rat retina. J Histochem Cytochem 49:87–96

Aliev G, Seyidova D, Lamb BT, Obrenovich ME, Siedlak SL, Vinters HV, Friedland RP, LaManna JC, Smith MA, Perry G (2003) Mitochondria and vascular lesions as a central target for the development of Alzheimer's disease and Alzheimer disease-like pathology in transgenic mice. Neurol Res 25:665–674

Armstrong D, Santangelo G, Connole E (1981) The distribution of peroxide regulating enzymes in the canine eye. Curr Eye Res 1:225–242

Atalla L, Fernandez MA, Rao NA (1987) Immunohistochemical localization of catalase in ocular tissue. Curr Eye Res 6:1181–1187

Bannai S (1986) Exchange of cystine and glutamate across plasma membrane of human fibroblasts. J Biol Chem 261:2256–2263

Barnett NL, Osborne NN (1995) Prolonged bilateral carotid artery occlusion induces electrophysiological and immunohistochemical changes to the rat retina without causing histological damage. Exp Eye Res 61:83–90

Bastianetto S (2002) Red wine consumption and brain aging. Nutrition 18:432–433

Bringmann A, Reichenbach A (2001) Role of Muller cells in retinal degenerations. Front Biosci 6:E72–92

Birinyi A, Parker D, Antal M, Shupliakov O (2001) Zinc co-localizes with GABA and glycine in synapses in the lamprey spinal cord. J Comp Neurol 433:208–221

Buettner GR (1993) The pecking order of free radicals and antioxidants: lipid peroxidation, alpha-tocopherol, and ascorbate. Arch Biochem Biophys 300:535–543

Buettner GR, Jurkiewicz BA (1996) Catalytic metals, ascorbate and free radicals: combinations to avoid. Radiat Res 145:532–541

Burke A, Fitzgerald GA (2003) Oxidative stress and smoking-induced vascular injury. Prog Cardiovasc Dis 46:79–90

Bush AI, Goldstein LE (2001) Specific metal-catalysed protein oxidation reactions in chronic degenerative disorders of aging: focus on Alzheimer's disease and age-related cataracts. Novartis Found Symp 235:26–38; 38–43

Cai J, Nelson KC, Wu M, Sternberg P Jr, Jones DP (2000) Oxidative damage and protection of the RPE. Prog Retin Eye Res 19:205–221

Chen CJ, Liao SL (2003) Zinc toxicity on neonatal cortical neurons: involvement of glutathione chelation. J Neurochem 85:443–453

Chen Y, Maret W (2001) Catalytic oxidation of zinc/sulfur coordination sites in proteins by selenium compounds. Antioxid Redox Signal 3:651–656

Cho E, Stampfer MJ, Seddon JM, Hung S, Spiegelman D, Rimm EB, Willett WC, Hankinson SE (2001) Prospective study of zinc intake and the risk of age-related macular degeneration. Ann Epidemiol 11:328–336

Cohen SM, Olin KL, Feuer WJ, Hjelmeland L, Keen CL, Morse LS (1994) Low glutathione reductase and peroxidase activity in age-related macular degeneration. Br J Ophthalmol 78:791–794

Crabb JW, Miyagi M, Gu X, Shadrach K, West KA, Sakaguchi H, Kamei M, Hasan A, Yan L, Rayborn ME, Salomon RG, Hollyfield JG (2002) Drusen proteome analysis: an approach to the etiology of age-related macular degeneration. Proc Natl Acad Sci USA 99:14682–14687

Curcio CA, Millican CL, Allen KA, Kalina RE (1993) Aging of the human photoreceptor mosaic: evidence for selective vulnerability of rods in central retina. Invest Ophthalmol Vis Sci 34:3278–3296

Curcio CA, Medeiros NE, Millican CL (1996) Photoreceptor loss in age-related macular degeneration. Invest Ophthalmol Vis Sci. 37:1236–1249

Delcourt C, Cristol JP, Leger CL, Descomps B, Papoz L (1999a) Associations of antioxidant enzymes with cataract and age-related macular degeneration. The POLA Study. Pathologies Oculaires Liees a l'Age. Ophthalmology 106:215–222

Delcourt C, Cristol JP, Tessier F, Leger CL, Descomps B, Papoz L (1999b) Age-related macular degeneration and antioxidant status in the POLA study. POLA Study Group. Pathologies Oculaires Liees a l'Age. Arch Ophthalmol 117:1384–1390

Derouiche A, Rauen T (1995) Coincidence of L-glutamate/L-aspartate transporter (GLAST) and glutamine synthetase (GS) immunoreactions in retinal glia: evidence for coupling of GLAST and GS in transmitter clearance. J Neurosci Res 42:131–143

Dunaief JL, Dentchev T, Ying GS, Milam AH (2002) The role of apoptosis in age-related macular degeneration. Arch Ophthalmol 120:1435–1442

Ebers GM, Stern L (1875) Papyrus Ebers. Facsimile with a partial translation. 2 vol

Erikson K, Aschner M (2002) Manganese causes differential regulation of glutamate transporter (GLAST), Taurine transporter and metallothionein in cultured rat astrocytes. Neurotoxicology 23:595–602

Evans JR (2002) Antioxidant vitamin and mineral supplements for age-related macular degeneration. Cochrane Database of Systematic Reviews (2): CD000254

Gelder NM van (1983) A central mechanism of action for taurine: osmoregulation, bivalent cations, and excitation threshold. Neurochem Res 8:687–699

Giardino I, Fard AK, Hatchell DL, Brownlee M (1998) Aminoguanidine inhibits reactive oxygen species formation, lipid peroxidation, and oxidant-induced apoptosis. Diabetes 47:1114–1120

Goldman SS (1990) Evidence that the gluconeogenic pathway is confined to an enriched Müller cell fraction derived from the amphibian retina. Exp Eye Res 50:213–218

Gordon WC, Casey DM, Lukiw WJ, Bazan NG (2002) DNA damage and repair in light-induced photoreceptor degeneration. Invest Ophthalmol Vis Sci 43:3511–3521

Goti D, Hammer A, Galla HJ, Malle E, Sattler W (2000). Uptake of lipoprotein-associated alpha-tocopherol by primary porcine brain capillary endothelial cells. J Neurochem 74:1374–1383

Gottschall-Pass KT, Grahn BH, Gorecki DK, Paterson PG (1997) Oscillatory potentials and light microscopic changes demonstrate an interaction between zinc and taurine in the developing rat retina. J Nutr 127:1206–1213

Grahn BH, Paterson PG, Gottschall-Pass KT, Zhang Z (2001) Zinc and the eye. J Am Coll Nutr 20:106–118

Han MH, Yang XL (1999) $Zn^{2+}$ differentially modulates kinetics of GABA(C) versus GABA(A) receptors in carp retinal bipolar cells. Neuroreport 10:2593–2597

Haug FM (1967) Electron microscopical localization of the zinc in hippocampal, mossy fiber synapses by a modified sulfide silver procedure. Histochemie 8:355–368

Hayes KC, Carey RE, Schmidt SY (1975) Retinal degeneration associated with taurine deficiency in the cat. Science 188:949–951

Hirata A, Kitaoka T, Ishigooka H, Ueno S (1991) Combined cytochemical detection of Müller-cell-specific enzyme activity and permeability tracers. Ophthalmologica 202:94–99

Huang PC, Gaitan AE, Hao Y, Petters RM, Wong F (1993) Cellular interactions implicated in the mechanism of photoreceptor degeneration in transgenic mice expressing a mutant rhodopsin gene. Proc Natl Acad Sci USA 90:8484–8488

Huster D, Reichenbach A, Reichelt W (2000) The glutathione content of retinal Müller (glial) cells: effect of pathological conditions. Neurochem Int 36:461–469

Hyman L, Neborsky R (2002) Risk factors for age-related macular degeneration: an update. Curr Opin Ophthalmol 13:171–175

Jacobson SG, Meadows NJ, Keeling PW, Mitchell WD, Thompson RP (1986) Rod mediated retinal dysfunction in cats with zinc depletion: comparison with taurine depletion. Clin Sci (Lond) 71:559–564

John SK, Smith JE, Aguirre GD, Milam AH (2000) Loss of cone molecular markers in rhodopsin-mutant human retinas with retinitis pigmentosa. Mol Vis 6:204–215

Kato S, Ishita S, Sugawara K, Mawatari K (1993) Cystine/glutamate antiporter expression in retinal Müller glial cells: implications for DL-alpha-aminoadipate toxicity. Neuroscience 57:473–482

Kannan R, Bao Y, Wang Y, Sarthy VP, Kaplowitz N (1999) Protection from oxidant injury by sodium-dependent GSH uptake in retinal Müller cells. Exp Eye Res 68:609–616

Kay AR (2003) Evidence for chelatable zinc in the extracellular space of the hippocampus, but little evidence for synaptic release of Zn. J Neurosci 23:6847–6855

Kedzierski W, Bok D, Travis GH (1998) Non-cell-autonomous photoreceptor degeneration in rds mutant mice mosaic for expression of a rescue transgene. J Neurosci 18:4076–4082

Kelly FJ (1998) Use of antioxidants in the prevention and treatment of disease. J Int Fed Clin Chem 10:21–23

Kim D, Joe CO, Han PL (2003) Extracellular and intracellular glutathione protects astrocytes from $Zn^{2+}$-induced cell death. Neuroreport 14:187–190

Kim IB, Kim KY, Joo CK, Lee MY, Oh SJ, Chung JW, Chun MH (1998) Reaction of Müller cells after increased intraocular pressure in the rat retina. Exp Brain Res 121:419–424

Kornzweig AL (1977) Changes in the choriocapillaris associated with senile macular degeneration. Ann Ophthalmol 9:753–756, 759–762

Kulms D, Schwarz T (2002) Molecular mechanisms involved in UV-induced apoptotic cell death. Skin Pharmacol Appl Skin Physiol 15:342–347

La Vail MM (1976) Survival of some photoreceptor cells in albino rats following long-term exposure to continuous light. Invest Ophthalmol Vis Sci 15:64–70

Leon A, Levick WR, Sarossy MG (1995) Lesion topography and new histological features in feline taurine deficiency retinopathy. Exp Eye Res 161:731–741

Liang FQ, Godley BF (2003) Oxidative stress-induced mitochondrial DNA damage in human retinal pigment epithelial cells: a possible mechanism for RPE aging and age-related macular degeneration. Exp Eye Res 76:397–403

Liu RM (2002) Down-regulation of gamma-glutamylcysteine synthetase regulatory subunit gene expression in rat brain tissue during aging. J Neurosci Res 68:344–351

Lynch JW, Jacques P, Pierce KD, Schofield PR (1998) Zinc potentiation of the glycine receptor chloride channel is mediated by allosteric pathways. J Neurochem 71:2159–2168

Makar TK, Nedergaard M, Preuss A, Gelbard AS, Perumal AS, Cooper AJ (1994) Vitamin E, ascorbate, glutathione, glutathione disulfide, and enzymes of glutathione metabolism in cultures of chick astrocytes and neurons: evidence that astrocytes play an important role in antioxidative processes in the brain. J Neurochem 62:45–53

Maret W (2003) Cellular zinc and redox states converge in the metallothionein/thionein pair. J Nutr 133:1460S–1462S

Meister A (1983) Selective modification of glutathione metabolism. Science 220:472–477

Mittag TW, Bayer AU, La Vail MM (1999) Light-induced retinal damage in mice carrying a mutated SOD I gene. Exp Eye Res 69:677–683

Muller-Moule P, Conklin PL, Niyogi KK (2002) Ascorbate deficiency can limit violaxanthin de-epoxidase activity in vivo. Plant Physiol 128:970–977

Newsome DA, Swartz M, Leone NC, Elston RC, Miller E (1988) Oral zinc in macular degeneration. Arch Ophthalmol 106:192–198

Nicolas MG, Fujiki K, Murayama K, Suzuki MT, Shindo N, Hotta Y, Iwata F, Fujimura T, Yoshikawa Y, Cho F, Kanai A (1996) Studies on the mechanism of early onset macular degeneration in cynomolgus monkeys. II. Suppression of metallothionein synthesis in the retina in oxidative stress. Exp Eye Res 62:399–408

Nishikawa S, Tamai M (2001) Müller cells in the human foveal region. Curr Eye Res 22:34–41

Nowak M, Swietochowska E, Wielkoszynski T, Marek B, Karpe J, Gorski J, Glogowska-Szelag J, Kos-Kudla B, Ostrowska Z (2003) Changes in blood antioxidants and several lipid peroxidation products in women with age-related macular degeneration. Eur J Ophthalmol 13:281–286

Ogawa T, Ohira A, Amemiya T, Kubo N, Sato H (2001) Superoxide dismutase in senescence-accelerated mouse retina. Histochem J 33:43–50

Ohia SE, Bagchi M, Stohs SJ (1994) Age-related oxidative damage in Long-Evans rat retina. Res Commun Mol Pathol Pharmacol 85:21–31

Ohira A, Tanito M, Kaidzu S, Kondo T (2003) Glutathione peroxidase induced in rat retinas to counteract photic injury. Invest Ophthalmol Vis Sci 44:1230–1236

Pasantes-Morales H, Cruz C (1985) Taurine and hypotaurine inhibit light-induced lipid peroxidation and protect rod outer segment structure. Brain Res 1330:154–157

Pasantes-Morales H, Dominguez L, Campomanes MA, Pacheco P (1986) Retinal degeneration induced by taurine deficiency in light-deprived cats. Exp Eye Res 43:55–60

Pecci L, Montefoschi G, Fontana M, Dupre S, Costa M, Cavallini D (2000) Hypotaurine and superoxide dismutase: protection of the enzyme against inactivation by hydrogen peroxide and peroxidation to taurine. Adv Exp Med Biol 483:163–168

Permyakov SE, Cherskaya AM, Wasserman LA, Khokhlova TI, Senin II, Zargarov AA, Zinchenko DV, Zernii EY, Lipkin VM, Philippov PP, Uversky VN, Permyakov EA (2003) Recoverin is a zinc-binding protein. J Proteome Res 2:51–57

Pfeiffer B, Grosche J, Reichenbach A, Hamprecht B (1994) Immunocytochemical demonstration of glycogen phosphorylase in Müller (glial) cells of the mammalian retina. Glia 12:62–67

Pow DV (2001) Visualising the activity of the cystine-glutamate antiporter in glial cells using antibodies to aminoadipic acid, a selectively transported substrate. Glia 34:27–38

Pow DV, Barnett NL (1999) Changing patterns of spatial buffering of glutamate in developing rat retinae are mediated by the Müller cell glutamate transporter GLAST. Cell Tissue Res 297:57–66

Pow DV, Crook DK (1995) Immunocytochemical evidence for the presence of high levels of reduced glutathione in radial glial cells and horizontal cells in the rabbit retina. Neurosci Lett 193:25–28

Pow DV, Robinson SR (1994) Glutamate in some retinal neurons is derived solely from glia. Neuroscience 60:355–366

Pow DV, Sullivan R, Reye P, Hermanussen S (2002) Localization of taurine transporters, taurine, and (3)H taurine accumulation in the rat retina, pituitary, and brain. Glia 37:153–168

Ramirez JM, Ramirez AI, Salazar JJ, Hoz R de, Trivino A (2001) Changes of astrocytes in retinal aging and age-related macular degeneration. Exp Eye Res 73:601–615

Reichenbach A, Stolzenburg JU, Eberhardt W, Chao TI, Dettmer D, Hertz L (1993) What do retinal Müller (glial) cells do for their neuronal 'small siblings'? J Chem Neuroanat 6:201–213

Reid GM, Tervit H (1999) Sudden infant death syndrome: oxidative stress. Med Hypotheses 52:577–580

Rothstein JD, Bristol LA, Hosler B, Brown RH Jr, Kuncl RW (1994) Chronic inhibition of superoxide dismutase produces apoptotic death of spinal neurons. Proc Natl Acad Sci USA 91:4155–4159

Samiec PS, Drews-Botsch C, Flagg EW, Kurtz JC, Sternberg P Jr, Reed RL, Jones DP (1998) Glutathione in human plasma: decline in association with aging, age-related macular degeneration, and diabetes. Free Radic Biol Med 24:699–704

Sandbach JM, Coscun PE, Grossniklaus HE, Kokoszka JE, Newman NJ, Wallace DC (2001) Ocular pathology in mitochondrial superoxide dismutase (SOD2)-deficient mice. Invest Ophthalmol Vis Sci 42:2173–2178

Sato H, Tamba M, Ishii T, Bannai S (1999) Cloning and expression of a plasma membrane cystine/glutamate exchange transporter composed of two distinct proteins. J Biol Chem 274:11 455–11 458

Schalch W (1992) Carotenoids in the retina – a review of their possible role in preventing or limiting damage caused by light and oxygen. EXS 62:280–298

Schmidt SY, Berson EL, Hayes KC (1976) Retinal degeneration in cats fed casein. I. Taurine deficiency. Invest Ophthalmol 15:47–52

Schutte M, Werner P (1998) Redistribution of glutathione in the ischemic rat retina. Neurosci Lett 246:53–56

Schweitzer D, Thamm E, Hammer M, Kraft J (2001) A new method for the measurement of oxygen saturation at the human ocular fundus. Int Ophthalmol 23:347–353

Simpkins N, Jankovic J (2003) Neuroprotection in Parkinson disease. Arch Intern Med 163:1650–1654

Soltaninassab SR, Sekhar KR, Meredith MJ, Freeman ML (2000) Multi-faceted regulation of gamma-glutamylcysteine synthetase. J Cell Physiol 182:163–170

Soong NW, Dang MH, Hinton DR, Arnheim N (1996) Mitochondrial DNA deletions are rare in the free radical-rich retinal environment. Neurobiol Aging 17:827–831

Specht S, Leffak M, Darrow RM, Organisciak DT (1999) Damage to rat retinal DNA induced in vivo by visible light. Photochem Photobiol 69:91–98

Spiridon M, Kamm D, Billups B, Mobbs P, Attwell D (1998) Modulation by zinc of the glutamate transporters in glial cells and cones isolated from the tiger salamander retina. J Physiol 506:363–376

Stone J, Maslim J, Valter-Kocsi K, Mervin K, Bowers F, Chu Y, Barnett N, Provis J, Lewis G, Fisher SK, Bisti S, Gargini C, Cervetto L, Merin S, Peer J (1999) Mechanisms of photoreceptor death and survival in mammalian retina. Prog Retin Eye Res 18:689–735

Sturman JA, Wen GY, Wisniewski HM, Hayes KC (1981) Histochemical localization of zinc in the feline tapetum. Effect of taurine depletion. Histochemistry 72: 341–350

Sullivan R, Penfold P, Pow DV (2003) Neuronal migration and glial remodeling in degenerating retinas of aged rats and in nonneovascular AMD. Invest Ophthalmol Vis Sci 44:856–865

Takano T, Sada K, Yamamura H (2002) Role of protein-tyrosine kinase syk in oxidative stress signaling in B cells. Antioxid Redox Signal 4:533–541

Tsai MJ, Chang YF, Schwarcz R, Brookes N (1996) Characterization of L-alpha-aminoadipic acid transport in cultured rat astrocytes. Brain Res 741:166–173

Tso MO, Zhang C, Abler AS, Chang CJ, Wong F, Chang GQ, Lam TT (1994) Apoptosis leads to photoreceptor degeneration in inherited retinal dystrophy of RCS rats. Invest Ophthalmol Vis Sci 35:2693–2699

White AA, Crawford KM, Patt CS, Lad PJ (1976) Activation of soluble guanylate cyclase from rat lung by incubation or by hydrogen peroxide. J Biol Chem 251: 7304–7312

Winkler BS, Boulton ME, Gottsch JD, Sternberg P (1999) Oxidative damage and age-related macular degeneration. Mol Vis 5:32

Wong F (1990) Visual pigments, blue cone monochromasy, and retinitis pigmentosa. Arch Ophthalmol 108: 935–936

Wu KH, Madigan MC, Billson FA, Penfold PL (2003) Differential expression of GFAP in early v late AMD: a quantitative analysis. Br J Ophthalmol 87:1159–1166

Wu L, Cao XY, Chen Y, Wu DZ (1994) Metabolic disturbance in age-related macular degeneration. Metab Pediatr Syst Ophthalmol 17:38–40

Yant LJ, Ran Q, Rao L, Van Remmen H, Shibatani T, Belter JG, Motta L, Richardson A, Prolla TA (2003) The selenoprotein GPX4 is essential for mouse development and protects from radiation and oxidative damage insults. Free Radic Biol Med 34:496–502

Yokota T, Uchihara T, Kumagai J, Shiojiri T, Pang JJ, Arita M, Arai H, Hayashi M, Kiyosawa M, Okeda R, Mizusawa H (2000) Postmortem study of ataxia with retinitis pigmentosa by mutation of the alpha-tocopherol transfer protein gene. J Neurol Neurosurg Psychiatr 68:521–525

Yokota T, Igarashi K, Uchihara T, Jishage K, Tomita H, Inaba A, Li Y, Arita M, Suzuki H, Mizusawa H, Arai H (2001) Delayed-onset ataxia in mice lacking alpha-tocopherol transfer protein: model for neuronal degeneration caused by chronic oxidative stress. Proc Natl Acad Sci USA 98:15185–15190

Young RW (1988) Solar radiation and age-related macular degeneration. Surv Ophthalmol 32:252–269

Zelko IN, Mariani TJ, Folz RJ (2002) Superoxide dismutase multigene family: a comparison of the CuZn-SOD (SOD1), Mn-SOD (SOD2), and EC-SOD (SOD3) gene structures, evolution, and expression. Free Radic Biol Med 33:337–349

# Photoreceptor Stability and Degeneration in Mammalian Retina: Lessons from the Edge

Jonathan Stone, Kyle Mervin, Natalie Walsh, Krisztina Valter, Jan M. Provis, Philip L. Penfold

## Contents

9.1 Introduction  149
9.2 Approach  150
9.3 Stages in Photoreceptor Degeneration at the Edge of the Retina  150
9.3.1 Postnatal Development of the Edge: Site of Early Stress and Degeneration  150
9.3.2 The Edge of the Retina Is Functionally Degraded  154
9.3.3 The Edge of the Retina Is Highly Stable in the Face of Acute Stress  155
9.3.4 The Adult Retina Aged 2 Years: Observations in the Marmoset  155
9.3.5 The Adult Retina Aged 20 Years: Observations in the Baboon  155
9.3.6 The Adult Retina Aged 30–70 Years: Observations in the Human  158
9.3.7 Evidence of Progression in Edge Degeneration in Humans  161
9.4 Discussion  161
9.4.1 The Edge of the Retina as a Model of Retinal Degeneration  162
9.4.2 Why Is the Protection of Photoreceptors Stress-Inducible?  162
9.4.3 Vulnerability to Oxidative Stress in Mice and Humans  163
9.4.4 The Link to AMD  163

References  164

## 9.1 Introduction

Surveys of the human retina have reported the degeneration of the anterior edge of the retina to be a common feature (Vrabec 1967; Gartner 1975; Gartner and Henkind 1981; Ahnelt 1998), sufficiently common in older humans to be regarded as "normal". Detailed observations of mouse and rat retina have suggested that these degenerative phenomena begin very early in postnatal life (Mervin and Stone 2002a, 2002b), with a period of localized photoreceptor death. Conversely, in experimental studies of photoreceptor degeneration in adult rodents, the edge of the retina has been described as highly resistant to degeneration, whether light- or mutation-induced (LaVail 1981; Valter et al. 1998; Bowers et al. 2001).

This paradox, that a chronically degenerative region of retina is resistant to acute stress, has been reproduced in studies of "pre-conditioning" of the retina, in which limited damage to the retina (including some photoreceptor death) upregulates protective mechanisms which make the surviving photoreceptors resistant to subsequent stress (Liu et al. 1998; Nir et al. 1999; Cao et al. 2001). The same paradox is evident in the impact of ambient light levels on photoreceptor structure and vulnerability (Penn and Anderson 1991). Bright ambient light kills some photoreceptors and makes the survivors resistant to acute stress. Penn and Anderson's data make clear that degenerative and stabilizing properties appeared in the same cell, photoreceptors exposed to stress showing degenerative membrane changes, yet proving highly resistant to light challenge.

This chapter summarises results of studies of degenerative phenomena at the edge of the retina. Taken together, these observations suggest that the edge of the retina is subject throughout life to a localized stress, which induces a progressive degenerative process. These edge-specific changes are part of the life history of the normal retina, and form part of the baseline against which the retinal degenerations take place.

## 9.2
## Approach

It was of interest to study the edge of the retina from neonatal development to senescence. For practical reasons, these observations were collected from different species. Neonatal development was studied in rodents, while the most striking phenomena of senescence were apparently only in adult human retina, after several decades of normal function. Segments of human retina were obtained through the Eye Bank of the Save Sight Institute, University of Sydney. Material was successfully obtained from individuals aged 62, 47 and 29 years. The eyes had been immersion-fixed in paraformaldehyde for up to several years. Medical records in each case contained no evidence of ocular pathology. Segments of two marmoset retinas were obtained by courtesy of Dr. Paul Martin, from animals studied in unrelated physiological experiments. One baboon eye was obtained, from an individual aged 20 years killed because of multiple organ failure which did not involve the eyes. Rodent material is summarized from recent published studies (Maslim et al. 1997; Bowers et al. 2001; Walsh et al. 2001; Mervin and Stone 2002a).

## 9.3
## Stages in Photoreceptor Degeneration at the Edge of the Retina

### 9.3.1
### Postnatal Development of the Edge: Site of Early Stress and Degeneration

The most detailed evidence available of the onset of stressed conditions at the edge of the retina comes from the mouse retina (Mervin and Stone 2002a, 2002b), in which a clustering of photoreceptor death at the edge of the retina has been observed in early postnatal material (Fig. 9.1A–C). Evidence of the cause of this death comes from the observation that the site of the death (the extreme edge) is also the site of an upregulation of two stress-inducible proteins, the trophic factor FGF-2 (green in Fig. 9.1D, E) and the structural protein GFAP in the processes of Müller cells (red in Fig. 9.1D). The locus of this early episode of photoreceptor death is a narrow (<100 µm; Fig. 9.1H) strip at the extreme edge of the retina. The episode is also transient; by adulthood, the rate of photoreceptor death at the edge is not measurably higher than in the midperiphery. Nevertheless, the edge-specific upregulation of GFAP and FGF-2 persists into adulthood (Fig. 9.1F, G), suggesting that the stress persists after the rate of death slows. Finally, the upregulation of both FGF-2 and GFAP is maximal at the edge and falls towards the center of the retina (Fig. 9.1I), suggesting that the stress acts at the extreme edge, and reduces with distance from the edge.

Death of edge photoreceptors is evident in mouse retina from P14 (Fig. 9.1A). Upregulation of FGF-2 in edge photoreceptors is not evident at P14 (Fig. 9.2A), but is detected at P16 (Fig. 9.2B) and increased thereafter (Fig. 9.2C–F). It thus seems likely that the edge-specific stress causes photoreceptor death and then, with a 1- to 2-day delay, the upregulation of stress-inducible factors.

Insight into the nature of the edge-specific stress come from an analysis (Mervin and Stone 2002b) of the influence of hypoxia on photoreceptor death. In most of the retina, hypoxia induces photoreceptor-specific death. At the edge at P14–18 (when the episode of edge-specific photoreceptor death is maximal), hypoxia has the opposite effect, reducing death. Mervin and Stone have suggested that hyperoxic toxicity is a factor in edge-specific stress, the hyperoxia arising because of oxygen flowing to the edge from the choriocapillaris just beyond the edge of the retina.

In the rat, as in the mouse, the adult retina shows evidence of edge of stress, and is structurally and neurotrophically distinct. At the edge of the Sprague-Dawley rat retina for example, stress-inducible trophic factors CNTF and FGF-2 are upregulated, the CNTF princi-

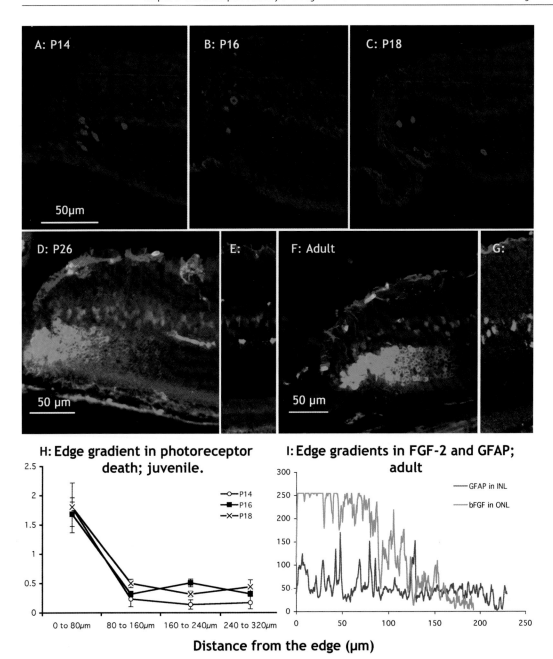

Fig. 9.1A–I. Evidence of stress at the edge of the neonatal C57BL/6J mouse retina. This strain is non-degenerative. A–C At the edge of the retina, TUNEL labelling demonstrates clustering of dying cells (red) at the extreme edge of the retina, and in the ONL. D–G By P26 (D, E) and in the adult (F, G), FGF-2 (green) is upregulated in the ONL at the edge (D, F), but not in mid-peripheral retina (E, G). GFAP (red) is present in all regions in astrocytes at the inner surface of the retina. At the edge, GFAP-labelling (red) is also seen in the radial processes of Müller cells. H Averaged over multiple sections of tissue, the cluster of dying cells is confined to the extreme edge of the retina, to a region <80 μm wide, at P14, P16 and P18. I Quantitation of the immunolabel signal in F confirms the visual impression that FGF-2-labelling is maximal at the edge and decreases towards the centre of the retina, following to low levels 200 μm from the edge. The *red* trace shows GFAP-labelling along a transect along the INL, crossing successive processes of Müller cells. GFAP-positive Müller cell processes show as peaks in the *red* line; they are confined to the most peripheral 150 μm of retina. (Data from Mervin and Stone 2002a)

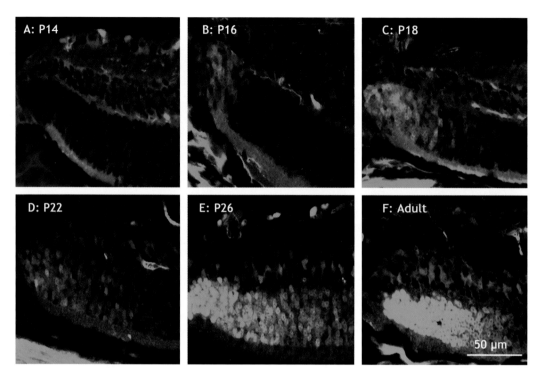

Fig. 9.2A–F. The development of the upregulation of FGF-2 in the ONL, in the developing C57BL/6J mouse retina. The upregulation can be detected at P16 and progresses over the following 10 days. The upregulation of FGF-2 thus seems to follow the occurrence of photoreceptor death at the edge, which is well developed at P14 (Fig. 9.1). (Data from Mervin and Stone 2002a)

Fig. 9.3A–F. Edge/midperiphery comparison for a rat retina (non-degenerative, Sprague-Dawley albino): A–D are from control retinas, raised in dim cyclic light; F, G are from a retina after exposure to bright continuous light. A, B FGF-2-immunolabelling in mid-peripheral retina is prominent only in Müller cell somas in the INL and in the nuclei of astrocytes at the inner surface (A). CNTF-labelling (*red*) is prominent only in astrocytes processes at the inner surface (A). At the edge (B), FGF-2 is prominent in the ONL and CNTF in Müller cells, cell processes extending across the retina. Note the overall thinning of the retina, including the ONL. C, D Immunolabelling (*red*) for CO (cytochrome oxidase) shows that in mid-peripheral retina (C) the enzyme is prominent in the inner segments (*lower two asterisks*) and in the OPL (*top asterisk*). Opsin (*green*) is prominent in outer segments (*arrow* in C). At the edge (D), the ONL has thinned to <50% of its normal thickness, and CO-labelling is less intense in both the outer segments (*lower asterisk*) and in the OPL (*upper asterisk*). The outer segments remain strongly opsin-positive but are markedly shortened (*arrow*). E, F Labelling for DNA (*blue*) and fragmenting DNA (*red*, using the TUNEL technique) in mid-peripheral and edge regions of a rat retina suffering light damage after exposure to bright continuous light (Bowers et al. 2001). The photoreceptors in the ONL undergo massive DNA fragmentation (*red*) in mid-peripheral retina (E), with some of the fragmented DNA appearing in Müller cells in the INL. The neat stacking of somas in the ONL, apparent in C, is broken down. At the edge (F), by contrast, the photoreceptors are light-damage resistant, only a few showing evidence of DNA fragmentation. The neat packing of ONL cells is maintained

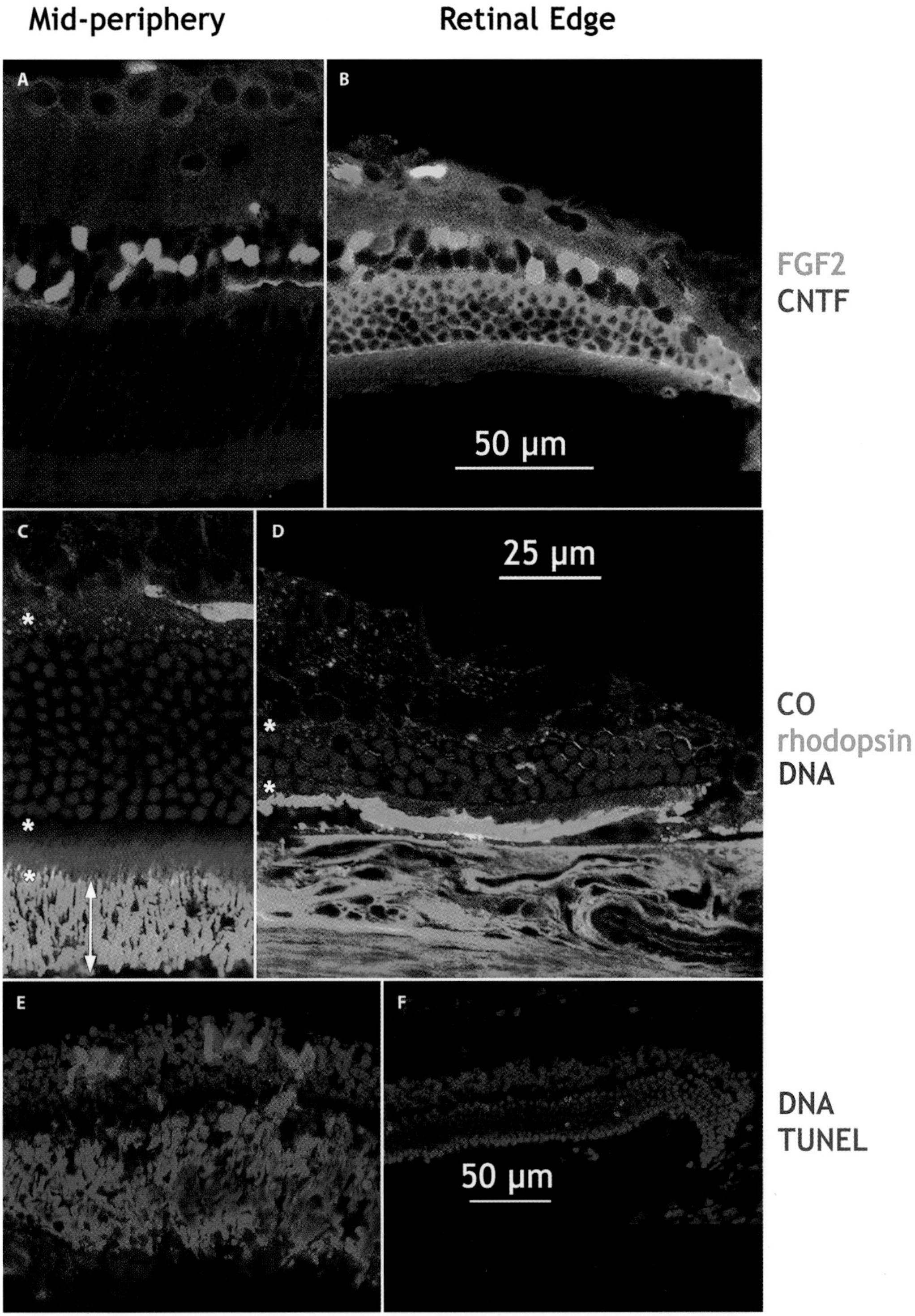

Chapter 9 Photoreceptor Stability and Degeneration in Mammalian Retina: Lessons from the Edge 153

**Mid-periphery** **Retinal Edge**

FGF2
CNTF

50 μm

25 μm

CO
rhodopsin
DNA

DNA
TUNEL

50 μm

pally in Müller cells, the FGF-2 in photoreceptor somas in the ONL (Walsh et al. 2001; Fig. 9.3A, B). The ONL at the edge thins to <50% of its thickness in central retina (compare Fig. 9.3A with B, C with D), suggesting significant cumulative loss of photoreceptors. The thinning of layers at the edge, and the upregulation of FGF-2 and CNTF, are shown quantitatively in Fig. 9.4. FGF-2 (green in Fig. 9.4A, B) is most prominently upregulated in the ONL (between the OPL and OLM in Fig. 9.4B), while CNTF (red in Fig. 9.4B) is most strongly upregulated at the inner surface of the retina, and in the OPL and OLM.

## 9.3.2
### The Edge of the Retina Is Functionally Degraded

It is a consistent feature in the rat (Fig. 9.3C, D) and other species (below) that, at its edge, the retina appears functionally degraded. This is evident when functional molecules are labelled. In Fig. 9.3C (mid-peripheral retina of the rat), the red label is immunolabelling for cytochrome oxidase (CO), the mitochondrial enzyme which sequesters oxygen for oxidative metabolism. The green label is immunolabelling for opsin, the protein of rhodopsin, and the blue is a DNA-specific dye. The ONL is 10–12 cells thick. CO is highly prominent in the outer part of the inner segments (between the two lower asterisks in Fig. 9.3C), and prominent also in the OPL (top asterisk in Fig. 9.3C), as blobs which correspond to the large mitochondria in the axon terminals of photoreceptors. Opsin is prominent in the outer segments (arrow in Fig. 9.3C). At the edge of the retina (Fig. 9.3D), the ONL is reduced to two to four cells, the labelling of CO in the OPL (upper asterisk in Fig. 9.3D) and in inner segments (lower asterisk in Fig. 9.3D) is less intense and the inner segments appear shorter, while opsin labelling of the OS is intense (Fig. 9.3D) but the OSs are relatively short.

Fig. 9.4A, B.
Detail of edge/mid-periphery differences in the expression of two stress-inducible protective factors, in a normal rat retina (from Fig. 9.3A, B). Each trace shows the strength of immunolabelling for FGF-2 (*green*) and CNTF (*red*), measured along transects which pass from the inner to the outer limiting membranes (ILM to ILM). The retina is thinner at the edge, and the strength of labelling is generally higher. In particular, FGF-2 labelling in the ONL (between the OPL and OLM) is many times higher in B than in A; and CNTF-labelling at the OLM and close to the ILM is several times stronger in B

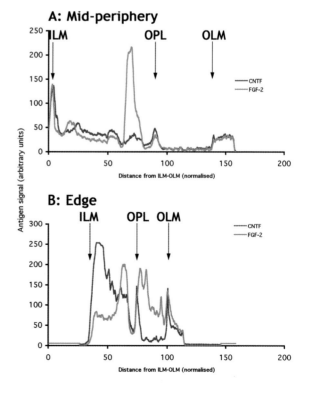

## 9.3.3
### The Edge of the Retina Is Highly Stable in the Face of Acute Stress

The edge/mid-peripheral difference in trophic factor expression and functional morphology is complemented by a striking difference in the vulnerability of photoreceptors to stress. In the Sprague-Dawley rat, exposure of the retina to bright continuous light (e.g. 2,000 lux for 24 h or 48 h) induces a photoreceptor-specific death. The severity of the death depends on the brightness, wavelength and duration of the exposure, and on the light history of the retina (reviewed by Organisciak et al. 1998). However, within one retina (thus with all the above factors constant), the severity of damage also varies with retinal location, being least at the retinal edge (Fig. 9.3E, F).

In the RCS rat, an inherited abnormality of a transport enzyme of the retinal pigment epithelium (D'Cruz et al. 2000) causes a failure of phagocytosis of the shed membrane of photoreceptor outer segments (LaVail et al. 1992). The outer segment membrane accumulates in the subretinal space, lengthening the diffusion path for oxygen to reach the photoreceptors. During the critical period of photoreceptor degeneration, RCS photoreceptors die at a high rate, driven partly by hypoxia (Valter et al. 1998). In the face of this genetically induced stress, the photoreceptors at the edge of the retina are relatively resistant, confirming earlier observations (LaVail and Battelle 1975).

In brief, photoreceptors at the edge are degraded morphologically and functionally by a chronic edge-specific stress, but contain or are surrounded by protective factors, and are highly resistant to damage.

## 9.3.4
### The Adult Retina Aged 2 Years: Observations in the Marmoset

The morphology of a 2-year-old marmoset retina near the anterior edge is shown in Fig. 9.5A–H. Centrally, the laminar structure of the retina seems well differentiated (Fig. 9.5D). Towards the edge (Fig. 9.5D–A), the retina thins, GFAP becomes more prominent in the radial processes of Müller cells (Fig. 9.5B) and FGF-2, which centrally is prominent only in the INL (Fig. 9.5D), is found in most surviving neurons (Fig. 9.5A). Centrally, the inner and outer segments of photoreceptors are well differentiated (Fig. 9.5H), with CO prominent in the inner segments (red) and opsin in the outer segments (green). Towards the edge (Fig. 9.5G, F, H), both inner and outer segments become shorter and less well separated. At the edge (Fig. 9.5E), neither inner nor outer segments are apparent; instead a few cells, probably surviving photoreceptors, appear rounded, with opsin prominent in their cytoplasm.

In a 7-year-old retina (Fig. 9.5I), the same trends are apparent, shown here at low power. Centrally (to the right in Fig. 9.5I), GFAP (red) is prominent in astrocytes at the inner surface of the retina. Towards the edge (arrow in Fig. 9.5I) the intensity of GFAP-labelling increases, both at the inner edge of the retina, and across its thickness. Centrally, FGF-2 (green in Fig. 9.5I) is prominent in Müller cells in the INL; towards the edge (arrow), FGF-2 is prominent in all cells present. The DNA label (blue in Fig. 9.5I) shows the cellular layers of the retina; two of the layers (INL and ONL) are seen at this magnification. At the edge (arrow in Fig. 9.5I) the separation between these two layers becomes indistinct.

## 9.3.5
### The Adult Retina Aged 20 Years: Observations in the Baboon

In this retina, again, there is a gradual breakdown of retinal structure over the peripheralmost few millimeters of retina. For example, the inner and outer segments of photoreceptors are neatly arranged in mid-peripheral retina (>10 mm from the edge, Fig. 9.6A), with opsin (green in Fig. 9.6A) detectable in the membranes of somas, showing a limited concentration in the inner parts of the inner segments and prominent in the outer segments. CO (red) is prominent in the outer parts of the inner segments. At 1 mm and 0.6 mm from the edge (Fig. 9.6A–C), this neat lamination progressive-

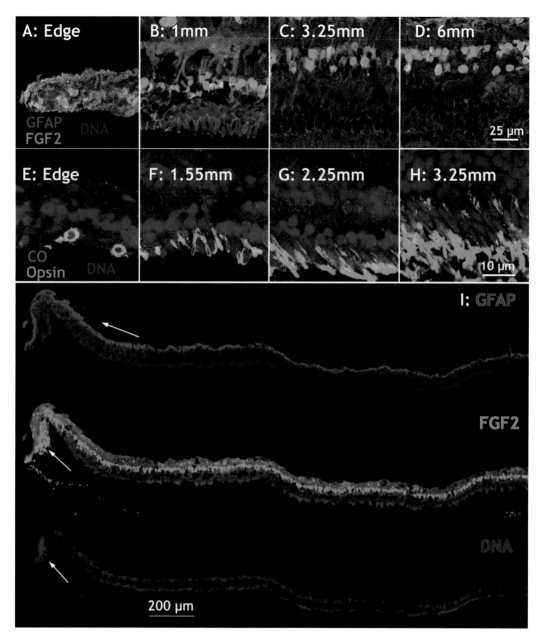

Fig. 9.5A–I. Degeneration at the edge of the marmoset retina. A–D From a section labelled for GFAP (*red*), FGF-2 (*green*) and DNA (*blue*), from a retina aged 2 - years. At the extreme edge (A) and 1 mm away (B), the retina is thin, and GFAP is prominent in astrocytes and Müller cells. FGF-2 is prominent in cells in the INL. More centrally (C, D) the retina is thicker, and GFAP less prominent in Müller cells. E–H From a section labelled for CO (*red*), opsin (*green*) and DNA (*blue*), from the retina aged 2 years. Centrally (H) CO is prominent in the inner segments of photoreceptors, and opsin in outer segments. Towards the edge (G–E) the separation of inner and outer segments becomes less clear. At the edge (E) inner and outer segments cannot be distinguished, and opsin is found in the cytoplasm of a few cells, presumably surviving photoreceptors. I Section at the edge of marmoset retina aged 7 years, labelled for GFAP, FGF-2 and DNA, with the fluophores shown separately. At lower power the upregulation of GFAP (*red*) is apparent at the edge (*arrow in top image*). The upregulation of FGF-2 (*green*) in photoreceptors is also apparent (*arrow in middle image*) and the breakdown of laminar structure is also seen (*arrow in lower image*)

ly degrades, and opsin-positive neurites grow abnormally into inner retina (Fig. 9.6C). At the extreme edge (Fig. 9.6D), inner and outer segment structure are totally lost, although a few opsin-positive structures remain. Some of these structures resemble cells, with opsin-positive cytoplasm. There is evidence of pigmentary invasion (arrow in Fig. 9.6G).

At lower power (Fig. 9.6E, F, H), it is apparent that the retina thins markedly towards the edge, to <50% of its central thickness (compare Fig. 9.6E, H). Centrally (Fig. 9.6H), GFAP (red) is confined to astrocytes at the inner surface of the retina, while, towards the edge, GFAP is expressed increasingly in the radial processes of Müller cells (Fig. 9.6E, F). The green label in

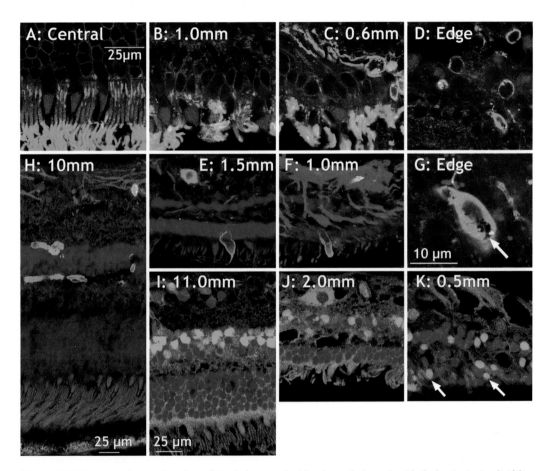

Fig. 9.6A–K. Degeneration at the edge of the baboon retina aged 20 years. A–D, G From a section labelled for CO (*red*) and opsin (*green*). Centrally (A), CO is prominent in the outer part of inner segments of rods and cones, and opsin in outer segments and, less markedly, in the inner parts of inner segments of rods. Towards the edge this separation of inner and outer segments breaks down (B). Some cells grow opsin-positive processes which extend into inner retina (C). At the edge (D), opsin is found only in the cytoplasm of isolated cells, presumably surviving photoreceptors. Pigmentary inclusions were seen in some cells, at the edge (*arrow* in G). E, F, H From a section labelled for GFAP (*red*), and for blood vessels (*green*), with the lectin G. simplicifolia. Centrally (H) the retina is thick, GFAP is confined to astrocytes at the inner edge of the retina and capillaries (*green*) extend to the outer aspect of the INL. Near the edge (E, F), the retina is markedly thinned, blood vessels are absent and the lectin labels occasional cone sheaths and cells in the ganglion cell layer. GFAP is present in Müller cells processes, becoming particularly prominent at the edge (F). I–K From a section labelled for GFAP (*red*), FGF-2 (*green* and DNA (*blue*) Again, towards the edge, the thinning of the retina is clear, FGF-2 becomes prominent in cells of the outer nuclear layer (*arrows* in K) and GFAP in the processes of Müller cells

Fig. 9.6E, F, H is a lectin (*G. simplicifolia*), which in central retina labels blood vessels (Fig. 9.6H), showing capillaries on the inner and outer aspects of the INL. Peripherally (Fig. 9.6E, F), no vessels are apparent; i.e. the retina is avascular and the lectin binding is abnormal, labelling neurons in the ganglion cell layer (example in Fig. 9.6E) and the sheaths of cones (Fig. 9.6E, F).

The thinning of the retina and upregulation of GFAP expression at the edge are also evident in Fig. 9.6I–K. In these images the green labelling is FGF-2. As in the species described above, FGF-2 is prominent in mid-peripheral retina in somas (probable Müller cells) in the INL and in the cytoplasm of ganglion cells. At the edge FGF-2 is also prominent in somas of photoreceptors in the outer nuclear layer (arrows in Fig. 9.6K).

Two features of this baboon retina may reflect its greater age. First, the extension of opsin-positive neurites into inner retina has been described in human retinas affected by long-standing retinitis pigmentosa (Li et al. 1995). Second, the presence, presumably by migration from the RPE, of pigment granules within the retina at the edge is a feature of the oldest human retina studied (below).

## 9.3.6
### The Adult Retina Aged 30–70 Years: Observations in the Human

The most spectacular "edge effects" are seen in the oldest retinas studied, from humans aged 29, 45 and 69 years. Throughout much of these retinas the layering structure is normal (Fig. 9.7A). GFAP is restricted to astrocytes at the inner surface, the three cellular layers of the retina are neatly intact and photoreceptors have ordered inner and outer segments, with CO concentrating in the inner segments and opsin in the outer segments (Fig. 9.7E).

Towards the edge of the retina (compare Fig. 9.7A with B, C with D), the retina thins. FGF-2 becomes prominent in cells in all retinal layers (Fig. 9.7B), and GFAP becomes prominent in Müller cells (Fig. 9.7D). The retinal vasculature (green in Fig. 9.7C) is absent from the edge region (Fig. 9.7D). In the 69-year retina, the neat segregation of CO-immunoreactive inner segments (red in Fig. 9.7E) from opsin-immunoreactive outer segments (green) persists to 7 mm from the edge (Fig. 9.7F), but is progressively degraded nearer the edge (Fig. 9.7G–I). At the absolute edge (Fig. 9.7I), inner and outer segment structure is totally lost. A similar sequence from the 45-year-old is shown in Fig. 9.7J–M. Again the structure of inner and outer segments breaks down towards the edge, and some opsin-immunoreactive cells, presumably photoreceptors, extended opsin-positive neurites into the inner retina, an abnormality previously noted in degenerating human retina (Li Y et al. 1995).

At low power, the edge region of the 69-year-old retina shows morphological changes which are quite spectacular (Fig. 9.8A). Vacuoles appear within the retina, and the layering of the retina becomes confused and then obliterated. Damage is most severe at the distal edge (at the right hand margin of Fig. 9.7A), and decreases with distance from the edge, consistent with the idea that the edge region is subject to a progressive edge-to-centre degeneration. For descriptive purposes, three zones can be distinguished in Fig. 9.8A. At the left of the image, the retina is near normal in structure and labelling. In the central part of the image is a zone of cystoid degeneration, within which retinal layering is disrupted. In this region some elements of normal structure remain. GFAP-labelled glial processes are prominent at the inner surface, and extend radially across the retina, along the "columns" of tissue which persist between the vacuoles (Fig. 9.8B). Among those glial processes neurone-like cells are present, which label for FGF-2 (green) and autofluoresce blue. At the outer surface (bottom of Fig. 9.8B), a few outer-segment-like processes remain. At the transition between the zones of "cystoid" and "complete" degeneration (shown at higher magnification in Fig. 9.8C), the GFAP-labelling stops abruptly, and only the vacuoles give indication that the most peripheral region was once part of neural retina.

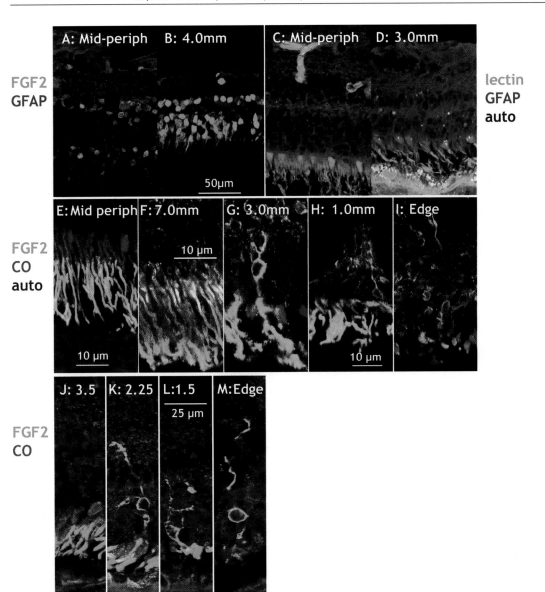

Fig. 9.7A–M. Degeneration at the edge of human retina. A, B From a 69-year retina. The retina thins markedly towards the edge (B, 4 mm from edge), and FGF-2 (*green*) is upregulated in cells of all layers. C, D From a 45-year retina. Again the retina thins towards the edge (D, approximately 3 mm from edge). GFAP (*red*) becomes more prominent, appearing in the radial processes of Müller cells. Vessels of the retinal circulation (*green* in C) are absent from the edge (D). E–I From a 69-year retina. CO (*red*) is prominent in inner segments and opsin (*green*) in outer segments. Inner and outer segments are well differentiated up to 7 mm from the edge (F). More peripherally, inner segments cannot be distinguished, outer segments shorten and opsin appears in cells of the outer nuclear layer. At the extreme edge (I), outer segments also cannot be distinguished. The *blue* in these images is autofluorescence. J–M From a 45-year retina, labelled as for E–I, but without the *blue* channel. Again, the inner and outer segments are nicely separate away from the edge (image at *left*). Close to the edge, inner and outer segments are lost, opsin appears in surviving cells of the ONL, and some opsin-positive cells extend neurites into the inner retina

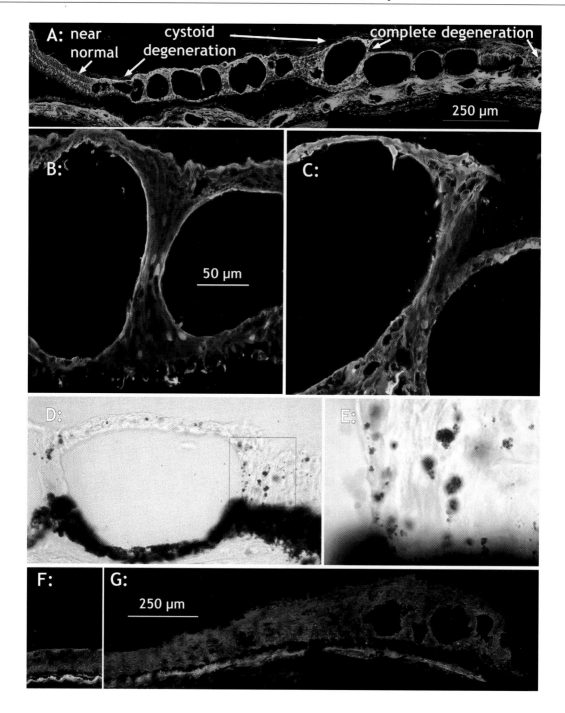

**Fig. 9.8. A** Edge of the 69-year retina, labelled for GFAP (*red*) and FGF-2 (*green*). *Blue* is autofluorescence. Cystoid degeneration is prominent over an extended region (2–3 mm). Most peripherally, the *red* labelling of GFAP in macroglia is lost, and the degeneration of the retina is complete. **B** In the cystoid degeneration region, GFAP-positive processes extend across the retina, between the vacuoles. FGF-2-positive cells persist among these glial processes and a few outer segments can be distinguished, at bottom. **C** Towards the edge, the *red* (GFAP) label stops and a cell-sparse matrix remains (*right*). **D, E** In the region of complete degeneration, pigmented cells are present in the surviving matrix, apparently migrating inwards from the retinal pigment epithelium (*bottom*). **E** shows at higher power the region outlined in D. **F, G** Edge of the 45-year retina, labelled as for A. Towards the midperiphery (F), retinal structure appears normal. At the edge the retina swells, vacuoles appear, and GFAP-labelling becomes intense. The cystoid degeneration is less extensive than at 69 years (A), and there was no region of complete degeneration

---

Under transmitted light (Fig. 9.8D,E), an additional abnormality of the zone of complete degeneration is apparent. Pigment clusters are present in the once-retinal tissue. At higher power (Fig. 9.8E) the clusters seem to form in radial strands, raising the possibility that they have migrated into the residual retinal structure, from the RPE (seen at bottom of both Fig. 9.8D and E).

### 9.3.7 Evidence of Progression in Edge Degeneration in Humans

The 29- and 45-year-old retinas also show cystoid degeneration at the edge (Fig. 9.8F), but less severely than at 69 years. There is, for example, no zone of "complete" degeneration at the extreme edge of either retina and no evidence of pigmentary invasion, and the cystoid disruption of retinal lamination is less extensive (extending ~0.8 mm from the edge, as compared to 1.5 mm in the 69-year retina). Finally, when the 29- and 45-year retinas are compared, there is evidence of progression of the degeneration between these ages. The extent of vacuolar degeneration is similar in the two retinas (~0.8 mm from the edge). At the cellular level, however, the INL and ONL are still recognizable at 29 years, the vacuoles forming in the OPL. At 45 years, the INL and ONL can not be distinguished, and the number of neurons present in this region of vacuolar degeneration is greatly reduced (data not shown).

## 9.4 Discussion

The evidence presented above that, in a range of mammalian species, the retina undergoes progressive degeneration from neonatal life to the seventh decade (in humans), confirms and expands earlier evidence of age-related cell loss in the normal retina. Gartner and Henkind (1981), for example, have noted, in a study of the edge of human retina, that "peripheral cystoid ... degeneration (is present) ... even as early as 1 month of age. It was more frequent after 3 years ... almost always present at age 8, and ... quite extensive after age 30". Further, systematic counts of photoreceptors in human retina as a function of age (Gao and Hollyfield 1992) show a progress of loss of photoreceptors throughout the retina, which is rapid in young adults, slowing but persisting into advanced age. The present observations document the special status of the edge of the retina in rodents (mouse and rat) and a range of primates (marmoset, baboon, human). The novel elements of the data reviewed include:

A. The degeneration of the edge is stress-induced, as indicated by the upregulation at the edge of stress-induced proteins (GFAP, CNTF, FGF-2)
B. The edge-specific stress includes hyperoxia (because hypoxia reduces cell death at the edge)
C. The damage to the retina at the edge includes cell death, thinning and eventual disruption of the layers of the retina, the formation of vacuoles (cystoid degeneration) and the shortening and loss of inner and outer segments

D. These effects are maximal at the edge and graded towards the center of the retina, suggesting that the stress acts at the edge of the retina

E. Where, in the human, retinal structure is entirely lost, there follows an invasion of the retina by pigmented cells, apparently from the retinal pigment epithelium

For technical reasons, the juvenile material has been obtained from rodents, and mature (>2-year-old) adult and senescent material from primates, so that the present observations are species-specific vignettes of the life history of the retinal edge. A full life-history of the edge is warranted for each of these species. There is, nevertheless, a common thread to these observations. In all species and at all ages, the edge of the retina appears subject to a localized chronic stress. Initially, in the neonate mouse, the stress causes an episode of rapid cell death, and then upregulates the expression of protective factors, with the result that the edge region, although slowly degenerating as a result (presumably) of chronic stress, is highly resistant to acute stress. Over decades, however, the degenerative process at the edge runs slowly to total disintegration of neuronal structure and pigmentary invasion of the degenerative site, and the process spreads centrally, from the edge towards the posterior pole of the eye.

### 9.4.1
### The Edge of the Retina as a Model of Retinal Degeneration

In analyzing retinas of normal animals, and of models of degeneration, in our own work on the RCS rat (Valter et al. 1998), it is evident that the edge region of retina is not distinctive in the mechanisms which operate there in response to stress. Upregulation of FGF-2, CNTF and GFAP, such as occurs at the edge, will occur throughout the retina, if the stress applies throughout the retina (Steinberg 1994; Cao et al. 2001) and will occur focally if the lesion is localized (Cao et al. 2001). Further, the upregulation of these factors is associated with an increase in photoreceptor resistance to damage. The pigmentary invasion of the retina, described here at the edge of the human retina, occurs throughout the retina when there is widespread photoreceptor degeneration. This is the signature pathology of the human retinal degenerations in retinitis pigmentosa.

Experimental data in the mouse (Mervin and Stone 2002b) suggest that hyperoxia is a factor in the chronic stress which causes edge-specific degeneration. We have argued previously (Stone et al. 1999) that hyperoxia may be a factor in the late stages of all retinal degenerations. For the investigator, therefore, the edge of normal retina is a useful model of retina-wide degeneration, a site at which the mechanisms of degeneration, and of resistance to degeneration, can be effectively studied in the retina is genetically normal, and has experienced normal levels of light retina.

### 9.4.2
### Why Is the Protection of Photoreceptors Stress-Inducible?

The retina's ability to protect its photoreceptors by upregulation of protective factors is well established. The factors most clearly established to be protective (FGF-2, CNTF, BDNF) are all stress-induced (Steinberg 1994; Wen et al. 1995; Bowers et al. 2001; Walsh and Stone 2001; Walsh et al. 2001). The upregulation of these factors in response to acute stress requires some hours (Walsh and Stone 2001). The rate of upregulation appears to be sufficient to protect the retina against many natural forms in increased light exposure, such as the brightness of summer, but is too slow to prevent photoreceptor degeneration in response to a more acute challenge, such as the switching on of a bright, continuous light. These observations raise a question: Why have these mechanisms of protection evolved as stress-induced? If they are available, why are they not just "switched on"?

One plausible answer to this question is suggested by the observation that the upregulation of protective factors (the evidence is clearest for FGF-2) is associated with a reduction in photoreceptor sensitivity (Gargini et al. 1999). The upregulation of protection may have evolved to be stress-inducible, so that sensitivity is maximized to the stress levels actually experienced. This maximization of sensitivity seems to involve, however, the risk of vulnerability to acute stress.

This point is of clinical as well as teleological interest, if the regulation of trophic factors is reversible, so that long-term reduction of stress (e.g. reducing ambient light exposure) may increase sensitivity. This could be of value to patients suffering vision loss due to photoreceptor loss. The possibility that the re-sensitized retina is also increasingly vulnerable to acute light stress would need to be understood, in managing light exposure over the long term.

### 9.4.3
### Vulnerability to Oxidative Stress in Mice and Humans

We have drawn data from different mammalian species, for pragmatic reasons, and species differences need to be considered when bringing these observations together. Mouse fibroblasts in vitro, for example, show senescent changes earlier than human fibroblast (Parrinello et al. 2003), and this early senescence seems associated with a higher vulnerability to oxidative stress. Put conversely, human fibroblasts may have evolved mechanisms to limit their vulnerability to oxidative stress, necessary for the long human lifespan.

Despite such differences, several themes of the present story appear to be common to rodents and primates. Regardless of species, photoreceptors are, among retinal cells, the most exposed to stress, and the most fragile in the face of environmental or genetic stress. Conversely, photoreceptors have also developed strong mechanisms of protection. The erosion of the edge of the retina, which begins in early postnatal life and progresses for the full human lifespan, shows one limit of the effectiveness of the retina's protective mechanisms. As far as we are aware, no other cell or region of the central nervous system shows a comparable "normal" degeneration, beginning as soon as tissue starts to function, and progressing throughout life.

### 9.4.4
### The Link to AMD

#### 9.4.4.1
#### Are Rods More Susceptible to Edge-Specific Stress?

In the mouse (Gresh et al. 2003) and human (Curcio et al. 1993, 1996; Adler et al. 1999; Jackson et al. 2002), rods are more vulnerable than cones to senescence. The numbers of both rods and cones decrease with age, with rod depletion occurring earlier, both in the retina as a whole and, in the human, in the cone-rich foveal region. The present evidence of an edge-specific degenerative process raises the question whether rods are also more susceptible in this form of degeneration. The present observations do not include the quantitative comparison needed to answer this question, but studies of the edge of human retina have shown that cones are relatively numerous at the edge (Williams 1991; Ahnelt 1998). Although previous authors have described the edge of human retina as "cone-enriched" rim, it is possible that the region has been selectively depleted of rods. One study (Mollon et al. 1998) sought psychophysical correlates of the relatively high cone density at the edge of the retina and was unable to demonstrate a correlate. The present evidence that the edge is degenerate, its photoreceptor population depleted and the functional morphology of the survivors degraded, confirm this evidence that the edge of the retina is unlikely to subserve a specialized visual function.

#### 9.4.4.2
#### Photoreceptor Depletion Destabilizes

The mammalian retina has evolved as a highly durable sensor, functioning for the full life of the organism, in humans for many decades. Nevertheless the retina is arguably the least stable part of the central nervous system, and loss of visual function is "normal" in older humans (Jackson et al. 1999). In recent years, evidence has accumulated that an important factor in maintaining retinal stability is the photoreceptor population. Put conversely, photoreceptor depletion destabilizes the photoreceptors which survive this depletion. Four lines of evidence support this proposition: First, clinically the rod/cone dystrophies and cone/rod dystrophies are rarely stable. They progress, raising the possibility that each photoreceptor class provides survival factors to the other (reviewed by Adler et al. 1999; Curcio et al. 2000). Second, experimental evidence is now available suggesting that a survival factor is provided to cones by rods, at least in mouse retina (Bateman et al. 1992; Mohand-Said et al. 1998; Adler et al. 1999). Third, rod loss is a precursor of foveal instability in humans; i.e. loss of rod function and rod numbers precedes AMD (Jackson et al. 2002). Fourth, photoreceptor depletion causes a rise in oxygen tension in outer retina (Yu et al. 2000), and hyperoxia is toxic to photoreceptors (Yamada et al. 1999). This is the oxygen toxicity hypothesis of (Stone et al. 1999).

In normal human retina, photoreceptor loss is measurable throughout life. The data of Gao and Hollyfield suggest, for example, a depletion rate of ~1000 rods/mm$^2$ per year in the third decade, declining to about half that value in the fifth to tenth decades. If these figures are extrapolated to the full area of human retina (~900 mm$^2$; Stone and Johnston 1981), they suggest a rod loss of 2,000/day in young adulthood, falling to 1,000/day in later life. These numbers appear disconcerting, but each retina has $1-2 \times 10^8$ photoreceptors, so that many should survive to the ninth decade of life. The functional effect of this normal photoreceptor loss can be measured as an age-related, decade-by-decade rise in dark-adapted thresholds, and a slowing of dark adaptation (Jackson 1999). The destabilizing effect of this photoreceptor loss has been recognized for many years in the loss of cone photoreceptors in rod-specific forms of retinitis pigmentosa (above). For normal retina, the destabilizing effect of photoreceptor depletion has been described only recently. We have noted a trend for non-familial forms of retinal degeneration to be associated with perinatal stress (Stone et al. 2001), which may cause early photoreceptor depletion. At the other end of human life, a trend has been reported (Jackson et al. 2002) for AMD to occur preferentially in retinas from which age-related loss of rods has been relatively high. Finally, we note here a trend for early depletion of photoreceptors, at the edge of the retina, to be associated with a life-long degenerative process, seen as a chronic, progressive degeneration of the edge of the retina. Clues to the management of the ageing retina may thus be gleaned from the systematic study of "the edge".

## References

Adler R, Curcio C, Hicks D, Price D, Wong F (1999) Cell death in age-related macular degeneration. Mol Vis 5:31

Ahnelt P (1998) The photoreceptor mosaic. Eye 2: 531–540

Bateman J, Klisak I, Kojis T, Mohandas T, Sparkes R, Li T (1992) Assignment of the beta-subunit of rod photoreceptor cGMP phosphodiesterase gene PDEB (homolog of the mouse rd gene) to human chromosome 4p16. Genomics 12:601–603

Bowers F, Valter K, Chan S, Walsh N, Maslim J, Stone J (2001) Effects of oxygen and bFGF on the vulnerability of photoreceptors to light damage. Invest Ophthalmol Vis Sci 42:804–815

Cao W, Li F, Steinberg RH, Lavail MM (2001) Development of normal and injury-induced gene expression of aFGF, bFGF, CNTF, BDNF, GFAP and IGF-I in the rat retina. Exp Eye Res 72:591–604

Curcio C, Millican C, Allen K, Kalina R (1993) Aging of the human photoreceptor mosaic: evidence for selective vulnerability of rods in central retina. Invest Ophthalmol Vis Sci 34:3278–3296

Curcio C, Medeiros N, Millican C (1996) Photoreceptor loss in age-related macular degeneration. Invest Ophthalmol Vis Sci 37:1236–1249

Curcio CA, Owsley C, Jackson GR (2000) Spare the rods, save the cones in aging and age-related maculopathy. Invest Ophthalmol Vis Sci 41:2015–2018

D'Cruz PM, Yasumura D, Weir J, Matthes MT, Abderrahim H, LaVail MM, Vollrath D (2000) Mutation of the receptor tyrosine kinase gene Mertk in the retinal dystrophic RCS rat. Hum Mol Genet 9:645–651

Gao H, Hollyfield JG (1992) Aging of the human retina: differential loss of neurons and retinal pigment epithelial cells. Invest Ophthalmol Vis Sci 33:1–17

Gargini C, Belfiore M, Bisti S, Cervetto L, Valter K, Stone J (1999) The impact of bFGF on photoreceptor function and morphology. Invest Ophthalmol Vis Sci 40: 2088–2099

Gartner J (1975) Fine structural changes of peripheral cystoid degenerations during life. Mod Probl Ophthalmol 15:98–102

Gartner S, Henkind P (1981) Lange's folds: a meaningful ocular artifact. Ophthalmology 88:1307–1310

Gresh J, Goletz PW, Crouch RK, Rohrer B (2003) Structure-function analysis of rods and cones in juvenile, adult, and aged C57bl/6 and Balb/c mice. Vis Neurosci 20:211–220

Jackson GR, Owsley C, McGwin G Jr (1999) Aging and dark adaptation. Vis Res 39:3975–3982

Jackson GR, Owsley C, Curcio CA (2002) Photoreceptor degeneration and dysfunction in aging and age-related maculopathy. Ageing Res Rev 1:381–396

LaVail MM (1981) Photoreceptor characteristics in congenic strains of RCS rats. Invest Ophthalmol Vis Sci 20:671–675

LaVail M, Battelle B (1975) Influence of eye pigmentation and light deprivation on inherited retinal dystrophy on the rat. Exp Eye Res 21:167–192

LaVail MM, Li L, Turner JE, Yasumura D (1992) Retinal pigment epithelial cell transplantation in rcs rats: normal metabolism in rescued photoreceptors. Exp Eye Res 55:555–562

Li Y, Kljavin I, Milam A (1995) Rod photoreceptor neurite sprouting in retinitis pigmentosa. J Neurosci 15:5429–5438

Liu C, Peng M, Laties A, Wen R (1998) Preconditioning with bright light evokes a protective response against light damage in the rat retina. J Neurosci 18:1337–1344

Maslim J, Valter K, Egensperger R, Hollander H, Stone J (1997) Tissue oxygen during a critical developmental period controls the death and survival of photoreceptors. Invest Ophthalmol Vis Sci 38:1667–1677

Mervin K, Stone J (2002a) Developmental death of photoreceptors in the C57BL/6J mouse: association with retinal function and self-protection. Exp Eye Res 75:703–713

Mervin K, Stone J (2002b) Regulation by oxygen of photoreceptor death in the developing and adult C57BL/6J mouse. Exp Eye Res 75:715–722

Mohand-Said S, Deudon-Combe A, Hicks D, Simonutti M, Forster V, Fintz AC, Leveillard T, Dreyfus H, Sahel JA (1998) Normal retina releases a diffusible factor stimulating cone survival in the retinal degeneration mouse. Proc Natl Acad Sci USA 95:8357–8362

Mollon J, Regan B, Bowmaker J (1998) What is the function of the cone-rich rim of the retina? Eye 12:548–552

Nir I, Liu C, Wen R (1999) Light treatment enhances photoreceptor survival in dystrophic retinas of Royal College of Surgeons rats. Invest Ophthalmol Vis Sci 40:2383–2390

Organisciak DT, Darrow RA, Darrow RA, Lininger LA (1998) Environmental light and age-related changes in retinal proteins. In: Photostasis and related phenomena. Plenum, New York, pp 79–92

Parrinello S, Samper E, Krtolica A, Goldstein J, Melov S, Campisi J (2003) Oxygen sensitivity severely limits the replicative lifespan of murine fibroblasts. Nat Cell Biol 5:741–747

Penn J, Anderson R (1991) Effects of light history on the rat retina. Prog Retin Res 11:75–98

Steinberg RH (1994) Survival factors in retinal degenerations. Curr Opin Neurobiol 4:515–524

Stone J, Johnston E (1981) The topography of primate retina: a study of the human, bushbaby, and new and old-world monkeys. J Comp Neurol 196:205–223

Stone J, Maslim J, Valter-Kocsi K, Mervin K, Bowers F, Chu Y, Barnett N, Provis J, Lewis G, Fisher S, Bisti S, Gargini C, Cervetto L, Merin S, Pe'er J (1999) Mechanism of photoreceptor death and survival. Prog Retin Eye Res 18:689–735

Stone J, Maslim J, Fawsi A, Lancaster P, Heckenlively J (2001) The role of perinatal stress in retinitis pigmentosa: evidence from surveys in Australia and the USA. Can J Ophthalmol 36:315–322

Valter K, Maslim J, Bowers F, Stone J (1998) Photoreceptor dystrophy in the RCS rat: roles of oxygen, debris and bFGF. Invest Ophthalmol Vis Sci 39:2427–2442

Vrabec F (1967) Neurohistology of cystoid degeneration of the peripheral human retina. Am J Ophthalmol 64:90–99

Walsh N, Stone J (2001) Timecourse of bFGF and CNTF expression in light-induced photoreceptor degeneration in the rat retina. In: New insights into retinal degenerative diseases. Kluwer, New York, pp 111–118

Walsh N, Valter K, Stone J (2001) Cellular and subcellular patterns of expression of bFGF and CNTF in the normal and light stressed adult rat retina. Exp Eye Res 72:495–501

Wen R, Song Y, Cheng T, Matthes M, Yasamura D, LaVail M, Steinberg R (1995) Injury-induced upregulation of bFGF and CNTF mRNAs in the rat retina. J Neurosci 15:7377–7385

Williams RW (1991) The human retina has a cone-enriched rim. Vis Neurosci 6:403–406

Yamada H, Yamada E, Hackett SF, Ozaki H, Okamoto N, Campochiaro PA (1999) Hyperoxia causes decreased expression of vascular endothelial growth factor and endothelial cell apoptosis in adult retina. J Cell Physiol 179:149–156

Yu DY, Cringle SJ, Su EN, Yu PK (2000) Intraretinal oxygen levels before and after photoreceptor loss in the RCS Rat. Invest Ophthalmol Vis Sci 41:3999–4006

Chapter 10

# Clinical Strategies for Diagnosis and Treatment of AMD: Implications from Research

Scott W. Cousins, Karl G. Csaky, Diego G. Espinosa-Heidmann

## Contents

10.1　Dry AMD　167
10.1.1　Definition of Dry AMD　167
10.1.2　Pathogenic Mechanisms for Drusen Formation　168
10.1.3　Established or Evaluated Treatments　170
10.1.4　Ongoing Trials: Multicenter Investigation of Rheopheresis for AMD (MIRA-1) Study Group　172
10.1.5　Future Research　172

10.2　Wet AMD　178
10.2.1　Definition of Neovascular AMD　178
10.2.2　Pathogenic Mechanisms for CNV Formation　180
10.2.3　Established or Evaluated Treatments for Neovascular AMD　184
10.2.4　Therapies Currently in Clinical Trial　185
10.2.5　Future Research　189

　　　References　191

## 10.1
## Dry AMD

### 10.1.1
### Definition of Dry AMD

Numerous biochemical and anatomical changes occur in Bruch's membrane (BrM) as part of aging retina in the absence of apparent retinal dysfunction, including collagenous thickening, calcification, and lipid infiltration. However, the accumulation of specific lipid-rich deposits under the RPE is a very prominent histopathologic feature of eyes with AMD (Young 1987; Green 1999). These deposits are generically termed "drusen" (clinical definition) when observed by physicians upon clinical examination of patients (Fig. 10.1; Zarbin 1998). However, histopathological examination allows three main types of sub-RPE deposits to be distinguished by location, thickness, and content: basal laminar deposits (BLD), basal linear deposits (BLinD), and nodular drusen (anatomic definition; van der Schaft et al. 1992; Green 1999). The RPE basement membrane seems to be the crucial dividing line in distinguishing between BLD and drusen or basal linear deposits. Drusen when detected clinically are more extensive BLD, BLinD, or nodular drusen.

BLD is the accumulation of amorphous material of intermediate electron density between the plasma membrane and the basement membrane of the RPE, often containing banded structures, patches of electron-dense fibrillar or granular material, and, occasionally, membranous debris (Kliffen et al. 1997). Basal linear deposits are diffuse, amorphous accumulations within the inner collagenous zone of BrM, external to RPE basement membrane, with similar content variations. Nodular drusen (anatomic definition) are discrete, dome-shaped deposits within the inner collagenous zone of BrM, often contiguous with basal linear deposits (Abdelsalam et al. 1999). In general, these deposits contain phospholipids, triglycerides, cholesterol, cholesterol esters, apolipoproteins, vitronectin, immunoglobulins, amyloid, complement, and many other poorly characterized components (see Chap. 6; Hageman and Mullins 1999; Hageman et al. 2001; Penfold et al. 2001; Anderson et al. 2002). Further, low-grade

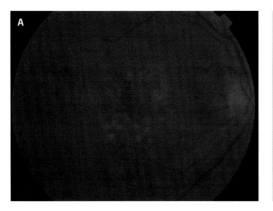
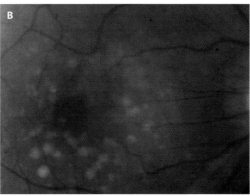

Fig. 10.1A, B. Color fundus photograph (A) and red-free photograph (B) showing predominantly large soft drusen and some subtle retinal pigment epithelial alterations. Drusen are generally the earliest sign of AMD, consisting of multiple discrete, round, slightly elevated, variable-sized sub-RPE deposits in the macula and elsewhere in the fundus of both eyes. Generally they are described clinically as hard or soft drusen, depending in their size and how well-defined their boundaries are

monocyte infiltration within the choriocapillaris is often present underlying areas of deposits (Penfold et al. 2001; Anderson et al. 2002).

The histological correlates of clinically observed drusen are controversial. Most expert agree that small hard drusen observed clinically consist mostly of nodular drusen histologically. However, no consensus exists among experts for the specific composition of clinically evident, large soft drusen, and definitive information is lacking that clarifies the specific contributions of each of the three ultrastructural deposit types. Some postulate that basal linear deposits are specific for AMD, whereas other believe only large nodular drusen define AMD (Curcio and Millican 1999; Curcio et al. 2000). Some investigators postulate that BLD are an aging change without disease significance, whereas others believe that they are an early manifestation of AMD and that they can evolve into linear deposits (van der Schaft et al. 1994). Nevertheless, many eyes with clinical AMD express all three deposit subtypes and, conversely, examples have been published in which conditions with clinical drusen-like deposits reveal histology dominated by thick variants of any one of the three subtypes (Penfold et al. 2001). Until better information is available, any research that clarifies the pathogenesis of any of the deposit subtypes is useful.

### 10.1.2
### Pathogenic Mechanisms for Drusen Formation

At least five different paradigms have been proposed to explain deposit formation in AMD. Many of these are detailed elsewhere in this book, but it is useful to summarize them here.

#### 10.1.2.1
#### Genetic Hypothesis

Twin and sibling studies clearly indicate that AMD has a genetic component, and two genetic models have been proposed. One model proposes that genetic mutations of protein(s) specific to retinal or RPE function cause the disease, similar to the pathogenesis of retinitis pigmentosa. This hypothesis has been supported by the identification of several specific mutations for various hereditary maculopathies, especially Stargardt's disease, Best's disease, and Sorsby's Fundus dystrophy (Felbor et al. 1997; Gorin et al. 1999; Yates and Moore 2000; Klaver and Allikmets 2003). To date, none of these genes seem to be associated with AMD (Yates and Moore 2000). Alternatively, mutations or polymorphisms in general metabolic pathways that interact with outer retinal function may serve as susceptibility cofactors in AMD (Chung and Lotery 2002; Hamdi and

Kenney 2003). The association of AMD with polymorphisms of apolipoprotein E supports this model (Klaver et al. 1998).

### 10.1.2.2
### Lysosomal Failure/Lipofuscin Hypothesis

This model proposes that an age-related failure in RPE lysosomal degradation of phagocytosed photoreceptor membranes is a primary cause of dry AMD (Dorey et al. 1989; von Ruckmann et al. 1998; Beatty et al. 2000). Initially, incomplete degradation of phagocytosed phospholipids and other materials lead to the accumulation of a potentially cytotoxic by-product, lipofuscin, within the RPE cytoplasm (Young 1987). Lipofuscin, which includes numerous distinct biochemical components, can be directly cytotoxic by disrupting lysosomal and cell membranes (Young 1987). It is also a powerful chromophore for visible light, resulting in the production of oxidants such as singlet oxygen upon blue light exposure (Beatty et al. 2000). According to this model, however, lysozomal failure ultimately causes failure to digest phagocytosed outer segments, which then leads to transcellular transport of undigested material, which is deposited under the RPE as drusen (Young 1987). Extensive basic scientific research has investigated the biochemical properties of lipofuscin and its components. However, very little data support a specific role of lysosomal failure as a mechanism for its accumulation, or for the secretion of sub-RPE deposits.

### 10.1.2.3
### Choroidal Hypoperfusion Hypothesis

Primary decrease in choroidal blood flow has been proposed to cause deposits in AMD, presumably by slowing removal of RPE-derived waste materials from Bruch's membrane and diminution of nutrients or oxygen delivered to the outer retina (Grunwald 1999; Lutty et al. 1999). Decreased choroidal blood flow has been observed in AMD eyes. Two structural changes that might contribute to hypoperfusion include decreased density or diameter of choriocapillaris lumens, leading to hypoperfusion of the outer retina, and increased scleral rigidity with age, leading to increased resistance to blood flow. Some have proposed that the submacular choroid is a vascular watershed area, suggesting a possible role for relative ischemia as a component in the pathogenesis of wet AMD (Grunwald 1999). However, no data are available to determine whether choroidal blood flow abnormalities are primary or reflect a change secondary to another pathogenic component of AMD.

### 10.1.2.4
### Barrier Hypothesis

This model postulates that an acquired defect in Bruch's membrane permeability impedes the flow of nutrients and oxygen from the choroid into the RPE, and waste products from RPE into the choroid. Marshall and colleagues have suggested that deposition of plasma-derived lipids and proteins may create the barrier (Moore et al. 1995; Cousins et al. 2002). Curcio and colleagues have proposed that lipids abnormally secreted by RPE may contribute to deposit formation (Curcio et al. 2002).

### 10.1.2.5
### RPE Injury Hypothesis

This model proposes that deposit formation is secondary to chronic, repetitive, but nonlethal RPE injury (Cousins et al. 2002). Two separate phenomena must be distinguished: the injury stimulus and the cellular response (Winkler et al. 1999; Beatty et al. 2000; Mullins et al. 2000; Johnson et al. 2001a, 2001b; Anderson et al. 2002). The most widely implicated injury stimuli are various oxidants, especially those induced by RPE exposure to visible light or those derived from endogenous metabolism (Beatty et al. 2000). Environmental oxidants derived from cigarette smoke, pollution, or industrial by-products have also been proposed to contribute to AMD (Cousins et al. 2003a). Two major targets of oxidants are cellular proteins and lipids within cell membranes (Davies 1995). More recently, inflammatory-derived injury stimuli have also become implicated, including oxidants, complement, immune complexes, and factors produced by macrophages or monocytes (Penfold et al. 2001). Irrespective of the injury, this model proposes that all stimuli result in a final common pathway of cellular responses that cause the actual deposits. Cellular responses that can lead to deposit formation include RPE cell membrane blebbing, dysregulation of extracellular matrix-modifying enzymes, synthesis of excess or abnormal collagens, and aberrant expression of cytokines or growth factors (Malorni et al. 1991; Malorni and Donelli 1992; Peten et al. 1992; Jacot et al. 1996; Belkhiri et al. 1997; Aumailley and Gayraud 1998; Strunnikova et al. 2001). Subsequent breakdown and resynthesis of basement membrane and Bruch's membrane extracellular matrix would cause a progression of BLD into Bruch's membrane, producing basal linear deposits and drusen.

### 10.1.2.6
### Progression of Drusen into Late Complications

Drusen are associated with progression into three main complications in AMD. First, they are directly associated with mild to moderate retinal degeneration and vision loss. Second, they are a prognostic marker for future progression into late dry AMD (geographic atrophy of the RPE and photoreceptors). And, third, drusen are the most important risk factor for the progression of dry AMD into wet AMD (Bressler et al. 1988; Zarbin 1998; Abdelsalam et al. 1999; Hageman and Mullins 1999). In various studies, greater size, confluence, and greater number of drusen have been associated with increasing rates of CNV (Pieramici and Bressler 1998). Unfortunately, no mechanistic model explains the conversion of dry AMD into wet AMD. It is not known whether drusen are the specific cause of progression or if they are mere epiphenomena indicating other, more widespread RPE metabolic dysfunction that results in cell death or induction of vascularization. In this regard, for example, the AREDS data indicate that vitamins reduce the risk of CNV, but without changing the appearance of drusen.

In support of drusen as the cause of progression is the observation that many deposits are associated with evidence of localized inflammation, including complement activation in drusen (Johnson et al. 2001a, 2001b) as well as evidence of monocytes or macrophage within the underlying choriocapillaris, often with insertion of processes into the drusen (Hageman et al. 2001; Penfold et al. 2001; Anderson et al. 2002). Conceivably, inflammatory mediators associated with drusen could induce the expression of angiogenic factors by RPE or directly damage choriocapillaris to trigger CNV (Penfold et al. 2001).

### 10.1.3
### Established or Evaluated Treatments

### 10.1.3.1
### Nutritional Supplementation

The role of nutritional supplementation in retinal degenerative diseases is a controversial topic (Beatty et al. 2000). Although supplements can affect the activity of various metabolic functions (such as metal-dependent lysosomal or matrix enzymes), the most prevalent hypothesis for the value of dietary supplements focuses on their antioxidant function (Beatty et al. 2000). In the retina, normal metabolic processes, exposure to blue light, and inflammation all generate potentially damaging, reactive oxygen intermediates (ROI), which can initiate oxidative damage to cell membranes, DNA, protein, and carbohydrate (Beatty et al. 2000). Although the retina contains numerous antioxidant systems that can inactivate ROI, it is hypothesized that these become deficient or overwhelmed with age (Winkler et al. 1999). Hypothetically, dietary supplementation may restore their function (Beatty et al. 2000). Seven randomized controlled trials have investigated the role of nutritional supplementation in AMD (Newsome et al. 1988; Kaiser et al. 1995; Stur et al. 1996; Teikari et al. 1998; AREDS 2001; Bartlett and Eperjesi 2003; Taylor et al. 2002). Three of these trials report a positive effect of nutritional supplementation in AMD – AREDS (vitamin C, vitamin E, beta-carotene, zinc; AREDS 2001), LAST (Lutein and antioxidants; Richer et al. 2002), and zinc in macular degeneration (Newsome et al. 1988) – while the other four show no effect – ATBC (vitamin E and beta-carotene; Teikari et al. 1998), VECAT (vitamin E; Taylor et al. 2002), Visaline (vitamin C, vitamin E, beta-carotene, buphenine) in the treatment of AMD (Kaiser et al. 1995), and Zinc in the second eye in AMD (Stur et al. 1996). Among these trials, the two most important are summarized below:

A. Age-related Eye Disease Study (AREDS): Laboratory and observational research data has suggested that antioxidant and/or zinc supplements may delay progression of AMD and vision loss (AREDS 2001a). These data inspired the AREDS, a double-masked clinical trial to evaluate the effect of high dose vitamins C and E, beta-carotene, and zinc supplements on AMD progression and visual acuity. The results indicate that the antioxidant-supplemented group with high risk of drusen develop significantly less conversion to late AMD, especially CNV, which is associated with expected preservation of vision (AREDS 2001). Curiously, vitamin E therapy fails to demonstrate any beneficial effect on the progression of drusen. Nevertheless, estimates suggest that if patients with extensive nonexudative dry AMD or advanced AMD in the fellow eye were to take AREDS supplements more than 300,000 cases of blindness from AMD could be prevented (Reuters press release, 11/11/03)

B. Lutein Antioxidant Supplementation Trial (LAST): Lutein and zeaxanthin are believed to protect the retina in two ways: First, they filter short wavelengths of light reducing the oxidative effects of blue light (Bartlett and Eperjesi 2003); secondly, carotenoids limit oxidant stress of tissue resulting from metabolism and light by quenching singlet oxygen and probably also peroxy radicals (Bartlett and Eperjesi 2003). The AREDS investigators state that lutein and zeaxanthin are considered for inclusion in the formulation but neither is available for manufacturing to a research formulation at AREDS's initiation. A separate trial (phase II) was started shortly after AREDS and preliminary results have reported a statistically significant concurrent improvement in glare recovery, contrast sensitivity, and distance/near visual acuity in patients with dry AMD (Richer et al. 2002).

### 10.1.3.2
### Laser Treatment of Drusen

Anecdotal observations by clinicians suggest that mild thermal laser treatment to the retina in eyes with drusen or pigment epithelial detachments is associated with drusen disappearance (Friberg 1999; Olk et al. 1999). Various protocols have been suggested, including the pattern of laser burn placement (the direct treatment of drusen or scattered "grid" treatment of the posterior pole), laser wavelength (green or red), or laser burn intensity (placing clinically visible burns or clinically unapparent, subthreshold laser burns; Ho 1999). Several clinical trials have been organized to test the efficacy of laser to drusen treatment in terms of visual stabilization and prevention of CNV:

A. CNV Prevention Trial (CNVPT): This clinical trial was initiated in 1998. This study is composed of two arms in which clinically apparent burns are placed in a grid fashion (Choroidal Neovascularization Prevention Trial Research Group 1998). The Bilateral Drusen Study arm has patients who have drusen in both eyes and one eye is selected for laser treatment, while the contralateral eye is observed. The Fellow Eye Study arm of this trial is composed of patients that have exudative AMD in one eye and drusen in the contralateral or fellow eye. Enrolment in these pilot studies has been suspended under recommendation of the Data and Safety Monitoring Committee, because there is a higher incidence of CNV within 12 months of study enrolment in laser-treated eyes in both arms of the trial than in observed eyes (Choroidal Neovascularization Prevention Trial Research Group 1998). The conclusion is that prophylactic laser treatment as applied in this trial caused an excess risk of choroidal neovascularization in the 1st year after treatment. By 30 months the incidence of CNV in the two treatment groups was the same. Additional follow-up has shown no statistical difference in visual acuity, contrast threshold, or incidence of geographic atrophy (Choroidal Neovascularization Prevention Trial Research Group 2003).

B. Prophylactic Treatment of AMD Trial (PTAMD): This trial has a similar design to the above laser trial, but utilizes diode red rather than green laser, and uses either clinically apparent or subthreshold burns in some arms of the trial. Two-year results of a randomized pilot study on the effectiveness and safety of infrared (810 nm) diode laser macular grid photocoagulation in patients with dry AMD suggest a reduction of drusen levels and improvements in visual acuity (Olk et al. 1999). However, preliminary unpublished data have suggested that prophylactic diode laser treatment in a fellow eye with CNV, using either clinically apparent or subthreshold laser burns, may promote CNV formation and visual loss at least in the short term (similar to the CNVPT results; Rodanant et al. 2002). For patients with bilateral drusen and no CNV, prophylactic subthreshold laser treatment is still ongoing and no safety issues have been raised on this arm of the study (Friberg and Musch 2002).

## 10.1.4
## Ongoing Trials: Multicenter Investigation of Rheopheresis for AMD (MIRA-1) Study Group

Rheopheresis is a hemodialysis technique using selective ultrafiltration to remove high molecular weight substances from plasma, including immune complexes, IgM, fibrinogen, LDL and VLDL cholesterol, von Willebrand factor, $\alpha$-2 macroglobulin, and vitronectin (Pulido 2002). Traditionally, it has been used for the treatment of hyperviscosity syndromes or related disorders caused by conditions with excess accumulation of plasma proteins such as multiple myeloma (Drew 2002). In the mid-1990s, this technology was applied to the treatment of dry AMD (Pulido 2002). The rationale for this treatment is based on the barrier hypothesis, supposing that accumulation of plasma substances can be diminished by rheopheresis (Pulido 2002); however, its mechanism of action is mostly speculative (Pulido 2002). A small, controlled randomized clinical trial to investigate the safety and efficacy of rheopheresis in patients with AMD (MAC-1 Trial) and another pilot study (IDE G970241) have demonstrated safety (Pulido 2002; Klingel et al. 2002). A placebo controlled phase-I trial (MIRA-1) has been initiated, to compare rheopheresis treatment with placebo control treatment in patients with dry AMD. The initial results of this trial (42 patients) have demonstrated a statistically significant and clinically relevant effect on visual acuity (main outcome measured) when compared with placebo for the 12-month interval reported (Pulido 2002). However, these results should be interpreted with caution since the control group actually lost more vision than has been seen in multiple larger clinical trials.

## 10.1.5
## Future Research

### 10.1.5.1
### Basic Scientific Research

Clearly the summary of knowledge for pathogenesis of dry AMD in this and other chapters indicates that much more research is imperative. Although biochemical and histological analysis of human AMD eyes is important, we also need to expand ongoing efforts to pursue basic scientific research into biochemical mechanisms for deposit formation. Biochemical mechanisms will provide potential therapeutic targets for drug inhibition of drusen formation. In this regard, ongoing research in our laboratory addressing the response to injury hypothesis illustrates how basic science can identify specific molecular targets for dry AMD.

Our research is based on the RPE injury model (Cousins et al. 2002). We hypothesize that progression of deposit formation in early AMD requires two sequential cell responses: initial RPE oxidant injury to cause extrusion of cell membrane "blebs" (see Fig. 10.2), which accumulate under the RPE as BLD; and subsequent RPE hormonal stimulation to increase synthesis of matrix metalloproteinases (MMPs), collagens, and other molecules responsible for basement membrane and BrM turnover, leading to admixture of blebs into BrM and formation of new basement membrane under the RPE (Strunnikova et al. 2003).

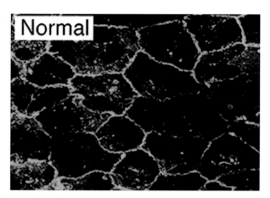
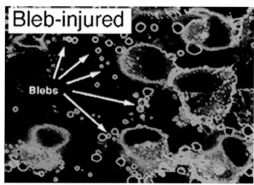

Fig. 10.2. Fluorescent microscope photographs of a spontaneously transformed human adult RPE cell line (ARPE-19), which was genetically modified by retroviral transduction with a construct containing green fluorescent protein-fernesylated r Ras (GFP-RPE) and attached to the inner leaflet of the plasma membrane. *Left panel* shows normal RPE cells in which their typically hexagonal shape can be seen in outline. Upon various kinds of oxidative injuries (menadione, MPO, hydroquinone), blebs (*right panel, white arrows*) can be readily detected and quantified by flow cytometry, Western blot, or ELISA

We have demonstrated that several different oxidants can induce blebbing, including macrophage-derived myeloperoxidase, as well as hydroquinone, an oxidant present in cigarette smoke, environmental pollution, and processed foods (Monroy et al. 2001; Cousins et al. 2003a). Repetitive oxidant injury can lead to accumulation of sub-RPE deposits when they are grown on special filter inserts (Strunnikova et al. 2003). As predicted by the RPE injury model, we can modify blebbing by two different strategies. First, blebbing can be lessened by interventions that modify the impact of oxidant injury, especially by the addition of vitamin E or lowering the polyunsaturated fatty acid content of the RPE cell membrane (Espinosa-Heidmann et al. 2004). Alternatively, it is possible to lessen blebbing by blocking the downstream cascades that trigger the ultimate cytoskeletal changes responsible for blebbing (Strunnikova et al. 2001). For example, interactions between heat-shock proteins and various mitogen-activated protein kinases are important in bleb formation, and our preliminary research indicates that pharmacologic strategies to inhibit these interactions diminishes blebbing (Strunnikova et al. 2001).

Based on the epidemiological association of worse AMD with female gender, hypertension, cigarette smoking and other environmental co-factors, we have evaluated the impact of molecular surrogates of these factors on blebbing and regulation of molecules responsible for the breakdown and synthesis of extracellular matrix (Smith et al. 1997; Hyman et al. 2000; Evans 2001). We have found that estrogens, hypertension-associated hormones such as angiotensin, macrophage-derived cytokines, and other relevant factors can have a powerful influence on the severity of blebbing and subsequent matrix accumulation in the laboratory. This approach will allow us to screen potential therapeutic agents.

### 10.1.5.2
### Preclinical Evaluation in Animal Models

Animal models are extremely useful in preclinical testing of theories of disease pathogenesis and therapeutic interventions. However, until recently, no animal models for dry AMD were available (see Chap. 11; Espinosa-Heidmann et al. 2004; Ambati et al. 2003). The potential usefulness of animal models is illustrated by recent data from our laboratory to evaluate mechanisms for gender differences and environmental toxins (Cousins et al. 2003a, 2003b).

Women with early menopause appear at risk for worse AMD (Cousins et al. 2003b). We reasoned that estrogen deficiency might contribute to the onset or severity of AMD in females. We used a mouse model in which mice fed a high-fat diet and briefly exposed to blue-green light develop significant sub-RPE deposits and mild Bruch's membrane (BrM) thickening (Cousins et al. 2003b). We sought to delineate the role of gender and estrogen status in this model.

Both male and female 16-month-old mice developed qualitatively similar basal laminar deposit morphology, but the severity in terms of thickness, continuity, and content was significantly greater in female mice (see Fig. 10.3). Aged female mice also demonstrated a trend toward more severe endothelial changes and increased BrM thickening compared with age-matched male mice. Middle-aged mice with estrogen deficiency induced by ovariectomy also developed more severe deposits compared with sham-operated controls (see Fig. 10.4). However, ovariectomized mice that received high-dose estrogen supplementation also developed significant deposits, although they had a thinner BrM compared with mice that were estrogen-deficient (see Fig. 10.5). We concluded that female gender in aged mice and estrogen deficiency in middle-aged mice appears to increase the severity of sub-RPE deposit formation. Estrogen deficiency may increase susceptibility to sub-RPE deposit formation by dysregulating turnover of BrM, contributing to collagenous thickening and endothelial changes. Estrogen supplementation at the dosages used in this study does not appear to protect against sub-RPE deposit formation (Cousins et al. 2003b).

Another use for animal models is to test the contribution of environmental toxins in deposit formation. Some investigators have suggested that the incidence and severity of AMD has increased among populations exposed to western lifestyle and urban environments (Pauleikhoff and Koch 1995; la Cour et al. 2002). One possible

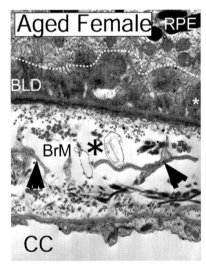 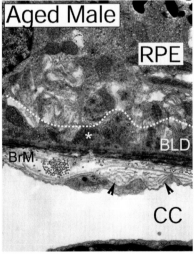

Fig. 10.3. Transmission electron microscopy of the outer retina and choroid from 16-month-old, fat-fed female or male mice exposed to blue-green light. *Left*: specimen from an aged female mouse revealed moderately thick sub-RPE deposits (*under white dotted line*) containing many banded structures (*white asterisk*) consistent with BLDs. BrM revealed marked collagenous thickening and scattered debris (*black asterisk*). Invasion by several cellular processes (*black arrows*) emanating from the CC was also evident. The endothelium revealed some loss of fenestrations with thickened basement membrane. *Right*: specimen from an aged male mouse revealed continuous moderately thick sub-RPE deposits (*under white dotted line*) containing banded structures (*white asterisk*) similar to those observed in the female. However, BrM was only mildly and irregularly thickened. The endothelial morphology appears slightly abnormal, with mild reduplication of the basement membrane (*arrowheads*), but evidence of severe injury or invasion was not observed. ×25,000. (*RPE* retinal pigment epithelium, *BLD* basal laminar deposits, *BrM* Bruch's membrane, *CC* choriocapillaris)

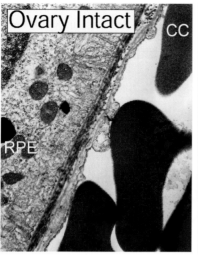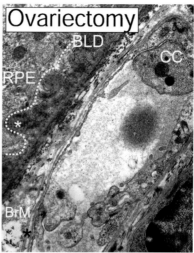

Fig. 10.4. Transmission electron microscopy of the outer retina and choroid from 9-month-old, fat-fed female mice exposed to blue-green light, undergoing sham surgery (*left*) or ovariectomy (*right*). *Left*: Specimen from a middle-aged sham-surgery (ovary intact) female mouse revealed no sub-RPE deposits and minimally thickened BrM. The CC appeared normal. *Right*: specimen from a middle-aged ovariectomized mouse revealed moderately continuous BLD (*under white dotted line*) with banded structures (*white asterisk*), irregular thickening of BrM, and normal choriocapillaris. Marked thickening and reduplication of the basement membrane (*black asterisk*) was evident. ×25,000. (*RPE* retinal pigment epithelium, *BLD* basal laminar deposits, *BrM* Bruch's membrane, *CC* choriocapillaris)

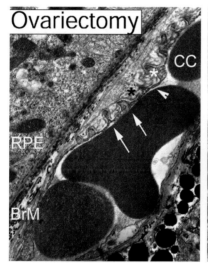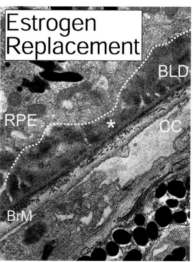

Fig. 10.5. Transmission electron microscopy of the outer retina and choroid from a 9-month-old, fat-fed female mice exposed to blue-green light, undergoing ovariectomy or ovariectomy with estrogen supplementation. *Left*: specimen from a middle-aged ovariectomized mouse revealed nodular BLDs and irregularly thickened BrM. The endothelium appeared hypertrophied (*white asterisk*) with irregular protrusions into BrM (*white arrows*), loss of fenestrations (*white arrowhead*), and reduplication of the basement membrane (*black asterisk*). No invasive processes were observed. *Right*: specimen from an ovariectomized mouse with estrogen replacement revealed moderately thick BLD (*under white dotted line*) of severity similar to the ovariectomized mice with banded structures (*white asterisk*). However, BrM appeared to be normally compact and the CC was normal. ×25,000. (*RPE* retinal pigment epithelium, *BLD* basal laminar deposits, *BrM* Bruch's membrane, *CC* choriocapillaris)

explanation is that toxic substances associated with western lifestyle might directly contribute to the formation of drusen and AMD (Hawkins et al. 1999; McCarty et al. 2001). Based on its biochemical properties, we reasoned that the oxidant hydroquinone, a prevalent component of cigarette smoke, automobile exhaust, and certain processed foods, might contribute to drusen pathogenesis. We evaluate the impact of oral feeding of this compound in mice (Cousins et al. 2003a).

Electron micrographs of HQ-exposed mice demonstrated that most mice examined revealed moderate deposits (Fig. 10.6). The sub-RPE changes were similar to the blue-light model, characterized by accumulation of moderately dense homogeneous material between the RPE and its basement membrane (Top, under white line; Cousins et al. 2002). Occasional blebs were observed, BrM was mildly thickened and banded structures were often present. This finding also indicated that two different oxidant stimuli, blue light and hydroquinone, can produce a common response in the RPE.

### 10.1.5.3
**Imaging Technologies**

The current standard technique to evaluate severity of dry AMD is fundus photography. The severity of drusen, pigmentary changes, and other features are usually stratified into four categories (Bird et al. 1995). Although reproducible, this method assumes that clinically apparent deposits on fundus photography are representative of the extent of deposit formation if the same eye were to be examined histologically. This assumption has never been evaluated.

Several imaging technologies are being developed that promise to provide additional information about the tissue changes associated with dry AMD. Among the best characterized are scanning laser ophthalmoscopy (SLO) and optical coherent tomography (OCT; Hee et al. 1996; Spraul et al. 1998; Ishiko et al. 2002). SLO can visualize lipofuscin-induced autofluorescence and pigmentary abnormalities. OCT can directly visualize the sub-RPE deposits, which can be quantified (see Fig. 10.7).

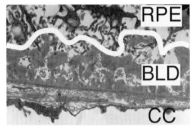
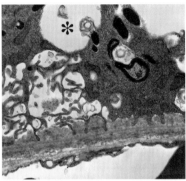

**Fig. 10.6.** Transmission electron microscopy of the outer retina and choroid from hydroquinone-exposed mice. *Top*: specimen from a hydroquinone diet revealed that sub-RPE changes were similar to the blue-light model, characterized by accumulation of moderately dense homogeneous material between the RPE and its basement membrane (*under white line*). *Middle*: specimen from a hydroquinone diet revealed similar changes as described above. Occasionally blebs were observed (*black asterisk*). *Bottom*: specimen from a hydroquinone diet revealed that banded structures were often present in the sub-RPE space while BrM was mildly thickened. These findings indicated that different oxidant stimuli (blue light and hydroquinone) can produce a common response in the RPE. ×25,000. (*RPE* retinal pigment epithelium, *BLD* basal laminar deposits, *CC* choriocapillaris)

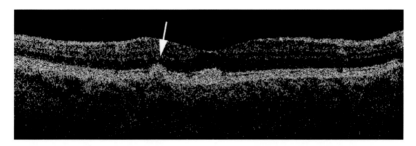

**Fig. 10.7.** Optical coherence tomography (*OCT*) of a human retina showing one of the newest imaging technologies that could provide additional information about tissue changes associated with dry AMD. The OCT image shows sub-RPE deposits, which can be spatially localized and quantified (i.e., *white arrow*)

Another new potential technology is scanning laser polarimetry, which measures the changes in the polarization properties of light as it passes through the retina. Deposits seem to induce increased depolarization in proportion to thickness (see Fig. 10.8). Other potential technologies in development include Raman spectroscopy of retinal carotenoid levels as well as new in vivo techniques to monitor real-time retinal metabolism.

Finally, improved technologies are becoming available to measure choroidal blood flow and to image choroidal vascular changes. Laser doppler flowmetry has demonstrated that 30% decreased blood flow is present in high-risk eyes (J. Grunwald, personal communication). Dynamic ICG indicates that some high-risk eyes with drusen demonstrate prominent increased vascularity under the macula, suggesting that intrachoroidal vascular remodeling may precede subretinal neovascularization (see Fig. 10.9; Hanutsaha et al. 1998). This observation may be useful to identify eyes at high risk for developing choroidal neovascularization.

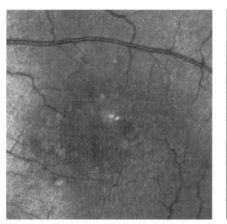
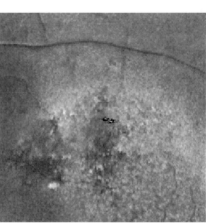
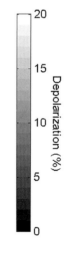

SLO Image         Depolarization Image

**Fig. 10.8.** Scanning laser polarimetry image of a human macula with dry AMD. *Left* panel shows standard red-free SLO image. *Right* panel shows extensive areas of "brightness" representing variable depolarization of light as it passed through clinically unapparent deposits

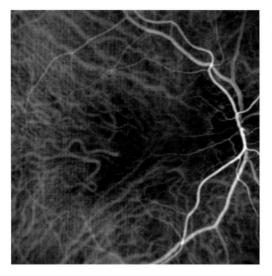 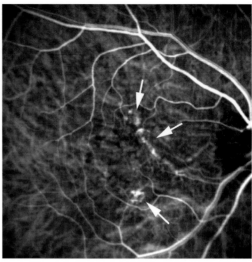

Fig. 10.9. Dynamic ICG images of normal (*left*) and dry AMD (*right*). The image from AMD demonstrates markedly increased vascularity underlying the macula (*white arrows*). This finding suggests intrachoroidal vascular remodeling, a precursor to subretinal neovascularization

### 10.1.5.4
### Biomarkers for Progression

The contribution (beneficial or harmful) of monocytes in AMD pathogenesis or progression is unknown. Theoretically, macrophages might mediate drusen resorption by scavenging and removing deposits (Killingsworth et al. 1990; Cousins and Csaky 2002). Conversely, macrophages may stimulate progression of drusen into CNV by releasing cytokines such as TNF-α or others into deposits, and by releasing factors that regulate CNV growth and severity (Killingsworth et al. 1990; Anderson et al. 2002; Grossniklaus et al. 2002).

Our laboratory has recently characterized TNF-α production by blood monocytes isolated from AMD patients and the results suggest that macrophage activation state, defined as TNF-α production, might serve as a predictor of risk for progression in patients with AMD (Cousins et al. 2004). The presence of inflammation in chronic diseases, irrespective of the cause, has been noted to serve as a risk factor for progression. For example, in atherosclerosis, high serum C-reactive protein levels is a strong predictor of myocardial infarction and is used as a biomarker to identify high-risk patients (Tracy 2003; Seddon et al. 2004). Similarly our research demonstrates that patients with AMD have a wide range in expression of TNF-α cytokine in culture, or messenger RNA (mRNA) in isolated monocytes. This high level of TNF-α mRNA correlated with a fivefold risk of neovascular AMD (Cousins et al. 2004). Implications of this data from analysis of monocyte TNF-α is of tremendous value because it might serve as a biomarker for risk of CNV formation in patients with early stages of AMD.

## 10.2
## Wet AMD

### 10.2.1
### Definition of Neovascular AMD

Wet or neovascular AMD is caused by choroidal neovascularization (CNV; Green 1999), the growth of pathologic new vessels from the choroid into the subretinal or sub-RPE location (Green 1999; Green and Harlan 1999). Clinically, CNV may appear as a greenish-gray lesion, often with exudative detachment of the neurosensory retina. A ring of hyperpigmentation may surround the lesion (Green and Harlan 1999; Maguire 2004). Although CNV are often conceptualized as well-defined capillary tubes, by light or electron microscopy, CNV demonstrate complex fibrovascular lesions (Green and Harlan 1999; Reynders et al. 2002; Hermans et al. 2003). Early intrachoroidal neovascularization is rarely detected clinically or histologically (Green and Harlan 1999). As choroidal neovascularization progresses, cells extend through the outer layer of Bruch's membrane and into the subretinal space usually associated with sub-RPE deposits (Green and Harlan 1999). In the majority of the cases, the vessels are located between the RPE and the remainder of Bruch's membrane (Gass 1987). Cellular components consist of many cell types in addition to endothelial cells (EC), including proliferating RPE, inflammatory cells (macrophages), vascular smooth muscle cells (VSMC), and poorly differentiated myofibroblastoid cells (Archer and Gardiner 1980; Lopez et al. 1996; Sarks et al. 1997; Green 1999). The importance of EC is self-evident, since they are required to form the lumens of perfused vascular tubes in the new vessels (Folkman and Shing 1992). However, the involvement of other cell types has been less well investigated.

Clinically, CNV are definitively identified by fluorescein angiography, a photographic imaging technique in which a fluorescent dye is injected intravenously, and then fluorescent dye accumulation is documented by photography using filter combinations to image fluorescent dye (see Fig. 10.10; Gass 1987). Although conceptually simple, the interpretation of angiography is surprising complex. The angiograph-

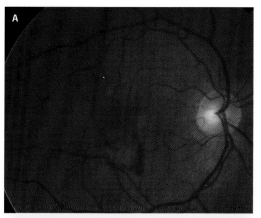
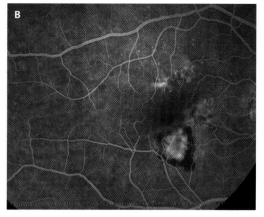
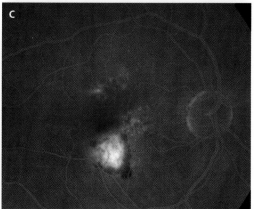

Fig. 10.10A–C. Color fundus photograph (A) and fluorescein angiogram (B, C) showing a choroidal neovascular (*CNV*) membrane, which characterizes the wet form of age-related macular degeneration (*AMD*). The color photograph in (A) shows a typical juxtafoveal CNV lesion (within 200 μm of the foveola) characterized by a greenish-gray elevated lesion with subretinal fluid and retinal hemorrhage. Angiographically an area of hyperfluorescence with well-demarcated boundaries on the early phase of fluorescein angiography characterizes classic CNV (B). In the latter phase of the angiogram, there is progressive dye leakage in the overlying subretinal space that typically obscures the boundaries of the CNV apparent in the early phase (C)

ic classification that is most widely utilized in clinical trials was developed by investigators in the Macular Photocoagulation Study during the early 1980s. In this classification, CNV may be described as classic, occult, or a mixture of both classic and occult (Macular Photocoagulation Study Group 1991c). In classic CNV the area of hyperfluorescence has well-demarcated boundaries on early phases of the angiogram, while the late phases are characterized as progressive dye leakage which eventually obscures the boundaries of the CNV. Two angiographic patterns are described as occult CNV; a fibrovascular pigment epithelial detachment which is an area of irregular elevation of the RPE that produces an ill-defined area of hyperfluorescence followed by stippled hyperfluorescence in the early phases and the late phases characterized by persistent leakage or staining of fluorescein. The other pattern that defines occult CNV is late leakage of undetermined source (Guyer et al. 1996; Maguire 2004).

## 10.2.2
## Pathogenic Mechanisms for CNV Formation

Different pathogenic mechanisms have been proposed to explain CNV formation in AMD. The most relevant paradigms are summarized below.

### 10.2.2.1
### Angiogenesis

Angiogenesis proposes that CNV develop in response to local overexpression of angiogenic growth factors (especially vascular endothelial growth factor or VEGF) produced by the RPE or other cells in the outer retina (Penfold et al. 1987; Schlingemann 2004; Witmer et al. 2003). Angiogenic factors then activate endothelial cells from the subjacent choriocapillaris to produce various degradatory enzymes, such as matrix metalloproteinase-2 and -9, allowing them to invade Bruch's membrane and the subretinal space (Steen et al. 1998; Hoffmann et al. 2002; Lambert et al. 2002; Holz and Miller 2003). Subsequent proliferation, migration, and then differentiation into a new vessel complex occurs.

The newly formed blood vessels leak plasma or whole blood. This paradigm has been extensively evaluated and has produced most of our current knowledge and therapies (Folkman 2003).

### 10.2.2.2
### Inflammation

Inflammation plays a role in many degenerative diseases such as atherosclerosis, Alzheimer's disease, glomerulosclerosis, rheumatoid arthritis, and pulmonary fibrosis (Gebicke-Haerter et al. 1996; Ross 1999; Walsh and Pearson 2001; Goncalves et al. 2003; Gupta and Pansari 2003; Ito and Ikeda 2003; Kramer et al. 2003; McGeer and McGeer 2003; Vizcarra 2003). The concept that inflammatory mechanisms play an important role in the pathogenesis of wet AMD has emerged as a paradigm shift in AMD (Hageman et al. 2001; Penfold et al. 2001; Johnson et al. 2001a, 2001b; Anderson et al. 2002; Grossniklaus et al. 2002). As discussed in Chap. 2, macrophages have been associated with drusen as well as with CNV (Anderson et al. 2002; Grossniklaus et al. 2002). Macrophages can contribute not only to the progression of drusen and induction of CNV (see above), but also in the regulation of the size and severity of CNV (Espinosa-Heidmann et al. 2003c; Sakurai et al. 2003). These observations and research studies implicate a potential pathogenic role for cytokines, chemical mediators, MMPs, mitogens, or angiogenic-factor release by macrophages from the choroid in promoting CNV (Grossniklaus et al. 2002). Furthermore, recent clinical trials support the use of anti-inflammatory agents as useful adjuncts in the treatment of CNV secondary to AMD (Ciulla et al. 2003; Spaide et al. 2003; Takahashi et al. 2003). This might be the best proof to support the hypothesis that inflammation plays an important pathogenic role in CNV (see Fig. 10.11).

### 10.2.2.3
### Repair to Injury

The response-to-injury hypothesis is a well-established concept in vascular pathology in many vascular beds, including atherosclerosis,

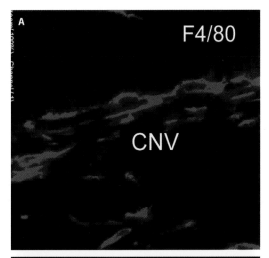

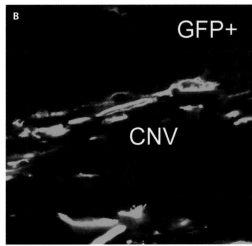

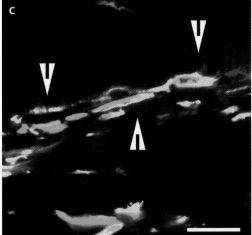

Fig. 10.11A–C. Confocal microscope photograph of a representative CNV lesion in mice with macrophages. Multiple groups have shown that CNV fibrovascular lesions are composed of multiple cells consisting in vascular smooth muscle cells, myofibroblastoid cells, proliferating retinal pigment epithelium cells, and inflammatory cells apart from the essential endothelial cells. Immunohistochemistry techniques were used to mark macrophages with F4/80 antibody (A). Similarly, to differentiate from resident macrophages after bone marrow transplantation experiments these cells show positivity for green fluorescent protein (GFP) markers confirming that they were recruited from the circulation after the animals were irradiated and reconstituted with donor GFP+ bone marrow (B). Colocalization was reflected by the presence of both markers (*yellow*) in the cells shown inside the CNV lesion (C, *white arrowheads*). Scale bar: 20 µm

transplant vasculopathy, post operative vasculopathies, and restenosis after angioplasty (Ross 1990, 1999; Azevedo et al. 2000; Lowe 2001). This paradigm proposes that, following an endothelial insult (inflammation, hemodynamic, physical), a fibroproliferative repair response is initiated, involving primarily vascular smooth muscle cells. These cells proliferate, producing growth factors and extracellular matrix-modifying proteins that ultimately result in a fibrovascular scar tissue. To our knowledge, this paradigm has not been applied to the pathogenesis of wet AMD. However, VSMC have been shown to invade Bruch's membrane with EC early in CNV formation (Sarks et al. 1997). By inference from the role of VSMC in other blood vessels, these cells might regulate EC proliferation and differentiation by production of specific growth factors, as well as synthesize extracellular matrix collagens and degradatory proteins (like matrix metalloproteinases) contributing to vascular remodeling and fibrosis (Abedi and Zachary 1995; Bundy et al. 2000).

#### 10.2.2.4
#### Vasculogenesis

Postnatal vasculogenesis is another new paradigm for the formation of pathological neovascularization in many different tissues (Murohara 2003). In contrast to angiogenesis (which proposes that the new vessel complex is derived from endothelial cells in an adjacent normal capillary), vasculogenesis proposes that the cellular components of the new vessels complex are derived from circulating vascular progenitors derived from bone marrow (Asahara et al. 1997, 1999; Isner and Asahara 1999). Bone marrow has been shown to contain stem and progenitor cells that demonstrate the capacity to enter the circulation, home into peripheral tissues, and differentiate into parenchymal tissues such as liver, heart, blood vessels, pancreas, muscle, and even neurons (Cornacchia et al. 2001; Grant et al. 2002; Laboratoryarge and Blau 2002; Religa et al. 2002; Ianus et al. 2003; Weimann et al. 2003). Evidence from several laboratories indicates that many cells in neovascularization or in other vascular reparative responses after injury are derived from circulating bone marrow progenitors (Bailey and Fleming 2003). It has been shown recently that progenitor cells might also, in part, contribute to CNV (see Fig. 10.12; Espinosa-Heidmann et al. 2003b; Sengupta et al. 2003). The physiological importance of this observation remains unclear, since no research has demonstrated that the recruited cells or the preexisting resident cells respond differently to angiogenic stimuli within the injured blood vessel. Some laboratories have proposed that bone marrow progenitors may be a source of cells to promote regeneration of damaged vessels in ischemic tissue (Harraz et al. 2001; Fujiyama et al. 2003). Alternatively, progenitors may also contribute to pathologic responses due to acquire abnormalities in function that cause them to contribute to age-related abnormalities in vascular responses to injury (Espinosa-Heidmann et al. 2003a).

### 10.2.2.5
**Environmental and Systemic Health Cofactors**

Epidemiologic studies indicate that both systemic health and environmental cofactors contribute to induction and regulation of severity (Klein 1999). Systemic health factors associated with wet AMD include aging, hypertension, cardiovascular diseases, gender, serum lipids and apolipoprotein E4 among the most important described to date (Snow and Seddon 1999; Hyman and Neborsky 2002; Ambati et al. 2003). Environmental factors may also play an important role for increased severity of CNV (Suner et al. 2004). The most important is cigarette smoking, which has been associated with a two- to fourfold increased incidence of neovascular AMD (Evans 2001). Other environmental factors with potential implication for AMD are light exposure and ionizing radiation, but the causal relationship between these and AMD has been difficult to prove (Beatty et al. 2000). Surprisingly, almost no research has been performed to evaluate the pathogenic contributions of these factors in wet AMD. Many potential therapeutic targets already available could be used if a causal relationship is found, which

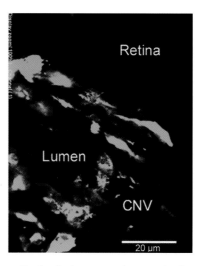

Fig. 10.12. Confocal microscope photograph of a vertical section of a CNV lesion in mice. Green fluorescent protein (GFP)-labeled cells can be seen surrounding a vascular lumen. GFP-labeled cells derived from bone marrow expressed markers for vascular smooth muscle cells (~40%), endothelial cells (~40%) or macrophages (~20%) confirming the contribution of vasculogenesis to experimental CNV. This alternative paradigm termed postnatal vasculogenesis states that the cellular components of the new vessel complex are derived in part from bone marrow-derived circulating vascular progenitors, which differentiate into mature endothelial cells or vascular smooth muscle cells

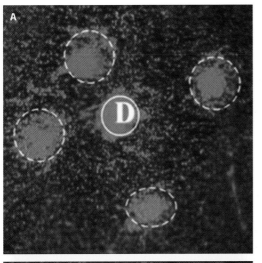
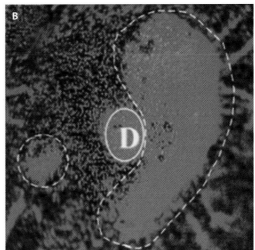
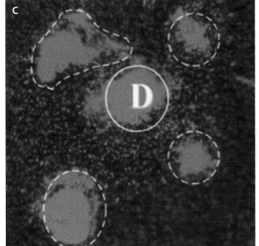

Fig. 10.13A–C. Flat-mount preparations of the posterior pole of mice eyes in which CNV lesions (*dashed white lines*) are depicted around the optic disc (A, control animals). Nicotine was shown to be important in promoting increased severity of CNV lesions by stimulating increased proliferation of vascular smooth muscle cells through a PDGF synergism (B, nicotine-treated animals). This effect of nicotine was blocked by concurrent administration of hexamethonium, a nicotinic receptor antagonist (C, nicotine + hexamethonium-treated animals). Nonneuronal nicotinic receptors may regulate size and severity of CNV, suggesting that cigarette smoke and nicotine replacement therapy may have adverse effects in patients with active CNV as well as be targets of potential future therapeutic intervention. (*D* optic disc)

would positively impact the pathogenesis of AMD. In this regard, for example, we have found that nicotine increases size and severity of experimental CNV, possibly by potentiating PDGF-mediated upregulation of proliferation of choroidal smooth muscle cells (Fig. 10.13). These results suggest that nonneuronal nicotinic receptor activation probably mediate some of the harmful effects of cigarette smoking and could be targets of potential future therapeutic intervention (Suner et al. 2004).

### 10.2.3
### Established or Evaluated Treatments for Neovascular AMD

#### 10.2.3.1
#### Thermal Laser Treatment

Shortly after the introduction of lasers into clinical ophthalmology practice in the late 1960s, clinicians established the idea of destroying CNV by thermal ablation with laser (Singerman 1988). This technique was formally evaluated in a series of clinical trials in the United States and Europe in the 1970s. Typical of these trials was the Macular Photocoagulation Study (MPS; Singerman 1988; Zimmer-Galler et al. 1995). The MPS data have demonstrated clear benefit of thermal laser photocoagulation in slowing vision loss for extrafoveal and juxtafoveal CNV, as well as for a small subgroup of subfoveal CNV (Macular Photocoagulation Study Group 1991a, 1991b, 1991c, 1994, 1996). However, only a small percentage of patients with CNV qualify for treatment by the MPS criteria (Macular Photocoagulation Study Group 1991c; Moisseiev et al. 1995). Further, the data indicate a high rate of recurrence, and many patients lost vision due to the detrimental effect of thermal damage to the photoreceptors (Mittra and Singerman 2002). Nevertheless, thermal laser ablation of CNV remains the gold standard by which other therapies must be evaluated.

#### 10.2.3.2
#### Photodynamic Therapy

Photodynamic therapy (PDT) is the combined use of low-energy light and a photosensitizing agent to induce vascular occlusion (Harding 2001; Lim 2002; Hunt and Margaron 2003; Wormald et al. 2003). The generation of free radicals and other oxidants by the light-activated photosensitizing agent results in damage to the blood vessel endothelium, leading to endothelial death and vascular thrombosis. A number of experimental agents have been evaluated in preclinical studies, but at least three drugs have been evaluated in clinical trials (Lim 2002). Verteporfin is the most successful and widely studied of these photosensitizing drugs (TAP Study Group and VIP Study Group 2002). The TAP study investigated the use of Verteporfin versus placebo in patients with subfoveal CNV secondary to AMD, while the VIP report 2 summarized the 1- and 2-year results of the Verteporfin clinical trial for occult CNV secondary to AMD (TAP Study Group 1999; Bressler 2001; VIP Study Group 2001). These studies show that CNV patients, particularly those with a predominantly classic component, have a reduced risk of moderate visual loss at 12 and 24 months (Lim 2002). While encouraging, this subgroup constitutes less than one-third of patients with neovascular AMD (Donati et al. 1999; Bressler 2001). More recently the VIP study has demonstrated some benefit for patients with pure occult CNV after 2 years of treatment with PDT (VIP Study Group 2001). However, the principal problem with PDT is controversial cost-effectiveness, especially the high cost of the drug, the need for many retreatments, and the continuing visual decline that most patients experience even with treatment. Several other photodynamic therapy trials are underway. Other photosensitizing agents with potential human applications for the treatment of CNV include motexafin lutetium (lutetium texaphyrin/LuTex), tin ethyl etiopurpurin (SnET2/Purlytin), and mono-L-aspartyl chlorin e6 (Npe6; Moshfeghi et al. 1998; Blumenkranz et al. 2000; Mori et al. 2001; Nakashizuka et al. 2001). The first two agents have undergone clinical trials for the treatment of wet AMD, but have not been approved for human use. The third has shown promising results in nonhuman primates.

#### 10.2.3.3
#### Radiation Therapy

Gamma irradiation is well known to inhibit cellular proliferation but also induces vascular injury (Archer and Gardiner 1994; Michalowski 1994). Based on the assumption that CNV represents actively dividing vascular tissue with increased sensitivity to ionizing radiation, several investigators have proposed the use of external-beam gamma irradiation for the treatment of CNV (Archer and Gardiner 1994; Valmaggia et al. 2002). In the AMD radiation study, patients were randomized to receive or not a fractionated radiation dose of 16 Gy divided over 8 days. A 1- to 2-year follow-up was completed and no significant differences were found for visual acuity in patients with subfoveal CNV who received radiation therapy vs. the sham radiation (Bellmann et al. 2003). Another multicenter study on radiation therapy in AMD showed significantly better contrast sensitivity but not vision after 2 years (Hart et al. 2002), while a prospective, double-masked, randomized clinical trial at 1 year follow-up showed neither beneficial nor harmful effects for subfoveal CNV complicating AMD (Marcus et al. 2001).

## 10.2.4
### Therapies Currently in Clinical Trial

#### 10.2.4.1
**Transpupillary Thermotherapy**

Transpupillary thermotherapy (TTT) is the use of infrared laser to induce mild, localized hyperthermia of the choroid and CNV complex (Subramanian and Reichel 2003). Infrared energy penetrates to the choroid and RPE, causing localized temperature rise while minimizing absorption of energy in the neurosensory retina. The rationale is that hyperthermia will induce endothelial cell damage that leads to thrombus formation and occlusion of neovessels (Subramanian and Reichel 2003; Rogers and Reichel 2001). Several retrospective case series show that TTT lessens leakage in most eyes and results in visual improvement in many cases (Reichel et al. 1999; Newsom et al. 2001; Rogers and Reichel 2001). The Transpupillary Thermotherapy for CNV trial (phase III) is currently in progress to compare the effectiveness of TTT with sham treatment for occult CNV (Subramanian and Reichel 2003).

#### 10.2.4.2
**Feeder-Vessel Laser**

Most laser therapies directly treat the new vessel capillary complex, which usually lies under the center of the macula. However, many CNV lesions consist of three distinct, neovascular anatomical components: the subretinal/sub-RPE capillary complex; an afferent "feeder" arteriole that connects the capillary to a larger choroidal artery; and an efferent "draining" venule that removes blood from the capillary into the choroidal venous system (Schneider et al. 1998; Shiraga et al. 1998; York et al. 2000). The hypothesis of feeder-vessel laser is that laser-induced occlusion of the afferent feeder arteriole, which is usually eccentric from the foveal center, will prevent perfusion of the capillary complex, thereby preventing exudation and leakage into the macular area (Shiraga et al. 1998). Also, loss of capillary perfusion is supposed to stimulate involution of the CNV. Until recently identification of these feeder vessels has been difficult, but new, dynamic imaging of the choroid with indocyanine green (ICG) dye has allowed for more precise detection of these vessels (York et al. 2000; Flower 2002; Yamamoto 2003). Then pulse-diode laser photocoagulation can be used to attempt closure of the vessels. Several case reports indicate the potential utility of this technique (Staurenghi et al. 1998; Desatnik et al. 2000). A phase I/II interventional study is underway to provide information on the feasibility of standardizing this procedure and estimating its potential efficacy (NIH clinical research studies – Protocol number 01-EI-0208).

#### 10.2.4.3
**Anti-inflammatory Therapy**

As discussed in detail in Chap. 3, CNV usually contain histopathologic evidence of inflammation, especially macrophages (Sarks et al. 1997; Grossniklaus et al. 2000; Penfold et al. 2001; Grossniklaus et al. 2002). These cells secrete numerous proangiogenic factors, including VEGF, prostaglandins, matrix metalloproteinases, and others (Oh et al. 1999; Grossniklaus et al. 2002). Based on these observations, two small, non-comparative case series were done to evaluate the effectiveness of intraocular injection of triamcinolone acetonide, both suggesting it was effective in lessening leakage from CNV (Navajas et al. 2003; Rechtman et al. 2003). A phase-I trial is being organized in the US (NIH clinical research studies – Protocol number 04-EI-0013).

By its intrinsic nature, PDT is a proinflammatory stimulus, associated with a posttreatment increase in macular edema and inflammation. Recently, investigators have suggested combined treatment of CNV with PDT and anti-inflammatory agents with intravitreal triamcinolone acetonide (noncomparative case series) and PDT has shown improvement in visual acuity and less-frequent requirement for retreatments in short-term follow-up (Spaide et al. 2003). Another multicenter, randomized, prospective phase-II clinical trial is underway of a cyclooygenase-2 (COX-2) inhibitor on the response to PDT in patients with subfoveal CNV secondary to AMD (NIH clinical research studies – Protocol number 02-EI-0257). Celecoxib is a COX-2 enzyme inhibitor (nonsteroidal anti-inflammatory agent) that acts by inhibiting prostaglandins expressed at inflammatory sites by leukocytes (polymorphonuclear neutrophils and monocytes/macrophages), as well as by activated mesenchymal cells (fibroblasts), which might as well potentiate the results of PDT treatment (Seibert et al. 1999; Deviere 2002). In addition, Celecoxib has been shown to have direct antiangiogenic effects through both increasing the levels of the potent angiostatic protein endostatin and reducing the levels of PDGF2-mediated VGEF secretion through the HIF1-pathway. Another multicenter phase-II clinical trial is studying an intravitreous implant (Retisert Implant) containing fluocinolone in isolation or in conjunction with PDT (Control Delivery Systems – Bausch and Lomb Pharmaceuticals – CDS FL 004).

### 10.2.4.4
**Broad-Acting Antiangiogenic Drugs**

As described above, angiogenesis involves a process of endothelial activation, invasion, proliferation, and differentiation (Folkman and Shing 1992; Diaz-Flores et al. 1994; Folkman 2003). Each stage represents a potential target for anti-angiogenesis therapy (Kieran and Billett 2001; Tosetti et al. 2002).

#### 10.2.4.4.1
*Antiangiogenic Steroids*

The Anecortave Acetate Clinical Study Group has initiated a trial to compare the clinical efficacy of anecortave acetate versus placebo in patients with primary or recurrent subfoveal CNV secondary to AMD (D'Amico et al. 2003). Anecortave acetate is a synthetic corticosteroid derivative specifically modified to eliminate its corticosteroid activity in vivo, eliminating side effects such as elevated intraocular pressure or accelerated cataract progression (Ciardella et al. 2002; D'Amico et al. 2003). It is a unique angiostatic agent that inhibits both urokinase-like plasminogen activator and matrix matalloproteinase-3, two enzymes necessary for vascular endothelial cell migration during blood vessel growth (Blei et al. 1993; DeFaller and Clark 2000; Penn et al. 2001; D'Amico et al. 2003). The 6- and 12-month analyses of clinical safety and efficacy have shown that anecortave acetate in a single dose of 15 mg as a posterior juxtascleral injection onto the posterior scleral surface is effective for preserving or improving vision and for inhibiting lesion growth in patients with subfoveal AMD (D'Amico et al. 2003; Slakter 2003). Its antiangiogenic activity has been proven effective as well in several neovascular models, including rabbit cornea neovascularization, hypoxic retinal neovascularization in rats and kitten, and murine uveal melanoma (BenEzra et al. 1997; Clark et al. 1999; McNatt et al. 1999; Penn et al. 2001). Topical ocular anecortave acetate significantly inhibits the regrowth of ocular fibrovascular membranes in patient with recurrent pterygium (DeFaller and Clark 2000). It is important to remark that no statistically significant difference is conferred by using anecortave acetate in conjunction with Verteporfin, but it is suggested that the combination maintains visual acuity better than PDT alone. Of the patients who received combined therapy, 78% had no significant loss of visual acuity compared with 67% for the group that was treated with PDT alone (Alcon Press Release Results for Anecortave Acetate Therapy for Wet Age-Related Macular Degeneration – ARVO 05/06/02). A phase-III pivotal trial is now recruiting patients with predominantly classic CNV who will be randomized to treatment with either anecortave acetate 15 mg or PDT with Verteporfin (Alcon Research- Study ID Number C-01-99). A clinical trial to evaluate anecortave for the prevention of CNV in high-risk eyes is currently being organized.

*10.2.4.4.2*
*Other Antiangiogenic Drugs*

Squalamine, a broad-spectrum aminosterol antibiotic originally isolated from the dogfish shark *Squalus acanthias* has been reported to inhibit tumor-induced angiogenesis and tumor growth (Moore et al. 1993; Sills et al. 1998; Teicher et al. 1998; Higgins et al. 2000; Rao et al. 2000). On a cellular level, squalamine inhibits growth factor-mediated endothelial cell proliferation and migration, including that of VEGF (Sills et al. 1998; Higgins et al. 2000). Higgins et al. have shown in a mouse model of oxygen-induced retinopathy that squalamine is effective in reducing retinal neovascularization (Higgins et al. 2000). A more recent publication from Genaidy et al. has shown that systemic but not intravitreally injected squalamine on iris neovascularization in monkeys is effective in inhibiting the development of iris neovascularization and causing partial regression of new vessels (Genaidy et al. 2002). A phase I/II clinical trial of squalamine antiangiogenic properties for the treatment of AMD recently emitted a press release in which there is evidence that squalamine produces a shrinkage of size of CNV lesions in some patients, and stabilization of lesions in others (Genaera Corporation press release – 10/07/03). Early trends for visual acuity shows some improvement of up to three lines of vision, in some patients, and stabilization of vision in all patients thus far in the 4 months after initiation of therapy. The study is evaluating visual acuity, ocular angiography, and fundus photography in 40 patients with each angiographic subtype of wet AMD, including classic, occult, and mixed CNV for the safety and efficacy of squalamine treatment.

Combretastatin A-4 phosphate (CA4P), derived from the bark of the South African willow tree, is a structural analogue of colchicine that binds tubulin and causes necrosis and shrinkage of tumors by damaging their blood vessels (Pettit et al. 1989; Woods et al. 1995; Griggs et al. 2002). Recently it has been shown that administration of this agent result in microthrombus formation of new vessels surrounding a hyperplastic thyroid, indicating that its effects are not limited to tumor vasculature (Griggs et al. 2001a). In a paper Nambu H et al. have suggested that CA4P suppresses the development of VEGF-induced neovascularization in the retina and both blocks development and promotes regression of CNV in a murine model of VEGF overexpression and mouse model of experimental CNV (Nambu et al. 2003). Approximately 20 patients will be undergoing a phase I/II safety study in patients with wet AMD approved by the FDA (Oxigene Protocol number: FBO-206). In addition to the wet AMD trial, CA4P is being studied in four clinical trials in cancer patients (Oxigene Protocol Numbers: UKR-104, UKCTC-207, ICC-2302, and PH1/092; Griggs et al. 2001b, 2002; Tozer et al. 2002).

**10.2.4.5**
**Targeted Molecular Therapy**

A different approach to the treatment of ocular neovascularization is antiangiogenic therapy with targeted molecular therapy. One of the potential targets for antiangiogenic therapy is VEGF.

*10.2.4.5.1*
*RhuFab (Lucentis)*

Growth factors, cytokines, or other types of signaling molecules regulate important cell functions in the disease process by activating specific receptors on target cells such as endothelium (Distler et al. 2003). Antibodies, which bind to specific growth factors, thereby preventing their activation of cellular receptors, have recently become important therapeutic agents (Mordenti et al. 1999). RhuFab is genetically modified anti-VEGF monoclonal antibody representing the small antigen-binding fragment of a recombinant humanized anti-VEGF monoclonal antibody (Krzystolik et al. 2002). Safety and efficacy data from several phase I/II studies using monthly intravitreal injections of Rhu-Fab indicate that the drug is well tolerated and highly effective in the treatment of neovascular AMD (Krzystolik et al. 2002). A significant number of patients have improved visual acuity. Confirmatory phase-III trials are underway.

*10.2.4.5.2*
*Macugen (Eyetech 001)*

Aptamers are oligonucleotides designed to bind to specific protein targets, thereby interfering with receptor binding or other functions (similar to antibodies; Aiello et al. 1995). Macugen is a polyethylene glycol (PEG)-conjugated oligonucleotide aptamer that binds to the major soluble human VEGF isoform, $VEGF_{165}$, with high specificity and affinity. Pegylation improves the pharmacokinetics by decreasing clearance from the vitreous (Drolet et al. 2000). When injected into the vitreous, the compound is typically conjugated to polyethylene glycol, which slows its disappearance (Eyetech Study Group 2002). Phase-II trials show good safety and tolerance of multiple intravitreal injections, with a trend toward improved vision (Eyetech Study Group 2003). The phase III trial has been completed, but the results are currently unavailable. While the aptamer only binds one isoform of VEGF, rhuFab has a broader spectrum of activity, binding all four known isomers of VEGF (Presta et al. 1997; Csaky 2003).

### 10.2.4.6
### Gene Therapy

Gene therapy is a family of technologies designed to transfer a specific gene into a specific target cell for the purposes of inducing the target to endogenously synthesize the protein product of the gene in question (Harjai et al. 2002). Typically, gene transfer is achieved by use of a disabled (i.e., replication-defective) virus as a vector containing the specified gene of choice, especially adenovirus, retrovirus, or adeno-associated virus (Burton et al. 2002; Lai et al. 2002). Each has their respective strengths and weaknesses as a vector. Also, other gene-transduction technologies such as oligonucleotides are being developed (Andersen et al. 2002).

Although replacement of genetically abnormal genes is the most ambitious goal of gene therapy, this technology can also be used to induce a target cell to produce sustained synthesis of a protein with beneficial properties in disease treatment. Recently it has been demonstrated that pigment epithelium-derived factor (PEDF) is a potent antiangiogenic agent for various models of neovascularization (Rasmussen et al. 2001). Periocular or intravitreal injection of an adenoviral vector encoding PEDF has been shown to inhibit choroidal neovascularization in a mouse model of experimental CNV (Mori et al. 2002; Renno et al. 2002; Gehlbach et al. 2003). Based on these data, a phase-I, single intravitreal administration of adenovirus PEDF.11D in humans is underway to test the inhibition of subfoveal CNV secondary to AMD (Rasmussen et al. 2001).

### 10.2.4.7
### Surgical Trials

*10.2.4.7.1*
*Surgical Extraction of CNV or Blood*

In the early 1990s, surgical removal of the CNV was introduced as an alternative therapy to laser photocoagulation for neovascular AMD (de Juan and Machemer 1988; Submacular Surgery Trials Pilot Study 2000a; Stone and Sternberg 2002). The rationale for the procedure was that, unlike laser photocoagulation, surgical removal of CNV might not destroy all central macular photoreceptors, and evacuation of blood may prevent toxicity from breakdown products of hemorrhage (Submacular Surgery Trials Pilot Study 2000a; Stone and Sternberg 2002). Pilot data from various case reports have suggested promising results, although recurrent CNV is a common complication (de Juan and Machemer 1988; Vander et al. 1991; Mandelcorn and Menezes 1993; Li and Gao 1995; Castellarin et al. 1998; Gandorfer et al. 1998). However, a committee appointed by the National Eye Institute of the National Institutes of Health in the US has recommended that surgery is not generally indicated until further research verifies the effectiveness and safety of the procedure (Submacular Surgery Trials Pilot Study 2000a, 2000b; Stone and Sternberg 2002). This recommendation led to the organization of the Submacular Surgery Trials (SST) study group, which has initiated a trio of multicenter, randomized clinical trials with the goal of determining whether surgical removal of subfoveal CNV stabilizes or improves vision more often than observation (NIH clinical research studies – Protocol number NEI-52). Three groups of patients are under study – Group B (blood), Group N (new CNV), and Group H (histoplasmosis/idiopathic; Stone and Sternberg 2002). The final results are pending. In 2001 the Swedish national survey of surgical excision for submacular CNV was published (Berglin et al. 2001). The study compared visual outcomes after surgical removal of subfoveal CNV between patients younger or older than 50 years of age. It was concluded that surgical removal of submacular CNV does not appear to improve visual acuity in patients older than 50 years of age (i.e., with AMD).

*10.2.4.7.2*
*Macular Translocation*

Surgical translocation of the macula is another surgical alternative to laser or drug treatment for wet AMD. In this technique, the neurosensory retina is surgically detached, then shifted away from the underlying subfoveal CNV to overly normal RPE (Au Eong et al. 2001; Fujii et al. 2002). The CNV can then be treated with thermal laser or other techniques. Two techniques have been developed. In the macular translocation surgery 360° (MTS360) technique, the entire retina is detached with 360° retinectomy (segmentation of the retina from its peripheral insertion) allowing macular rotation 5–20° from its original location (Terasaki 2001; Toth and Freedman 2001). The procedure is not without significant risks for intraoperative and postoperative complications such as retinal detachment, macular pucker, increased lens opacity in the phakic eyes, diplopia, and tilted image (Toth and Freedman 2001; Aisenbrey et al. 2002; Fujii et al. 2002). Also, the vitreous cavity is usually filled with silicone oil to tamponade the retina in place, which often requires surgical removal or can lead to associated toxicity (Kampik and Gandorfer 2000). Subsequent extraocular muscles movement is usually required to realign the eye. However, recent case reports indicate that, in the hands of experienced surgeons, MTS360 can achieve excellent results in selected cases (Toth and Freedman 2001; Aisenbrey et al. 2002).

A less-extensive technique involves limited macular translocation (LMTS; de Juan and Fujii 2001; Fujii et al. 2001b, 2002). It consists of partial-thickness scleral resections near the equator at either the superotemporal or the inferotemporal quadrant followed by a near-total retinal detachment (de Juan et al. 1998; de Juan and Vander 1999; de Juan and Fujii 2001). The resected sclera edges are sutured, causing shortening of the sclera with subsequent reattachment of the retina, resulting in translocation of the fovea to an area overlying nonfoveal RPE and choroid. Preliminary case reports indicate the technique seems to be effective in selected cases (de Juan and Vander 1999; Ho 2000; Fujii et al. 2001a; Ohji et al. 2001; Sullivan et al. 2002; Chang et al. 2003). Currently a phase I/II clinical trial is under way comparing this surgical technique with PDT in eyes with subfoveal CNV secondary to AMD.

## 10.2.5
## Future Research

### 10.2.5.1
### Basic Scientific and Preclinical Research

As outlined above, many recent, exciting developments are occurring in clinical trials for the treatment of wet AMD. To a great extent, this progress is the result of outstanding basic scientific and preclinical research. In particular, excellent research data are available that address the paradigm of angiogenesis for neovascular AMD with in vitro systems and various animal models. However, less well developed are data supporting the potential contributions for the other paradigms. As described above, ongoing work from our and other laboratories suggests how research in these other paradigms of CNV pathogenesis may result in improved knowledge and new therapeutic ideas.

### 10.2.5.2
### Prevention of Vision Loss

CNV causes the most severe cases of vision loss in AMD (Green 1999), but minimal research is available to define the specific mechanism for retinal dysfunction. At least three possibilities explain vision loss: permanent retinal dysfunction due to death of photoreceptors (Young 1987; Curcio et al. 1996); retinal dysfunction due to leakage (Donati et al. 1999; Bressler 2001; Lim 2002); or dysfunction due to some other property, such as disruption of the bipolar-photoreceptor synapse (Caicedo et al. 2003). The latter two are potentially reversible. In fact, the phase-II trials with Lucentis described above indicate that blockade of VEGF and improvement of retinal leakage results in improved vision in some patients, proving reversible vision loss.

Recent work in our laboratory has suggested another mechanism for preventable or reversible vision loss in CNV. We hypothesize that the following sequences of events are induced in the retina overlying CNV. First, blood-derived macrophages invade the retina at onset of CNV, which infiltrate into the plexiform layers and co-localize with Muller cells (Fig. 10.14). Then, Muller cells become activated by mediators produced by infiltrating macrophages. Muller cell activation results in subsequent loss of neurotrophic factor production and other changes in function. Finally, the changes in the activated Muller cells result in disruption of the bipolar photoreceptor synapse (Fig. 10.15; Caicedo 2003; Caicedo et al. 2003). Much more work needs to be done in order to clarify and fully understand the pathogenesis that leads to vision loss in AMD.

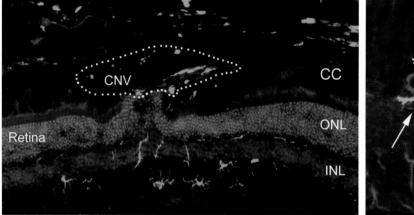
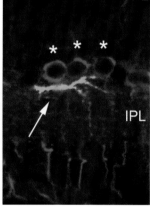

Fig. 10.14. Fluorescent microscope photograph of a vertical section of a choroidal neovascular (CNV) lesion and retina under CNV. *Left*: Macrophages (*green*) invade the retina under CNV and predominantly localize to the outer and inner borders of the inner nuclear layer were they can be seen in close association with glial cells, such as Müller cells. Adjacent regions of the retina away from CNV are practically devoid of macrophages. *Right*: Confocal microscope photograph of a vertical section of a retina under a CNV lesion showing activated Muller cells (*white asterisks*) in relation to a recruited circulating macrophage (*white arrow*). (*CC* choriocapillaris, *ONL* outer nuclear layer, *OPL* outer plexiform layer, *INL* inner nuclear layer, *IPL* inner plexiform layer)

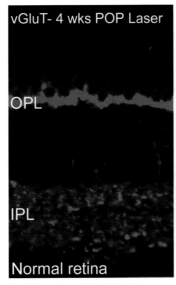
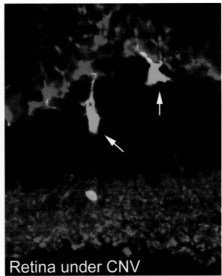

Fig. 10.15. Fluorescent microscope photograph of a vertical section of a retina under a choroidal neovascular (*CNV*) membrane. *Left*: Control retina showing a normal distribution of a synaptic marker (vGluT, *red*). *Right*: Retina under CNV 4 weeks after experimental induction of CNV lesion. There is an abnormal redistribution of the synaptic marker at the level of the outer plexiform layer where macrophages (*green/white arrows*) are clearly associated. Inner plexiform layer is preserved in comparison with the control. (*VGluT* vesicular glutamate transporter, *OPL* outer plexiform layer, *IPL* inner plexiform layer)

## References

Abdelsalam A, Del Priore L, Zarbin MA (1999) Drusen in age-related macular degeneration: pathogenesis, natural course, and laser photocoagulation-induced regression. Surv Ophthalmol 44:1–29

Abedi H, Zachary I (1995) Signalling mechanisms in the regulation of vascular cell migration. Cardiovasc Res 30:544–556

Aiello LP, Pierce EA, Foley ED, Takagi H, Chen H, Riddle L, Ferrara N, King GL, Smith LE (1995) Suppression of retinal neovascularization in vivo by inhibition of vascular endothelial growth factor (VEGF) using soluble VEGF-receptor chimeric proteins. Proc Natl Acad Sci USA 92:10457–10461

Aisenbrey S, Lafaut BA, Szurman P, Grisanti S, Luke C, Krott R, Thumann G, Fricke J, Neugebauer A, Hilgers RD, Esser P, Walter P, Bartz-Schmidt KU (2002) Macular translocation with 360 degrees retinotomy for exudative age-related macular degeneration. Arch Ophthalmol 120:451–459

Ambati J, Ambati BK, Yoo SH, Ianchulev S, Adamis AP (2003) Age-related macular degeneration: etiology, pathogenesis, and therapeutic strategies. Surv Ophthalmol 48:257–293

Andersen MS, Sorensen CB, Bolund L, Jensen TG (2002) Mechanisms underlying targeted gene correction using chimeric RNA/DNA and single-stranded DNA oligonucleotides. J Mol Med 80:770–781

Anderson DH, Mullins RF, Hageman GS, Johnson LV (2002) A role for local inflammation in the formation of drusen in the aging eye. Am J Ophthalmol 134:411–431

Archer DB, Gardiner TA (1980) Experimental subretinal neovascularization. Trans Ophthalmol Soc UK 100:363–368

Archer DB, Gardiner TA (1994) Ionizing radiation and the retina. Curr Opin Ophthalmol 5:59–65

AREDS (2001) A randomized, placebo-controlled, clinical trial of high-dose supplementation with vitamins C and E and beta carotene for age-related cataract and vision loss: AREDS report no. 9. Arch Ophthalmol 119:1417–1436

Asahara T, Murohara T, Sullivan A, Silver M, van der Zee R, Li T, Witzenbichler B, Schatteman G, Isner JM (1997) Isolation of putative progenitor endothelial cells for angiogenesis. Science 275:964–967

Asahara T, Takahashi T, Masuda H, Kalka C, Chen D, Iwaguro H, Inai Y, Silver M, Isner JM (1999) VEGF contributes to postnatal neovascularization by mobilizing bone marrow-derived endothelial progenitor cells. EMBO J 18:3964–3972

Au Eong KG, Pieramici DJ, Fujii GY, Ng EW, Humayun MS, Maia M, Harlan JB Jr, Schachat AP, Beatty S, Toth CA, Thomas MA, Lewis H, Eckardt C, Tano Y, de Juan E (2001) Macular translocation: unifying concepts, terminology, and classification. Am J Ophthalmol 131:244–253

Aumailley M, Gayraud B (1998) Structure and biological activity of the extracellular matrix. J Mol Med 76: 253–265

Azevedo LC, Pedro MA, Souza LC, Souza HP de, Janiszewski M, Luz PL da, Laurindo FR (2000) Oxidative stress as a signaling mechanism of the vascular response to injury: the redox hypothesis of restenosis. Cardiovasc Res 47:436–445

Bailey AS, Fleming WH (2003) Converging roads: evidence for an adult hemangioblast. Exp Hematol 31: 987–993

Bartlett H, Eperjesi F (2003) Age-related macular degeneration and nutritional supplementation: a review of randomised controlled trials. Ophthalmic Physiol Opt 23:383–399

Beatty S, Koh H, Phil M, Henson D, Boulton M (2000) The role of oxidative stress in the pathogenesis of age-related macular degeneration. Surv Ophthalmol 45:115–134

Belkhiri A, Richards C, Whaley M, McQueen SA, Orr FW (1997) Increased expression of activated matrix metalloproteinase-2 by human endothelial cells after sublethal $H_2O_2$ exposure. Laboratory Invest 77: 533–539

Bellmann C, Unnebrink K, Rubin GS, Miller D, Holz FG (2003) Visual acuity and contrast sensitivity in patients with neovascular age-related macular degeneration. Results from the Radiation Therapy for Age-Related Macular Degeneration (RAD-) Study. Graefes Arch Clin Exp Ophthalmol 241:968–974

BenEzra D, Griffin BW, Maftzir G, Sharif NA, Clark AF (1997) Topical formulations of novel angiostatic steroids inhibit rabbit corneal neovascularization. Invest Ophthalmol Vis Sci 38:1954–1962

Berglin L, Algvere P, Olivestedt G, Crafoord S, Stenkula S, Hansson LJ, Tomic Z, Kvanta A, Seregard S (2001) The Swedish national survey of surgical excision for submacular choroidal neovascularization (CNV). Acta Ophthalmol Scand 79:580–584

Bird AC, Bressler NM, Bressler SB, Chisholm IH, Coscas G, Davis MD, Jong PT de, Klaver CC, Klein BE, Klein R (1995) An international classification and grading system for age-related maculopathy and age-related macular degeneration. The International ARM Epidemiological Study Group. Surv Ophthalmol 39: 367–374

Blei F, Wilson EL, Mignatti P, Rifkin DB (1993) Mechanism of action of angiostatic steroids: suppression of plasminogen activator activity via stimulation of plasminogen activator inhibitor synthesis. J Cell Physiol 155:568–578

Blumenkranz MS, Woodburn KW, Qing F, Verdooner S, Kessel D, Miller R (2000) Lutetium texaphyrin (Lu-Tex): a potential new agent for ocular fundus angiography and photodynamic therapy. Am J Ophthalmol 129:353–362

Bressler NM (2001) Photodynamic therapy of subfoveal choroidal neovascularization in age-related macular degeneration with verteporfin: two-year results of 2 randomized clinical trials-tap report 2. Arch Ophthalmol 119:198–207

Bressler NM, Bressler SB, Fine SL (1988) Age-related macular degeneration. Surv Ophthalmol 32:375–413

Bundy RE, Marczin N, Birks EF, Chester AH, Yacoub MH (2000) Transplant atherosclerosis: role of phenotypic modulation of vascular smooth muscle by nitric oxide. Gen Pharmacol 34:73–84

Burton EA, Fink DJ, Glorioso JC (2002) Gene delivery using herpes simplex virus vectors. DNA Cell Biol 21:915–936

Caicedo A (2003) Photoreceptor synapses are disrupted after experimental choroidal neovascularization. Soc Neurosci 816:3 (abstract)

Caicedo A, Espinosa-Heidmann DG, Cousins SW (2003) Glutamate transporters at photoreceptor synapses are downregulated after experimental choroidal neovascularization. Invest Ophthalmol Vis Sci E (Abstract 3935)

Castellarin AA, Nasir MA, Sugino IK, Zarbin MA (1998) Clinicopathological correlation of primary and recurrent choroidal neovascularisation following surgical excision in age related macular degeneration. Br J Ophthalmol 82:480–487

Chang AA, Tan W, Beaumont PE, Zeldovich A (2003) Limited macular translocation for subfoveal choroidal neovascularization in age-related macular degeneration. Clin Exp Ophthalmol 31:103–109

Choroidal Neovascularization Prevention Trial Research Group (1998) Choroidal neovascularization in the Choroidal Neovascularization Prevention Trial. Ophthalmology 105:1364–1372

Choroidal Neovascularization Prevention Trial Research Group (2003) Laser treatment in fellow eyes with large drusen: updated findings from a pilot randomized clinical trial. Ophthalmology 110:971–978

Chung M, Lotery AJ (2002) Genetics update of macular diseases. Ophthalmol Clin North Am 15:459–465

Ciardella AP, Donsoff IM, Guyer DR, Adamis A, Yannuzzi LA (2002) Antiangiogenesis agents. Ophthalmol Clin North Am 15:453–458

Ciulla TA, Criswell MH, Danis RP, Fronheiser M, Yuan P, Cox TA, Csaky KG, Robinson MR (2003) Choroidal neovascular membrane inhibition in a laser treated rat model with intraocular sustained release triamcinolone acetonide microimplants. Br J Ophthalmol 87:1032–1037

Clark AF, Mellon J, Li XY, Ma D, Leher H, Apte R, Alizadeh H, Hegde S, McLenaghan A, Mayhew E, D'Orazio TJ, Niederkorn JY (1999) Inhibition of intraocular tumor growth by topical application of the angiostatic steroid anecortave acetate. Invest Ophthalmol Vis Sci 40:2158–2162

Cornacchia F, Fornoni A, Plati AR, Thomas A, Wang Y, Inverardi L, Striker LJ, Striker GE (2001) Glomerulosclerosis is transmitted by bone marrow-derived mesangial cell progenitors. J Clin Invest 108: 1649–1656

Cour M la, Kiilgaard JF, Nissen MH (2002) Age-related macular degeneration: epidemiology and optimal treatment. Drugs Aging 19:101–133

Cousins SW, Csaky KG (2002) Immunology of age-related macular degeneration. In: Lim JI (ed.) Age-related macular degeneration. Marcel Decker, New York, pp 27–65

Cousins SW, Espinosa-Heidmann DG, Alexandridou A, Sall J, Dubovy S, Csaky K (2002) The role of aging, high fat diet and blue light exposure in an experimental mouse model for basal laminar deposit formation. Exp Eye Res 75:543–553

Cousins SW, Espinosa-Heidmann DG, Marin-Castano ME (2003a) Smoking-related oxidants and RPE injury in vitro and in vivo. Invest Ophthalmol Vis Sci E (Abstract 1619)

Cousins SW, Marin-Castano ME, Espinosa-Heidmann DG, Alexandridou A, Striker L, Elliot S (2003b) Female gender, estrogen loss, and Sub-RPE deposit formation in aged mice. Invest Ophthalmol Vis Sci 44: 1221–1229

Cousins SW, Espinosa-Heidmann DG, Csaky KG (2004) Macrophage function in patients with age-related macular degeneration (AMD): A biomarker for progression? Arch Ophthalmol 22:1013–1018

Csaky K (2003) Anti-vascular endothelial growth factor therapy for neovascular age-related macular degeneration: promises and pitfalls. Ophthalmology 110: 879–881

Curcio CA, Millican CL (1999) Basal linear deposit and large drusen are specific for early age-related maculopathy. Arch Ophthalmol 117:329–339

Curcio CA, Medeiros NE, Millican CL (1996) Photoreceptor loss in age-related macular degeneration. Invest Ophthalmol Vis Sci 37:1236–1249

Curcio CA, Owsley C, Jackson GR (2000) Spare the rods, save the cones in aging and age-related maculopathy. Invest Ophthalmol Vis Sci 41:2015–2018

Curcio CA, Bradley K, Guidry C, et al. (2002) A local source for esterified cholesterol (EC) in human Bruch's membrane (BrM). Invest Ophthalmol Vis Sci E (Abstract 862)

D'Amico DJ, Goldberg MF, Hudson H, Jerdan JA, Krueger S, Luna S, Robertson SM, Russell S, Singerman L, Slakter JS, Sullivan EK, Yannuzzi L, Zilliox P (2003) Anecortave acetate as monotherapy for the treatment of subfoveal lesions in patients with exudative age-related macular degeneration (AMD): interim (month 6) analysis of clinical safety and efficacy. Retina 23:14–23

Davies KJ (1995) Oxidative stress: the paradox of aerobic life. Biochem Soc Symp 61:1–31

DeFaller JM, Clark AF (2000) A new pharmacological treatment for angiogenesis. In: Taylor HR (ed.) Pterygium. Kugler, The Hague, The Netherlands, pp 159–181

Desatnik H, Treister G, Alhalel A, Krupsky S, Moisseiev J (2000) ICGA-guided laser photocoagulation of feeder vessels of choroidal neovascular membranes in age-related macular degeneration. Indocyanine green angiography. Retina 20:143–150

Deviere J (2002) Do selective cyclo-oxygenase inhibitors eliminate the adverse events associated with nonsteroidal anti-inflammatory drug therapy? Eur J Gastroenterol Hepatol 14 (Suppl 1):S29–33

Diaz-Flores L, Gutierrez R, Varela H (1994) Angiogenesis: an update. Histol Histopathol 9:807–843

Distler JH, Hirth A, Kurowska-Stolarska M, Gay RE, Gay S, Distler O (2003) Angiogenic and angiostatic factors in the molecular control of angiogenesis. Q J Nucl Med 47:149–161

Donati G, Kapetanios AD, Pournaras CJ (1999) Principles of treatment of choroidal neovascularization with photodynamic therapy in age-related macular degeneration. Semin Ophthalmol 14:2–10

Dorey CK, Wu G, Ebenstein D, Garsd A, Weiter JJ (1989) Cell loss in the aging retina. Relationship to lipofuscin accumulation and macular degeneration. Invest Ophthalmol Vis Sci 30:1691–1699

Drew MJ (2002) Plasmapheresis in the dysproteinemias. Ther Apher 6:45–52

Drolet DW, Nelson J, Tucker CE, Zack PM, Nixon K, Bolin R, Judkins MB, Farmer JA, Wolf JL, Gill SC, Bendele RA (2000) Pharmacokinetics and safety of an anti-vascular endothelial growth factor aptamer (NX1838) following injection into the vitreous humor of rhesus monkeys. Pharm Res 17:1503–1510

Espinosa-Heidmann DG, Caicedo A, Cousins SW (2003a) Adult bone marrow derived progenitor cells contribute to choroidal neovascularization and modulate the severity. Invest Ophthalmol Vis Sci E (Abstract 3936)

Espinosa-Heidmann DG, Caicedo A, Hernandez EP, Csaky KG, Cousins SW (2003b) Bone marrow-derived progenitor cells contribute to experimental choroidal neovascularization. Invest Ophthalmol Vis Sci 44:4914–4919

Espinosa-Heidmann DG, Suner IJ, Hernandez EP, Monroy D, Csaky KG, Cousins SW (2003c) Macrophage depletion diminishes lesion size and severity in experimental choroidal neovascularization. Invest Ophthalmol Vis Sci 44:3586–3592

Espinosa-Heidmann DG, Sall J, Hernandez EP, Cousins SW (2004) Basal laminar deposit formation in APO B100 transgenic mice: complex interactions between dietary fat, blue light, and vitamin E. Invest Ophthalmol Vis Sci 45:260–266

Evans JR (2001) Risk factors for age-related macular degeneration. Prog Retin Eye Res 20:227–253

Eyetech Study Group (2002) Preclinical and phase 1A clinical evaluation of an anti-VEGF pegylated aptamer (EYE001) for the treatment of exudative age-related macular degeneration. Retina 22:143–152

Eyetech Study Group (2003) Anti-vascular endothelial growth factor therapy for subfoveal choroidal neovascularization secondary to age-related macular degeneration: phase II study results. Ophthalmology 110:979–986

Felbor U, Benkwitz C, Klein ML, Greenberg J, Gregory CY, Weber BH (1997) Sorsby fundus dystrophy: re-evaluation of variable expressivity in patients carrying a TIMP3 founder mutation. Arch Ophthalmol 115:1569–1571

Flower RW (2002) Optimizing treatment of choroidal neovascularization feeder vessels associated with age-related macular degeneration. Am J Ophthalmol 134:228–239

Folkman J (2003) Fundamental concepts of the angiogenic process. Curr Mol Med 3:643–651

Folkman J, Shing Y (1992) Angiogenesis. J Biol Chem 267:10931–10934

Friberg TR (1999) Laser photocoagulation of eyes with drusen: will It help? Semin Ophthalmol 14:45–50

Friberg TR, Musch D (2002) Prophylactic treatment of age-related macular degeneration (PTAMD): update on the clinical trial. Invest Ophthalmol Vis Sci E (Abstract 2904)

Fujii GY, Juan E Jr de, Au Eong KG, Harlan JB Jr (2001a) Effective nasal limited macular translocation. Am J Ophthalmol 132:124–126

Fujii GY, Juan E de, Thomas MA, Pieramici DJ, Humayun MS, Au Eong KG (2001b) Limited macular translocation for the management of subfoveal retinal pigment epithelial loss after submacular surgery. Am J Ophthalmol 131:272–275

Fujii GY, Au Eong KG, Humayun MS, Juan E Jr de (2002) Limited macular translocation: current concepts. Ophthalmol Clin North Am 15:425–436

Fujiyama S, Amano K, Uehira K, Yoshida M, Nishiwaki Y, Nozawa Y, Jin D, Takai S, Miyazaki M, Egashira K, Imada T, Iwasaka T, Matsubara H (2003) Bone marrow monocyte lineage cells adhere on injured endothelium in a monocyte chemoattractant protein-1-dependent manner and accelerate reendothelialization as endothelial progenitor cells. Circ Res 93:980–989

Gandorfer A, Scheider A, Gundisch O, Kampik A (1998) [Recurrent choroid neovascularization in age-related macular degeneration. Fluorescein angiographic morphology after surgical membranectomy]. Ophthalmologe 95:408–412

Gass JD (1987) Stereoscopic atlas of macular diseases. Mosby, St. Louis

Gebicke-Haerter PJ, Van Calker D, Norenberg W, Illes P (1996) Molecular mechanisms of microglial activation. A. Implications for regeneration and neurodegenerative diseases. Neurochem Int 29:1–12

Gehlbach P, Demetriades AM, Yamamoto S, Deering T, Duh EJ, Yang HS, Cingolani C, Lai H, Wei L, Campochiaro PA (2003) Periocular injection of an adenoviral vector encoding pigment epithelium-derived factor inhibits choroidal neovascularization. Gene Ther 10:637–646

Genaidy M, Kazi AA, Peyman GA, Passos-Machado E, Farahat HG, Williams JI, Holroyd KJ, Blake DA (2002) Effect of squalamine on iris neovascularization in monkeys. Retina 22:772–778

Goncalves AR, Fujihara CK, Mattar AL, Malheiros DM, Noronha Ide L, de Nucci G, Zatz R (2004) Renal expression of COX-2, ANG II, and AT1 receptor in remnant kidney: strong renoprotection by therapy with losartan and a nonsteroidal anti-inflammatory. Am J Physiol Renal Physiol 286:F945–954 (Epub 2003 Dec 16)

Gorin MB, Breitner JC, Jong PT de, Hageman GS, Klaver CC, Kuehn MH, Seddon JM (1999) The genetics of age-related macular degeneration. Mol Vis 5:29

Grant MB, May WS, Caballero S, Brown GA, Guthrie SM, Mames RN, Byrne BJ, Vaught T, Spoerri PE, Peck AB, Scott EW (2002) Adult hematopoietic stem cells provide functional hemangioblast activity during retinal neovascularization. Nat Med 8:607–612

Green WR (1999) Histopathology of age-related macular degeneration. Mol Vis 5:27

Green WR, Harlan JB (1999) Histopathologic Features. In: Berger JW, Fine SL, Maguire MG (eds) Age-related macular degeneration. Mosby, St. Louis, pp 81–154

Griggs J, Hesketh R, Smith GA, Brindle KM, Metcalfe JC, Thomas GA, Williams ED (2001a) Combretastatin-A4 disrupts neovascular development in non-neoplastic tissue. Br J Cancer 84:832–835

Griggs J, Metcalfe JC, Hesketh R (2001b) Targeting tumour vasculature: the development of combretastatin A4. Lancet Oncol 2:82–87

Griggs J, Skepper JN, Smith GA, Brindle KM, Metcalfe JC, Hesketh R (2002) Inhibition of proliferative retinopathy by the anti-vascular agent combretastatin-A4. Am J Pathol 160:1097–1103

Grossniklaus HE, Cingle KA, Yoon YD, Ketkar N, L'Hernault N, Brown S (2000) Correlation of histologic 2-dimensional reconstruction and confocal scanning laser microscopic imaging of choroidal neovascularization in eyes with age-related maculopathy. Arch Ophthalmol 118:625–629

Grossniklaus HE, Ling JX, Wallace TM, Dithmar S, Lawson DH, Cohen C, Elner VM, Elner SG, Sternberg P Jr (2002) Macrophage and retinal pigment epithelium expression of angiogenic cytokines in choroidal neovascularization. Mol Vis 8:119–126

Grunwald JE (1999) Choroidal blood flow. In: Berger JW, Fine SL, Maguire MG (eds) Age-related macular degeneration. Mosby, St. Louis, pp 167–172

Gupta A, Pansari K (2003) Inflammation and Alzheimer's disease. Int J Clin Pract 57:36–39

Guyer DR, Yannuzzi LA, Slakter JS, Sorenson JA, Hanutsaha P, Spaide RF, Schwartz SG, Hirschfeld JM, Orlock DA (1996) Classification of choroidal neovascularization by digital indocyanine green videoangiography. Ophthalmology 103:2054–2060

Hageman GS, Mullins RF (1999) Molecular composition of drusen as related to substructural phenotype. Mol Vis 5:28

Hageman GS, Luthert PJ, Victor Chong NH, Johnson LV, Anderson DH, Mullins RF (2001) An integrated hypothesis that considers drusen as biomarkers of immune-mediated processes at the RPE-Bruch's membrane interface in aging and age-related macular degeneration. Prog Retin Eye Res 20:705–732

Hamdi HK, Kenney C (2003) Age-related macular degeneration: a new viewpoint. Front Biosci 8:e305–e314

Hanutsaha P, Guyer DR, Yannuzzi LA, Naing A, Slakter JS, Sorenson JS, Spaide RF, Freund KB, Feinsod M, Orlock DA (1998) Indocyanine-green videoangiography of drusen as a possible predictive indicator of exudative maculopathy. Ophthalmology 105:1632–1636

Harding S (2001) Photodynamic therapy in the treatment of subfoveal choroidal neovascularisation. Eye 15:407–412

Harjai KJ, Chowdhury P, Grines CL (2002) Therapeutic angiogenesis: a fantastic new adventure. J Interv Cardiol 15:223–229

Harraz M, Jiao C, Hanlon HD, Hartley RS, Schatteman GC (2001) CD34 – blood-derived human endothelial cell progenitors. Stem Cells 19:304–312

Hart PM, Chakravarthy U, Mackenzie G, Chisholm IH, Bird AC, Stevenson MR, Owens SL, Hall V, Houston RF, McCulloch DW, Plowman N (2002) Visual outcomes in the subfoveal radiotherapy study: a randomized controlled trial of teletherapy for age-related macular degeneration. Arch Ophthalmol 120:1029–1038

Hawkins BS, Bird A, Klein R, West SK (1999) Epidemiology of age-related macular degeneration. Mol Vis 5:26

Hee MR, Baumal CR, Puliafito CA, Duker JS, Reichel E, Wilkins JR, Coker JG, Schuman JS, Swanson EA, Fujimoto JG (1996) Optical coherence tomography of age-related macular degeneration and choroidal neovascularization. Ophthalmology 103:1260–1270

Hermans P, Lommatzsch A, Bomfeld N, Pauleikhoff D (2003) [Angiographic-histological correlation of late exudative age-related macular degeneration]. Ophthalmologe 100:378–383

Higgins RD, Sanders RJ, Yan Y, Zasloff M, Williams JI (2000) Squalamine improves retinal neovascularization. Invest Ophthalmol Vis Sci 41:1507–1512

Ho AC (1999) Laser treatment in eyes with drusen. Curr Opin Ophthalmol 10:204–208

Ho CL (2000) Macular translocation – an innovative treatment for macular degenerative diseases. Changgeng Yi Xue Za Zhi 23:672–680

Hoffmann S, Friedrichs U, Eichler W, Rosenthal A, Wiedemann P (2002) Advanced glycation end products induce choroidal endothelial cell proliferation, matrix metalloproteinase-2 and VEGF upregulation in vitro. Graefes Arch Clin Exp Ophthalmol 240:996–1002

Holz FG, Miller DW (2003) [Pharmacological therapy for age-related macular degeneration. Current developments and perspectives]. Ophthalmologe 100:97–103

Hunt DW, Margaron P (2003) Status of therapies in development for the treatment of age-related macular degeneration. IDrugs 6:464–469

Hyman L, Neborsky R (2002) Risk factors for age-related macular degeneration: an update. Curr Opin Ophthalmol 13:171–175

Hyman L, Schachat AP, He Q, Leske MC (2000) Hypertension, cardiovascular disease, and age-related macular degeneration. Age-Related Macular Degeneration Risk Factors Study Group. Arch Ophthalmol 118:351–358

Ianus A, Holz GG, Theise ND, Hussain MA (2003) In vivo derivation of glucose-competent pancreatic endocrine cells from bone marrow without evidence of cell fusion. J Clin Invest 111:843–850

Ishiko S, Akiba J, Horikawa Y, Yoshida A (2002) Detection of drusen in the fellow eye of Japanese patients with age-related macular degeneration using scanning laser ophthalmoscopy. Ophthalmology 109:2165–2169

Isner JM, Asahara T (1999) Angiogenesis and vasculogenesis as therapeutic strategies for postnatal neovascularization. J Clin Invest 103:1231–1236

Ito T, Ikeda U (2003) Inflammatory cytokines and cardiovascular disease. Curr Drug Targets Inflamm Allergy 2:257–265

Jacot TA, Striker GE, Stetler-Stevenson M, Striker LJ (1996) Mesangial cells from transgenic mice with progressive glomerulosclerosis exhibit stable, phenotypic changes including undetectable MMP-9 and increased type IV collagen. Laboratory Invest 75:791–799

Johnson LV, Leitner WP, Staples MK, Anderson DH (2001a) Complement activation and inflammatory processes in Drusen formation and age related macular degeneration. Exp Eye Res 73:887–896

Johnson LV, Ozaki, S, Staples MK, Erikson, PA and Anderson, DH (2001b) A potential role for immune complex pathogenesis in drusen formation. Exp Eye Res 70:441–449

Juan E Jr de, Fujii GY (2001) Limited macular translocation. Eye 15:413–423

Juan E Jr de, Machemer R (1988) Vitreous surgery for hemorrhagic and fibrous complications of age-related macular degeneration. Am J Ophthalmol 105:25–29

Juan E Jr de, Vander JF (1999) Effective macular translocation without scleral imbrication. Am J Ophthalmol 128:380–382

Juan E Jr de, Loewenstein A, Bressler NM, Alexander J (1998) Translocation of the retina for management of subfoveal choroidal neovascularization II: a preliminary report in humans. Am J Ophthalmol 125:635–646

Kaiser HJ, Flammer J, Stumpfig D, Hendrickson P (1995) Visaline in the treatment of age-related macular degeneration: a pilot study. Ophthalmologica 209:302–305

Kampik A, Gandorfer A (2000) Silicone oil removal strategies. Semin Ophthalmol 15:88–91

Kieran MW, Billett A (2001) Antiangiogenesis therapy. Current and future agents. Hematol Oncol Clin North Am 15:835–851, viii

Killingsworth MC, Sarks JP, Sarks SH (1990) Macrophages related to Bruch's membrane in age-related macular degeneration. Eye 4:613–621

Klaver CC, Allikmets R (2003) Genetics of macular dystrophies and implications for age-related macular degeneration. Dev Ophthalmol 37:155–169

Klaver CC, Kliffen M, Duijn CM van, Hofman A, Cruts M, Grobbee DE, Broeckhoven C van, Jong PT de (1998) Genetic association of apolipoprotein E with age-related macular degeneration. Am J Hum Genet 63:200–206

Klein EA (1999) Epidemiology. In: Berger JW, Fine SL, Maguire MG (eds) Age-related macular disease. Mosby, St. Louis, pp 31–55

Kliffen M, Schaft TL van der, Mooy CM, Jong PT de (1997) Morphologic changes in age-related maculopathy. Microsc Res Tech 36:106–122

Klingel R, Fassbender C, Fischer I, Hattenbach L, Gumbel H, Pulido J, Koch F (2002) Rheopheresis for age-related macular degeneration: a novel indication for therapeutic apheresis in ophthalmology. Ther Apher 6:271–281

Kramer I, Wibulswas A, Croft D, Genot E (2003) Rheumatoid arthritis: targeting the proliferative fibroblasts. Prog Cell Cycle Res 5:59–70

Krzystolik MG, Afshari MA, Adamis AP, Gaudreault J, Gragoudas ES, Michaud NA, Li W, Connolly E, O"-Neill CA, Miller JW (2002) Prevention of experimental choroidal neovascularization with intravitreal anti-vascular endothelial growth factor antibody fragment. Arch Ophthalmol 120:338–346

Labarge MA, Blau HM (2002) Biological progression from adult bone marrow to mononucleate muscle stem cell to multinucleate muscle fiber in response to injury. Cell 111:589–601

Lai CM, Lai YK, Rakoczy PE (2002) Adenovirus and adeno-associated virus vectors. DNA Cell Biol 21:895–913

Lambert V, Munaut C, Jost M, Noel A, Werb Z, Foidart JM, Rakic JM (2002) Matrix metalloproteinase-9 contributes to choroidal neovascularization. Am J Pathol 161:1247–1253

Li G, Gao R (1995) [Vitrectomy for treatment of vitreous hemorrhage associated with age-related macular degeneration]. Zhonghua Yan Ke Za Zhi 31:262–263

Lim JI (2002) Photodynamic therapy for choroidal neovascular disease: photosensitizers and clinical trials. Ophthalmol Clin North Am 15:473–478, vii

Lopez PF, Sippy BD, Lambert HM, Thach AB, Hinton DR (1996) Transdifferentiated retinal pigment epithelial cells are immunoreactive for vascular endothelial growth factor in surgically excised age-related macular degeneration-related choroidal neovascular membranes. Invest Ophthalmol Vis Sci 37:855–868

Lowe GD (2001) The relationship between infection, inflammation, and cardiovascular disease: an overview. Ann Periodontol 6:1–8

Lutty G, Grunwald J, Majji AB, Uyama M, Yoneya S (1999) Changes in choriocapillaris and retinal pigment epithelium in age-related macular degeneration. Mol Vis 5:35

Macular Photocoagulation Study Group (1991a) Laser photocoagulation of subfoveal neovascular lesions in age-related macular degeneration. Results of a randomized clinical trial. Macular Photocoagulation Study Group. Arch Ophthalmol 109:1220–1231

Macular Photocoagulation Study Group (1991b) Laser photocoagulation of subfoveal recurrent neovascular lesions in age-related macular degeneration. Results of a randomized clinical trial. Macular Photocoagulation Study Group. Arch Ophthalmol 109:1232–1241

Macular Photocoagulation Study Group (1991c) Subfoveal neovascular lesions in age-related macular degeneration. Guidelines for evaluation and treatment in the macular photocoagulation study. Arch Ophthalmol 109:1242–1257

Macular Photocoagulation Study Group (1994) Laser photocoagulation for juxtafoveal choroidal neovascularization. Five-year results from randomized clinical trials. Macular Photocoagulation Study Group. Arch Ophthalmol 112:500–509

Macular Photocoagulation Study Group (1996) Occult choroidal neovascularization. Influence on visual outcome in patients with age-related macular degeneration. Arch Ophthalmol 114:400–412

Maguire MG (2004) Natural history. In: Berger JW, Fine SL, Maguire MG (eds) Age-related macular degeneration. Mosby, St. Louis, pp 16–30

Malorni W, Donelli G (1992) Cell death. General features and morphological aspects. Ann NY Acad Sci 663:218–233

Malorni W, Iosi F, Mirabelli F, Bellomo G (1991) Cytoskeleton as a target in menadione-induced oxidative stress in cultured mammalian cells: alterations underlying surface bleb formation. Chem Biol Interact 80:217–236

Mandelcorn MS, Menezes AV (1993) Surgical removal of subretinal hemorrhage and choroidal neovascular membranes in acute hemorrhagic age-related macular degeneration. Can J Ophthalmol 28:19–23

Marcus DM, Sheils W, Johnson MH, McIntosh SB, Leibach DB, Maguire A, Alexander J, Samy CN (2001) External beam irradiation of subfoveal choroidal neovascularization complicating age-related macular degeneration: one-year results of a prospective, double-masked, randomized clinical trial. Arch Ophthalmol 119:171–180

McCarty CA, Mukesh BN, Fu CL, Mitchell P, Wang JJ, Taylor HR (2001) Risk factors for age-related maculopathy: the Visual Impairment Project. Arch Ophthalmol 119:1455–1462

McGeer EG, McGeer PL (2003) Inflammatory processes in Alzheimer's disease. Prog Neuropsychopharmacol Biol Psychiatry 27:741–749

McNatt LG, Weimer L, Yanni J, Clark AF (1999) Angiostatic activity of steroids in the chick embryo CAM and rabbit cornea models of neovascularization. J Ocul Pharmacol Ther 15:413–423

Michalowski AS (1994) On radiation damage to normal tissues and its treatment. II. Anti-inflammatory drugs. Acta Oncol 33:139–157

Mittra RA, Singerman LJ (2002) Recent advances in the management of age-related macular degeneration. Optom Vis Sci 79:218–224

Moisseiev J, Alhalel A, Masuri R, Treister G (1995) The impact of the macular photocoagulation study results on the treatment of exudative age-related macular degeneration. Arch Ophthalmol 113:185–189

Monroy D, Marin-Castano ME, Striker LJ, et al. (2001) RPE expression of matrix metalloproteinase (MMP-2) and MCP-1 after different kinds of injury. Invest Ophthalmol Vis Sci 42(4):4060

Moore DJ, Hussain AA, Marshall J (1995) Age-related variation in the hydraulic conductivity of Bruch's membrane. Invest Ophthalmol Vis Sci 36:1290–1297

Moore KS, Wehrli S, Roder H, Rogers M, Forrest JN Jr, McCrimmon D, Zasloff M (1993) Squalamine: an aminosterol antibiotic from the shark. Proc Natl Acad Sci USA 90:1354–1358

Mordenti J, Cuthbertson RA, Ferrara N, Thomsen K, Berleau L, Licko V, Allen PC, Valverde CR, Meng YG, Fei DT, Fourre KM, Ryan AM (1999) Comparisons of the intraocular tissue distribution, pharmacokinetics, and safety of $^{125}$I-labeled full-length and Fab antibodies in rhesus monkeys following intravitreal administration. Toxicol Pathol 27:536–544

Mori K, Yoneya S, Anzail K, Kabasawa S, Sodeyama T, Peyman GA, Moshfeghi DM (2001) Photodynamic therapy of experimental choroidal neovascularization with a hydrophilic photosensitizer: mono-L-aspartyl chlorine6. Retina 21:499–508

Mori K, Gehlbach P, Ando A, McVey D, Wei L, Campochiaro PA (2002) Regression of ocular neovascularization in response to increased expression of pigment epithelium-derived factor. Invest Ophthalmol Vis Sci 43:2428–2434

Moshfeghi DM, Peyman GA, Moshfeghi AA, Khoobehi B, Primbs GB, Crean DH (1998) Ocular vascular thrombosis following tin ethyl etiopurpurin (SnET2) photodynamic therapy: time dependencies. Ophthalmic Surg Lasers 29:663–668

Mullins RF, Russell SR, Anderson DH, Hageman GS (2000) Drusen associated with aging and age-related macular degeneration contain proteins common to extracellular deposits associated with atherosclerosis, elastosis, amyloidosis, and dense deposit disease. FASEB J 14:835–846

Murohara T (2003) Angiogenesis and vasculogenesis for therapeutic neovascularization. Nagoya J Med Sci 66:1–7

Nakashizuka T, Mori K, Hayashi N, Anzail K, Kanail K, Yoneya S, Moshfeghi DM, Peyman GA (2001) Retreatment effect of NPe6 photodynamic therapy on the normal primate macula. Retina 21:493–498

Nambu H, Nambu R, Melia M, Campochiaro PA (2003) Combretastatin A-4 phosphate suppresses development and induces regression of choroidal neovascularization. Invest Ophthalmol Vis Sci 44:3650–3655

Navajas EV, Costa RA, Farah ME, Cardillo JA, Bonomo PP (2003) Indocyanine green-mediated photothrombosis combined with intravitreal triamcinolone for the treatment of choroidal neovascularization in serpiginous choroiditis. Eye 17:563–566

Newsom RS, McAlister JC, Saeed M, McHugh JD (2001) Transpupillary thermotherapy (TTT) for the treatment of choroidal neovascularisation. Br J Ophthalmol 85:173–178

Newsome DA, Swartz M, Leone NC, Elston RC, Miller E (1988) Oral zinc in macular degeneration. Arch Ophthalmol 106:192–198

Oh H, Takagi H, Takagi C, Suzuma K, Otani A, Ishida K, Matsumura M, Ogura Y, Honda Y (1999) The potential angiogenic role of macrophages in the formation of choroidal neovascular membranes. Invest Ophthalmol Vis Sci 40:1891–1898

Ohji M, Fujikado T, Kusaka S, Hayashi A, Hosohata J, Ikuno Y, Sawa M, Kubota A, Hashida N, Tano Y (2001) Comparison of three techniques of foveal translocation in patients with subfoveal choroidal neovascularization resulting from age-related macular degeneration. Am J Ophthalmol 132:888–896

Olk RJ, Friberg TR, Stickney KL, Akduman L, Wong KL, Chen MC, Levy MH, Garcia CA, Morse LS (1999) Therapeutic benefits of infrared (810-nm) diode laser macular grid photocoagulation in prophylactic treatment of nonexudative age-related macular degeneration: two-year results of a randomized pilot study. Ophthalmology 106:2082–2090

Pauleikhoff D, Koch JM (1995) Prevalence of age-related macular degeneration. Curr Opin Ophthalmol 6:51–56

Penfold PL, Killingsworth MC, Sarks SH (1985) Senile macular degeneration: the involvement of immunocompetent cells. Graefes Arch Clin Exp Ophthalmol 223:69–76

Penfold PL, Provis JM, Billson FA (1987) Age-related macular degeneration: ultrastructural studies of the relationship of leucocytes to angiogenesis. Graefes Arch Clin Exp Ophthalmol 225:70–76

Penfold PL, Madigan MC, Gillies MC, Provis JM (2001) Immunological and aetiological aspects of macular degeneration. Prog Retin Eye Res 20:385–414

Penn JS, Rajaratnam VS, Collier RJ, Clark AF (2001) The effect of an angiostatic steroid on neovascularization in a rat model of retinopathy of prematurity. Invest Ophthalmol Vis Sci 42:283–290

Peten EP, Garcia-Perez A, Terada Y, Woodrow D, Martin BM, Striker GE, Striker LJ (1992) Age-related changes in alpha 1- and alpha 2-chain type IV collagen mRNAs in adult mouse glomeruli: competitive PCR. Am J Physiol 263:F951–F957

Pettit GR, Singh SB, Hamel E, Lin CM, Alberts DS, Garcia-Kendall D (1989) Isolation and structure of the strong cell growth and tubulin inhibitor combretastatin A-4. Experientia 45:209–211

Pieramici DJ, Bressler SB (1998) Age-related macular degeneration and risk factors for the development of choroidal neovascularization in the fellow eye. Curr Opin Ophthalmol 9:38–46

Presta LG, Chen H, O"Connor SJ, Chisholm V, Meng YG, Krummen L, Winkler M, Ferrara N (1997) Humanization of an anti-vascular endothelial growth factor monoclonal antibody for the therapy of solid tumors and other disorders. Cancer Res 57:4593–4599

Pulido JS (2002) Multicenter prospective, randomized, double-masked, placebo-controlled study of rheopheresis to treat nonexudative age-related macular degeneration: interim analysis. Trans Am Ophthalmol Soc 100:85–106; discussion 106–107

Rao MN, Shinnar AE, Noecker LA, Chao TL, Feibush B, Snyder B, Sharkansky I, Sarkahian A, Zhang X, Jones SR, Kinney WA, Zasloff M (2000) Aminosterols from the dogfish shark *Squalus acanthias*. J Nat Prod 63:631–635

Rasmussen H, Chu KW, Campochiaro P, Gehlbach PL, Haller JA, Handa JT, Nguyen QD, Sung JU (2001) Clinical protocol. An open-label, phase I, single administration, dose-escalation study of ADGV-PEDF.11D (ADPEDF) in neovascular age-related macular degeneration (AMD). Hum Gene Ther 12:2029–2032

Rechtman E, Allen VD, Danis RP, Pratt LM, Harris A, Speicher MA (2003) Intravitreal triamcinolone for choroidal neovascularization in ocular histoplasmosis syndrome. Am J Ophthalmol 136:739–741

Reichel E, Berrocal AM, Ip M, Kroll AJ, Desai V, Duker JS, Puliafito CA (1999) Transpupillary thermotherapy of occult subfoveal choroidal neovascularization in patients with age-related macular degeneration. Ophthalmology 106:1908–1914

Religa P, Bojakowski K, Maksymowicz M, Bojakowska M, Sirsjo A, Gaciong Z, Olszewski W, Hedin U, Thyberg J (2002) Smooth-muscle progenitor cells of bone marrow origin contribute to the development of neointimal thickenings in rat aortic allografts and injured rat carotid arteries. Transplantation 74:1310–1315

Renno RZ, Youssri AI, Michaud N, Gragoudas ES, Miller JW (2002) Expression of pigment epithelium-derived factor in experimental choroidal neovascularization. Invest Ophthalmol Vis Sci 43:1574–1580

Reynders S, Lafaut BA, Aisenbrey S, Broecke CV, Lucke K, Walter P, Kirchhof B, Bartz-Schmidt KU (2002) Clinicopathologic correlation in hemorrhagic age-related macular degeneration. Graefes Arch Clin Exp Ophthalmol 240:279–285

Richer SP, Stiles W, Statkute L, et al. (2002) The Lutein Antioxidant Supplementation Trial. Invest Ophthalmol Vis Sci E (Abstract 2542)

Rodanant N, Freeman WR, Musch DC, et al. (2002) Predictors of choroidal neovascularization in eyes treated with subthreshold diode grid laser therapy to prevent choroidal neovascularization. Results from the unilateral arm of PTAMD Study. Invest Ophthalmol Vis Sci E (Abstract 1211)

Rogers AH, Reichel E (2001) Transpupillary thermotherapy of subfoveal occult choroidal neovascularization. Curr Opin Ophthalmol 12:212–215

Ross R (1990) Mechanisms of atherosclerosis – a review. Adv Nephrol Necker Hosp 19:79–86

Ross R (1999) Atherosclerosis is an inflammatory disease. Am Heart J 138:S419–S420

Ruckmann A von, Schmidt KG, Fitzke FW, Bird AC, Jacobi KW (1998) [Dynamics of accumulation and degradation of lipofuscin in retinal pigment epithelium in senile macular degeneration]. Klin Monatsbl Augenheilkd 213:32–37

Sakurai E, Anand A, Ambati BK, Rooijen N van, Ambati J (2003) Macrophage depletion inhibits experimental choroidal neovascularization. Invest Ophthalmol Vis Sci 44:3578–3585

Sarks JP, Sarks SH, Killingsworth MC (1997) Morphology of early choroidal neovascularisation in age-related macular degeneration: correlation with activity. Eye 11:515–522

Schaft TL van der, Mooy CM, Bruijn WC de, Oron FG, Mulder PG, Jong PT de (1992) Histologic features of the early stages of age-related macular degeneration. A statistical analysis. Ophthalmology 99:278–286

Schaft TL van der, Mooy CM, Bruijn WC de, Bosman FT, Jong PT de (1994) Immunohistochemical light and electron microscopy of basal laminar deposit. Graefes Arch Clin Exp Ophthalmol 232:40–46

Schlingemann RO (2004) Role of growth factors and the wound healing response in age-related macular degeneration. Graefes Arch Clin Exp Ophthalmol 242:91–101

Schneider S, Greven CM, Green WR (1998) Photocoagulation of well-defined choroidal neovascularization in age-related macular degeneration: clinicopathologic correlation. Retina 18:242–250

Seddon JM, Gensler G, Milton RC, Klein ML, Rifai N (2004) Association between C-reactive protein and age-related macular degeneration. J Am Med Assoc 291:704–710

Seibert K, Lefkowith J, Tripp C, Isakson P, Needleman P (1999) COX-2 inhibitors – is there cause for concern? Nat Med 5:621–622

Sengupta N, Caballero S, Mames RN, Butler JM, Scott EW, Grant MB (2003) The role of adult bone marrow-derived stem cells in choroidal neovascularization. Invest Ophthalmol Vis Sci 44:4908–4913

Shiraga F, Ojima Y, Matsuo T, Takasu I, Matsuo N (1998) Feeder vessel photocoagulation of subfoveal choroidal neovascularization secondary to age-related macular degeneration. Ophthalmology 105:662–669

Sills AK Jr, Williams JI, Tyler BM, Epstein DS, Sipos EP, Davis JD, McLane MP, Pitchford S, Cheshire K, Gannon FH, Kinney WA, Chao TL, Donowitz M, Laterra J, Zasloff M, Brem H (1998) Squalamine inhibits angiogenesis and solid tumor growth in vivo and perturbs embryonic vasculature. Cancer Res 58:2784–2792

Singerman LJ (1988) Current management of choroidal neovascularization. Ann Ophthalmol 20:415–420, 423

Slakter JS (2003) Anecortave acetate as monotherapy for treatment of subfoveal neovascularization in age-related macular degeneration: twelve-month clinical outcomes. Ophthalmology 110:2372–2383

Smith W, Mitchell P, Wang JJ (1997) Gender, oestrogen, hormone replacement and age-related macular degeneration: results from the Blue Mountains Eye Study. Aust NZ J Ophthalmol 25 (Suppl 1):S13–S15

Snow KK, Seddon JM (1999) Do age-related macular degeneration and cardiovascular disease share common antecedents? Ophthalmic Epidemiol 6:125–143

Spaide RF, Sorenson J, Maranan L (2003) Combined photodynamic therapy with verteporfin and intravitreal triamcinolone acetonide for choroidal neovascularization. Ophthalmology 110:1517–1525

Spraul CW, Lang GE, Lang GK (1998) [Value of optical coherence tomography in diagnosis of age-related macular degeneration. Correlation of fluorescein angiography and OCT findings]. Klin Monatsbl Augenheilkd 212:141–148

Staurenghi G, Orzalesi N, La Capria A, Aschero M (1998) Laser treatment of feeder vessels in subfoveal choroidal neovascular membranes: a revisitation using dynamic indocyanine green angiography. Ophthalmology 105:2297–2305

Steen B, Sejersen S, Berglin L, Seregard S, Kvanta A (1998) Matrix metalloproteinases and metalloproteinase inhibitors in choroidal neovascular membranes. Invest Ophthalmol Vis Sci 39:2194–2200

Stone TW, Sternberg P Jr (2002) Submacular surgery trials update. Ophthalmol Clin North Am 15:479–488

Strunnikova N, Baffi J, Gonzalez A, Silk W, Cousins SW, Csaky KG (2001) Regulated heat shock protein 27 expression in human retinal pigment epithelium. Invest Ophthalmol Vis Sci 42:2130–2138

Strunnikova NV, Baffi J, Zhang C, et al. (2003) Cellular and molecular responses to sublethal oxidative injury in retinal pigment epithelial cells (RPE). Invest Ophthalmol Vis Sci E (Abstract 3145)

Stur M, Tittl M, Reitner A, Meisinger V (1996) Oral zinc and the second eye in age-related macular degeneration. Invest Ophthalmol Vis Sci 37:1225–1235

Submacular Surgery Trials Pilot Study (2000a) Submacular surgery trials randomized pilot trial of laser photocoagulation versus surgery for recurrent choroidal neovascularization secondary to age-related macular degeneration: 1. Ophthalmic outcomes submacular surgery trials pilot study report number 1. Am J Ophthalmol 130:387–407

Submacular Surgery Trials Pilot Study (2000b) Submacular surgery trials randomized pilot trial of laser photocoagulation versus surgery for recurrent choroidal neovascularization secondary to age-related macular degeneration: 2. Quality of life outcomes submacular surgery trials pilot study report number 2. Am J Ophthalmol 130:408–418

Subramanian ML, Reichel E (2003) Current indications of transpupillary thermotherapy for the treatment of posterior segment diseases. Curr Opin Ophthalmol 14:155–158

Sullivan P, Filsecker L, Sears J (2002) Limited macular translocation with scleral retraction suture. Br J Ophthalmol 86:434–439

Suner IJ, Espinosa-Heidmann DG, Marin-Castano ME, Hernandez EP, Pereira-Simon S, Cousins SW (2004) Nicotine increases size and severity of experimental choroidal neovascularization. Invest Ophthalmol Vis Sci 45:311–317

Takahashi K, Saishin Y, Saishin Y, Mori K, Ando A, Yamamoto S, Oshima Y, Nambu H, Melia MB, Bingaman DP, Campochiaro PA (2003) Topical nepafenac inhibits ocular neovascularization. Invest Ophthalmol Vis Sci 44:409–415

TAP Study Group (1999) Photodynamic therapy of subfoveal choroidal neovascularization in age-related macular degeneration with verteporfin: one-year results of 2 randomized clinical trials – TAP report. Treatment of age-related macular degeneration with photodynamic therapy (TAP) Study Group. Arch Ophthalmol 117:1329–1345

TAP Study Group and VIP Study Group (Verteporfin Roundtable 2000 and 2001 Participants) (2002) Guidelines for using verteporfin (visudyne) in photodynamic therapy to treat choroidal neovascularization due to age-related macular degeneration and other causes. Retina 22:6–18

Taylor HR, Tikellis G, Robman LD, McCarty CA, McNeil JJ (2002) vitamin E supplementation and macular degeneration: randomised controlled trial. BMJ 325:11

Teicher BA, Williams JI, Takeuchi H, Ara G, Herbst RS, Buxton D (1998) Potential of the aminosterol, squalamine in combination therapy in the rat 13,762 mammary carcinoma and the murine Lewis lung carcinoma. Anticancer Res 18:2567–2573

Teikari JM, Laatikainen L, Virtamo J, Haukka J, Rautalahti M, Liesto K, Albanes D, Taylor P, Heinonen OP (1998) Six-year supplementation with alpha-tocopherol and beta-carotene and age-related maculopathy. Acta Ophthalmol Scand 76:224–229

Terasaki H (2001) Rescue of retinal function by macular translocation surgery in age-related macular degeneration and other diseases with subfoveal choroidal neovascularization. Nagoya J Med Sci 64:1–9

Tosetti F, Ferrari N, De Flora S, Albini A (2002) Angioprevention": angiogenesis is a common and key target for cancer chemopreventive agents. FASEB J 16:2–14

Toth CA, Freedman SF (2001) Macular translocation with 360-degree peripheral retinectomy impact of technique and surgical experience on visual outcomes. Retina 21:293–303

Tozer GM, Kanthou C, Parkins CS, Hill SA (2002) The biology of the combretastatins as tumour vascular targeting agents. Int J Exp Pathol 83:21–38

Tracy RP (2003) Inflammation, the metabolic syndrome and cardiovascular risk. Int J Clin Pract Suppl 10–17

Valmaggia C, Ries G, Ballinari P (2002) Radiotherapy for subfoveal choroidal neovascularization in age-related macular degeneration: a randomized clinical trial. Am J Ophthalmol 133:521–529

Vander JF, Federman JL, Greven C, Slusher MM, Gabel VP (1991) Surgical removal of massive subretinal hemorrhage associated with age-related macular degeneration. Ophthalmology 98:23–27

VIP (Verteporfin in Photodynamic Therapy) Study Group (2001) Verteporfin therapy of subfoveal choroidal neovascularization in age-related macular degeneration: two-year results of a randomized clinical trial including lesions with occult with no classic choroidal neovascularization–verteporfin in photodynamic therapy report 2. Am J Ophthalmol 131: 541–560

Vizcarra C (2003) New perspectives and emerging therapies for immune-mediated inflammatory disorders. J Infus Nurs 26:319–325

Walsh DA, Pearson CI (2001) Angiogenesis in the pathogenesis of inflammatory joint and lung diseases. Arthritis Res 3:147–153

Weimann JM, Charlton CA, Brazelton TR, Hackman RC, Blau HM (2003) Contribution of transplanted bone marrow cells to Purkinje neurons in human adult brains. Proc Natl Acad Sci USA 100:2088–2093

Winkler BS, Boulton ME, Gottsch JD, Sternberg P (1999) Oxidative damage and age-related macular degeneration. Mol Vis 5:32

Witmer AN, Vrensen GF, Van Noorden CJ, Schlingemann RO (2003) Vascular endothelial growth factors and angiogenesis in eye disease. Prog Retin Eye Res 22:1–29

Woods JA, Hadfield JA, Pettit GR, Fox BW, McGown AT (1995) The interaction with tubulin of a series of stilbenes based on combretastatin A-4. Br J Cancer 71: 705–711

Wormald R, Evans J, Smeeth L, Henshaw K (2003) Photodynamic therapy for neovascular age-related macular degeneration. Cochrane Database Syst Rev CD002030

Yamamoto Y (2003) Measurement of blood flow velocity in feeder vessels of choroidal neovascularization by a scanning laser ophthalmoscope and image analysis system. Jpn J Ophthalmol 47:53–58

Yates JR, Moore AT (2000) Genetic susceptibility to age related macular degeneration. J Med Genet 37:83–87

York J, Glaser B, Murphy R (2000) High-speed ICG. Used for pinpoint laser treatment of feeder vessels in wet AMD. J Ophthalmic Nurs Technol 19:66–67

Young RW (1987) Pathophysiology of age-related macular degeneration. Surv Ophthalmol 31:291–306

Zarbin MA (1998) Age-related macular degeneration: review of pathogenesis. Eur J Ophthalmol 8:199–206

Zimmer-Galler IE, Bressler NM, Bressler SB (1995) Treatment of choroidal neovascularization: updated information from recent macular photocoagulation study group reports. Int Ophthalmol Clin 35:37–57

# Subject Index

**A**
angiogenesis 180
animal models 113
– baboon 155
– marmoset 155
– monkey 113
– mouse 114, 116, 150, 173
– rabbit 114
– rat 114, 116, 139, 150
antioxidants 124, 126, 173
apolipoprotein E (APOE) 72
ascorbic acid 138
astrocytes 12
atrophic macular degeneration (see also geographic atrophy) 37, 167
autoantibody 35

**B**
basal
– laminar deposit 167
– linear deposit 167
Beaver Dam Eye Study 53, 80, 82
Best disease 68, 70, 117
bipolar cell 9
blood-retinal barrier 26, 30
bone marrow progenitor 182
Bruch's membrane 54, 55, 167, 169

**C**
cell death (see also photoreceptor loss) 140
choroidal neovascularization 37, 38
choroidal new vessels
– growth factor-induced 114
– laser-induced 114, 115
clinical trial 46
– anti-angiogenic drugs 186
– anti-inflammatory 185
– gene therapy 188
– laser 184, 185, 171
– photodynamic therapy 184
– radiation 184
– rheopheresis 172
– surgery 188
– transpupillary thermotherapy 185
– VEGF 187
cone 4, 18
cystoid degeneration 160

**D**
degeneration
– atrophic macular 37, 167
– cystoid 160
– disciform 48
– dry macular 37
– neovascular macular 37, 178
– peripheral 150, 151, 157, 158
– wet macular 37
disciform degeneration 48
DNA damage 125
Down's syndrome 131
Doyne honeycomb dystrophy 71
drusen 31, 34, 37, 70, 125, 167, 170
– biogenesis 32
dry macular degeneration 37

**E**
early onset maculopathy 67, 69
electroretinogram 134
estrogen deficiency 174, 175

**F**
Fibulin family 70, 71
fovea 1, 2
– acuity 26
– anatomy 1
– avascular zone 5
– bipolar cells 17
– blood vessels 12, 19, 25
– comparative anatomy 14–17
– cone density 26, 47
– development 25
– dimensions 9
– ganglion cells 17
– glia 12
– position 3
– rods 47
foveola 2
free radicals 125

**G**
ganglion cell 6, 9
– density 6
– distribution 11
– midget 17
genetic study
– family 64, 66
– identification of genes 65
– linkage analysis 65, 66
– sibling 64, 168
– twins 64, 168
geographic atrophy 36, 48
giant cells 33
glutamate 141
– toxicity 144
– transporters 143
glutathione 128
GSH-GSSG cycle 129

**H**
Human Genome Project 63

**I**
ICAM-1 27
immunological factor 117
incidence 79, 82
inflammation 86, 178, 180
– CD45 40
– ICAM-1 30
– MHC II 30
inherited maculopathy 68

**L**
L/M cone, relative numbers 16
laser-induced choroidal new vessels 114, 115
lipid 125
lipofuscin 53, 169

**M**
macrophages 190
macula 1–3, 25
macular
– degeneration
– – atrophic 37
– – dry 37
– – incidence 79
– – neovascular 37
– – population studies 80
– – prevalence 79, 80
– – wet 37
– pigment 52, 138
maculopathy
– early onset 67, 69
– inherited 67

Mallatia Leventinese dystrophy 71
manganese 137
metallothionein 136
microarray 73
microglia 12, 29, 33
microphages 33
midget pathway 10
- M/L cones 11
mitochondria 154
mitochondrial DNA 140
morbidity 83
mouse models 117
Müller cells 12, 13, 129–131, 134, 135, 142, 143, 154, 158, 190
- metabolic functions 141

## N

neovascular macular degeneration 37, 178
neuroprotection 149, 151–153, 155, 163
nutritional supplementation 170

## O

opsin 5
oxidative damage 123

## P

Paraoxonase gene 73
PEDF 116
peripheral degeneration 150, 151, 157, 158
photoreceptor 138
- cone survival 48
- cones 6, 8
- dark adaptation 54, 55
- density 7, 48
- distribution at fovea 3
- dysfunction 51, 52, 54
- early changes 46
- loss 46, 47, 52, 150, 163, 164
- M/L cones 7
- number 4
- rods 6, 8
- - loss 47, 49
- S cones 6, 7
- scotopic sensitivity 55
- sensitivity 51, 53, 54

- stress 150, 162
- types 4
physiological stress 39
pigmentary disturbance 30
population study 80, 103
prevalence 79, 80, 105
- age 194
- ethnicity 81
- gender 104
- in China 103
- regional 105, 108
proteomics 74
public health 83

## R

radial drusen 71
redox reaction 123
retinal
- degeneration slow gene (RDS) 67
- imaging
- - fluorescein angiography 179
- - optical coherent tomography 176
- - scanning laser ophthalmoscopy 176
- - scanning laser polarimetry 177
- pigmented epithelium 13, 53, 54, 128, 169
- - loss 53
- - transport 56
- vascularization 38
- vasculature 158
retinoid cycle 56, 57
risk factors 82
- age 82, 83, 105, 109
- alcohol 87, 110
- arthritis 86
- atherosclerosis 84
- blood pressure 83, 111
- caffeine 88
- cataract surgery 91
- diabetes 86
- diet 89
- ethnicity 109
- gout 86
- hormone replacement therapy 88
- hyperopia 90

- in China 107
- inheritance 82, 109
- iris color 90
- light exposure 89, 109
- myopia 90
- obesity 86
- occupation 92, 109
- physical activity 88
- pulmonary disease 85, 111
- smoking 87, 109, 125
- socioeconomic 92
- UV radiation 89
- X-rays 90
rod 4, 18

## S

S cone pathway 10
scorbic acid 132
selenium 137
serology 35
single nucleotide polymorphism 73
Sorsby's fundus dystrophy 70, 117
Stargardt's disease 69, 117
superoxide 131

## T

taurine 133, 135
triamcinolone acetonide 27, 185

## V

vasculogenesis 181
VEGF 116
visual
- acuity
- - China (AMD) 107
- cortex 19
vitamin
- A 56
- E 132, 133

## W

wet macular degeneration 37
WHO 103

## Z

zinc 136